W9-ABO-712

# Student Interactive Workbook for Starr, Evers, and Starr's

# BIOLOGY: CONCEPTS AND APPLICATIONS

*Seventh Edition*

Michael Windelspecht
*Appalachian State University*

Amy Fenster
*Virginia Western Community College*
*Hollins University*

John D. Jackson
*North Hennepin Community College*

Jane B. Taylor
*Northern Virginia Community College*

THOMSON
™
BROOKS/COLE

Australia • Brazil • Canada • Mexico • Singapore • Spain • United Kingdom • United States

**Student Interactive Workbook for Starr, Evers, and Starr's Biology: Concepts and Applications**
Seventh Edition
Michael Windelspecht, Amy Fenster, John D. Jackson, Jane B. Taylor

© 2008 Thomson Brooks/Cole, a part of The Thomson Corporation. Thomson, the Star logo, and Brooks/Cole are trademarks used herein under license.

ALL RIGHTS RESERVED. No part of this work covered by the copyright hereon may be reproduced or used in any form or by any means—graphic, electronic, or mechanical, including photocopying, recording, taping, Web distribution, information storage and retrieval systems, or in any other manner—without the written permission of the publisher.

Printed in the United States of America

1 2 3 4 5 6 7  11 10 09 08 07

Printer: Thomson/West
Cover Image: Stephan Dalton/Minden Pictures

Library of Congress Control Number: 2007939724

ISBN-13: 978-0-495-11971-5
ISBN-10: 0-495-11971-7

**Thomson Higher Education**
**10 Davis Drive**
**Belmont, CA 94002-3098**
**USA**

For more information about our products,
contact us at:
**Thomson Learning Academic Resource Center**
**1-800-423-0563**
For permission to use material from this text or
product, submit a request online at
**http://www.thomsonrights.com.**
Any additional questions about permissions can be
submitted by email to **thomsonrights@thomson.com.**

# CONTENTS

# Credits and Sources

**Chapter 4**
p.39: Lisa Starr

**Chapter 5**
p.52: PDB files from NYU Scientific Visualization Lab.
p.53: After: David H. MacLennan, William J. Rice and N. Michael Green, "The Mechanism of Ca$^{2+}$ Transport by Sarco (Endo) Plasmic Reticulum Ca$^{2+}$-ATPases." *JBC* Volume 272, Number 46, Issue of November 14, 1997, pp.28815–28818.
p.55: Raychel Ciemma

**Chapter 13**
p.125: Gary Head

**Chapter 15**
p.140: Courtesy of © Genelex Corp.

**Chapter 16**
p.152: From left, © Hans Reinhard/Bruce Coleman, Inc.; © Phillip Colla Photography; © Randy Wells/Corbis; © Cousteau Society/The Image Bank/Getty Images; © Robert Dowling/Corbis

**Chapter 18**
pp.168–169: Raychell Ciemma and Precision Graphics, Inc.

**Chapter 19**
p.173: Lisa Starr

**Chapter 20**
p.184: © Dr. David Phillips/Visuals Unlimited
p.185: (a) © Sinclair Stammers/Photo Researchers, Inc.; (c) © London School of Hygiene & Tropical Medicine/Photo Researchers, Inc.; (e) Micrograph Steven L'Hernaults

**Chapter 22**
p.205: Left, Garry T. Cole, University of Texas, Austin/BPS; Right, © Eye of Science/Photo Researchers, Inc.; Art, After T. Rost, et al., *Botany*, Wiley, 1979.

**Chapter 23**
p.213: (c) © Andrew Syred/SPL/Photo Researchers, Inc.
p.215: Palay/Beaubois

**Chapter 24**
p.223: Above, © Patrick J. Lynch/Photo Researchers, Inc.
p.224: Gary Head
p.225: from E. Solomon, L. Berg, and D.W. Martin, *Biology,* Seventh Edition, Thomson Brooks/Cole.

**Chapter 26**
p.239: Raychel Ciemma
p.241: Art, Lisa Starr; Photograph, © Andrew Syred/Photo Researchers, Inc.
p.242: (1–6) Robert and Linda Mitchell Photography
p.242: (7–13) Art, Raychel Ciemma
p.243: (14) © Carolina Biological Supply Company
p.243: (15) Mike Clayton, University of Wisconsin, Botany Department
p.243: (24–25) N. Cattlin, Photo Researchers, Inc.
p.243: (26–31) Art, Raychel Ciemma
p.243: (32–35) C.E. Jeffree, et al., Planta, 172(1): 20–37, 1987. Reprinted by Permission of C.E. Jeffree and Springer-Verlag
p.243: (36–37) © Jeremy Bargess/SPL/Photo Researchers, Inc.
p.245: (1–11) After Salisbury and Ross, *Plant Physiology,* Fourth Edition, Wadsworth
p.245: (1–5, 5–6) Michael Clayton, University of Wisconsin, Botany Department
p.245: (12–13) Raychel Ciemma
p.246: (1–6 and 10) Omikron/Photo Researchers, Inc.
p.246: (7–9) H.A. Core, W.A. Cote, and A.C. Day, *Wood Structure and Identification,* Second Edition, University Press, 1979.

**Chapter 27**
p.251: (9–12) Micrograph Chuck Brown

**Chapter 28**
p.258: Raychel Ciemma
p.259: (left) © John McAnulty/CORBIS; (right) © Robert Essel NYC/CORBIS
p.261: Raychel Ciemma

**Chapter 29**
p.277: Robert Demarest

**Chapter 30**
p.283: Kevin Somerville
p.287: Robert Demarest

p.289: Robert Demarest
p.292: © Colin Chumbley/Science
    Source/Photo Researchers, Inc.

**Chapter 31**
p.298: Robert Demarest
p.300: Robert Demarest

**Chapter 33**
p.315: Raychel Ciemma
p.319: Raychel Ciemma

**Chapter 34**
p.329: Kevin Somerville
p.330: Raychel Ciemma
p.332: left, © Biophoto Associates/
    Photo Researchers, Inc.

**Chapter 35**
p.343: Left, © NIBSC/Photo Researchers, Inc.;
    (a–e) After Stephen Wolfe, *Molecular Biology
    of the Cell*, Wadsworth, 1993.

**Chapter 36**
p.349: Kevin Somerville

**Chapter 38**
p.370 (top): Robert Demarest
p.370 (bottom): Precision Graphics, Inc.

**Chapter 39**
p.381: Ed Reschke

**Chapter 40**
p.403: Preface, Inc. and Precision Graphics, Inc.

# PREFACE

*Tell me and I will forget, show me and I might remember, involve me and I will understand.*
—Chinese Proverb

This insightful proverb outlines three levels of learning, each successively more effective than the preceding one. The writer of the proverb understood that we humans learn most efficiently when we *involve* ourselves in the material to be learned. This study guide is like a tutor; when used properly, it increases the efficiency of your study periods. The guide's interactive exercises actively involve you in the most important terms and central ideas of your text. Specific tasks ask you to recall key concepts and terms and apply them to life; they test your understanding of the facts and indicate items to reexamine or clarify. Your performance on these tasks will help you estimate your performance on in-class tests of similar material. Most important, however, this biology study guide and text together help you make informed decisions about matters that affect not only your own well-being, but also that of your environment. In the years to come, human survival on planet Earth will demand that individual, national, and global decisions be based on an informed biological background.

## HOW TO USE THIS STUDY GUIDE

Each chapter of this guide begins with a title and a brief introduction regarding the content of the chapter. This is followed by a set of bulleted focal points. The purpose of the focal points is to direct you to key figures, tables, or material in the chapter. For easy reference to an answer or definition, each question and term in this comprehensive study guide is accompanied by the appropriate text page(s), and appears in the form: [p.352], for example. The Interactive Exercises begin with a list of Selected Words (other than boldfaced terms) chosen by the authors as those that are most likely to enhance your understanding of the material. In the text chapters, these words appear in italics, quotation marks, or normal type. Next, a list of Boldfaced Terms from the text appears. These terms are essential to your understanding of each text section. Space is provided by each term for you to formulate a definition in *your own words*. After the terms is a series of different types of exercises that may include Completion, Short Answer, True–False, Matching, Choice, Dichotomous Choice, Identification, Problems, Labeling, Sequence, Crossword Puzzle, Concept Maps, and Complete the Table. Dispersed throughout the Interactive Exercises are references to the TNOW Web site. These online exercises and assignments have been chosen to provide interactive versions of the printed material in this manual.

A Self-Quiz immediately follows the Interactive Exercises. This quiz is composed primarily of multiple-choice questions. Any wrong answers in the Self-Quiz indicate portions of the text you need to reexamine. A series of Chapter Objectives/Review Questions follows each Self-Quiz. These are tasks that you should be able to accomplish if you have understood the assigned reading in the text. Following the Chapter Objectives/Review Questions section is a Chapter Summary. This is a fill-in-the-blank version of the Key Concepts material at the beginning of the chapter.

Each study guide chapter concludes with the Integrating and Applying Key Concepts section. It invites you to try your hand at applying major text concepts to situations in which a single pat answer does not necessarily work—and so none is usually provided in the chapter answer section. Your text generally will provide enough clues to get you started on an answer, but this part is intended to stimulate your thought and provoke lively group discussions.

*A person's mind, once stretched by a new idea, can never return to its original dimension.*
—Oliver Wendell Holmes

## THE STRUCTURE OF THIS STUDY GUIDE

This overview shows how each chapter in the study guide is organized.

Chapter Number ————————➤

# 12

Chapter Title ————————➤

# DNA STRUCTURE AND FUNCTION

Introduction ————————➤

### INTRODUCTION

The DNA molecule has a history full of scientific intrigue, experimentation, and even scandal. While the molecule itself was discovered in the 1800s, it was not until the 1940s that researchers became certain that DNA was the heritable genetic material. Then, the race was on to discover the structure and replication scheme of this all-important molecule. This chapter describes the experiments that led to the elucidation of the function of DNA and the scandal which resulted in the discovery of the structure of this molecule. The chapter concludes with a look at DNA technology with respect not to how far we have come in our understanding, but to where this understanding may take us.

Focal Points ————————➤

### FOCAL POINTS

- Figure 12.3 [p.188] animates the experiment by Fredrick Griffith in which he discovered "the transforming principle."
- Figure 12.5 [p.190] illustrates the four nucleotide bases in the DNA molecule.
- Figure 12.6 [p.191] shows an animation of the double helix molecule of DNA.
- Figure 12.9 [p.193] depicts the formation of a newly synthesized strand during replication.
- Table 12.1 [p.194] compares various cloning methods.

Interactive Exercises ————————➤

The Interactive Exercises are divided into numbered sections by titles of main headings and corresponding page references to encourage your *interaction* with the text. This study guide is especially helpful in that each term or question is accompanied by a reference to the text page(s) where that definition or answer may be found. Each section begins with important author-selected or boldfaced terms and continues with various types of challenging exercises that require constant interaction with the text, which helps you to truly learn the important chapter information.

Self-Quiz ————————➤

These multiple-choice questions (or other testing devices) sample important blocks of text information to test your comprehension of the entire chapter.

Chapter Objectives/————————➤
Review Questions

This combination of relative objectives to be met and questions to be answered highlights what you need to learn in each chapter.

Chapter Summary ————————➤

A review of the Key Concepts at the start of the chapter.

Integrating and ————————➤
Applying Key Concepts

Applications of text material to questions for which there may be more than one correct answer.

Answers to Interactive ————————➤
Exercises and Self-Quizzes

Answers for all Interactive Exercises, Self-Quizzes, and Chapter Summaries can be found at the back of this study guide, organized by chapter and title and by the main headings with their page references.

Use the answers section wisely by thinking of them as a check on your progress. That is, always come up with your own answers first, then check them against those provided. Doing okay? Great—onto the next section. Not doing so well? Then it is time to review both the text and the study guide, and ask questions in class.

Learning is hard work, but its unending pursuit reaps indisputable personal enjoyment and satisfaction, and lays the stepping stones to success in one's life.

*Every day we spend without learning something is lost.*
—Beethoven

*An investment in knowledge pays the best interest.*
—Benjamin Franklin

# 1

## INVITATION TO BIOLOGY

### INTRODUCTION

This first chapter of the text serves as the introduction to many of the main concepts that are explored during the course.

### FOCAL POINTS

- Figure 1.3 [pp.4–5] illustrates the levels of organization in nature.
- Figure 1.4 [p.6] illustrates the difference between the flow of energy and nutrients in an ecosystem.
- Sections 1.6 and 1.7 [pp.12–15] explain how scientists explore nature using the scientific method.

## Interactive Exercises

*Note:* In the answer sections of this book, a specific molecule is most often indicated by its abbreviation. For example, adenosine triphosphate is ATP.

*Lost Worlds and Other Wonders* [pp.2–3]

### 1.1. LIFE'S LEVELS OF ORGANIZATION [pp.4–5]

*Selected Words*

In addition to the boldfaced terms, discussed in the next paragraph, the text features other important terms essential to understanding the assigned material. "Selected Words" lists these terms, which appear in the text in italics, in quotation marks, and occasionally in normal type.

*Life* [p.4], *nonlife* [p.4]

*Boldfaced Terms*

The boldfaced terms are particularly important. Write a definition for each term in your own words without looking at the text. Next, compare your definition with that given in the chapter or in the text glossary. If your definition seems inadequate, allow some time to pass and repeat this procedure until you can define each term quickly.

[p.4] nature _____

_____

[p.4] atoms _____

_____

[p.4] molecules _____

_____

[p.4] cell _____

_____

[pp.4–5] organism _____

_____

[p.5] population _____

_____

[p.5] community _____

_____

[p.5] ecosystem _____

_____

[p.5] biosphere _____

_____

[p.5] emergent properties_____

_____

## Matching

Match each of the following terms with its correct description. [pp.4–5]

1. _____ atom
2. _____ organ system
3. _____ organ
4. _____ cell
5. _____ organism
6. _____ community
7. _____ biosphere
8. _____ molecule
9. _____ tissue
10. _____ population
11. _____ ecosystem

a. structural unit of two or more tissues interacting in some task
b. all of the regions of the Earth's water, crust, and atmosphere that hold organisms
c. smallest unit that can live and reproduce on its own or as part of a multicelled organism
d. organs interacting physically and/or chemically in some task
e. two or more joined atoms of the same or different elements
f. all populations of all species occupying a specified area
g. smallest unit of an element that still retains the element's properties
h. a community that is interacting with its physical environment
i. individual made of different types of cells
j. a unit of cells interacting in some task
k. group of individuals of the same species in a given area

## Concept Map

In the concept map below, provide the missing terms associated with the numbers in parentheses. [pp.4–5]

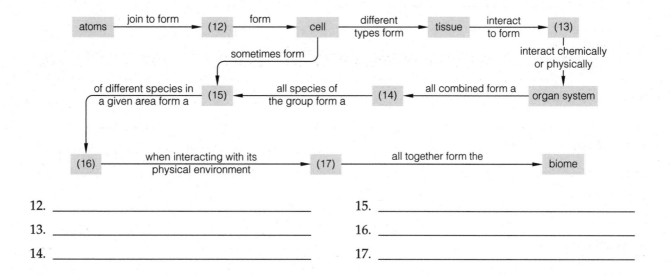

12. _____

13. _____

14. _____

15. _____

16. _____

17. _____

*TNOW:* For an animation of this exercise, refer to Chapter 1: Art Labeling; Levels of Organization.

## 1.2. OVERVIEW OF LIFE'S UNITY [pp.6–7]

## 1.3. IF SO MUCH UNITY, WHY SO MANY SPECIES? [pp.8–9]

## 1.4. AN EVOLUTIONARY VIEW OF DIVERSITY. [p.10]

### Selected Words

Decomposers [p.6], *internal* environment [pp.6–7], eukaryotic [p.8], adaptive [p.10], *artificial* selection [p.10], *natural* selection [p.10], "selective agents" [p.10]

### Boldfaced Terms

[p.6] energy _____

_____

[p.6] nutrient _____

_____

[p.6] producers _____

_____

[p.6] consumers _____

_____

[p.6] receptor _____

_____

[p.7] homeostasis _____

_____

[p.7] DNA _____

_____

[p.7] inheritance _____

_____

[p.7] reproduction _____

_____

[p.7] development _____

_____

[p.8] species _____

_____

[p.8] genus _____

_____

[p.8] bacteria _____

_____

[p.8] archaeans _____

_____

[p.8] protists _____

_____

[p.8] fungi _____

_____

[p.8] plants _____

_____

[p.9] animals _____

_____

[p.10] mutations _____

_____

[p.10] natural selection _____

_____

[p.10] evolution _____

_____

## Matching

Match each of the following definitions with the correct term. [pp.6–7]

1. _____ the capacity to do work                                                          a. consumers
2. _____ structure that detects a specific form of stimulation                            b. homeostasis
3. _____ a nucleic acid that is the signature molecule of life                            c. producers
4. _____ organisms that have the capacity to make their own food                          d. development
5. _____ mechanism by which parents transmit DNA to offspring                             e. inheritance
6. _____ transmission of genetic information from parents to offspring                    f. energy
7. _____ transformation of cells into an adult organism                                   g. nutrient
8. _____ maintenance of internal conditions within a range suitable for life              h. receptor
9. _____ organisms that must get their energy from other organisms                        i. DNA
10. _____ an atom or molecule that has an essential role in growth and survival           j. reproduction

## Choice

For each of the following, choose from one of the following answers. [pp.6–7]

    a. energy     b. nutrient     c. both a and b

11. _____ photosynthesis captures this for use by the producers
12. _____ cycled between producers and consumers
13. _____ flows through the world of life in one direction
14. _____ decomposers return this to the environment for the producers
15. _____ consumers must get this from the producers

## Matching

Match each of the following kingdom or domain to its correct description. [pp.8–9]

16. _____ protists       a. the simplest of the eukaryotic organisms
17. _____ animals        b. multicelled, eukaryotic decomposers
18. _____ fungi          c. the most ancient of the single-celled prokaryotes
19. _____ bacteria       d. the prokaryotes that are most closely related to the eukaryotes
20. _____ plants         e. photosynthetic, multicelled eukaryotes
21. _____ archaeans      f. multicelled and eukaryotic consumers

## Choice

For each of the following, choose from one of the following terms. [p.10]

    a. evolution by natural selection     b. evolution by artificial selection

22. _____ pigeon breeding
23. _____ a favoring of adaptive traits in nature
24. _____ the selection of one trait over another in contrived, manipulated conditions
25. _____ environmentally influenced, inheritable changes occurring in a population over time
26. _____ Darwin viewed this as his model for understanding change over time.

27. _____ breeders represent the "selective agent"

28. _____ predators represent the "selective agent"

## 1.5. CRITICAL THINKING AND SCIENCE [p.11]

## 1.6. HOW SCIENCE WORKS [pp.12–13]

### Selected Words

*Kriticos* [p.11], "the scientific method" [p.12], "theory" [p.13]

### Boldfaced Terms

[p.11] critical thinking _____

_____

[p.11] science _____

_____

[p.11] supernatural _____

_____

[p.12] hypothesis _____

_____

[p.12] prediction _____

_____

[p.12] test _____

_____

[p.12] model _____

_____

[p.12] scientific theory _____

_____

### Matching

Match each of the following terms with its correct definition.

1. _____ supernatural [p.11]
2. _____ scientific theory [p.12]
3. _____ critical thinking [p.11]
4. _____ model [p.12]
5. _____ science [p.11]
6. _____ prediction [p.12]
7. _____ hypothesis [p.12]
8. _____ test [p.12]

a. a hypothesis that has been repeatedly verified by tests and is used to make additional predictions
b. an analogous system for testing a hypothesis
c. either systematic observations or experiments
d. a statement of a condition that should exist if a hypothesis is valid
e. a testable answer to a question
f. the judging of information before accepting it
g. the systematic study of nature
h. anything that is beyond nature

## Concept Map

Provide the missing word(s) for each number in parentheses in the following concept map of scientific inquiry. [p.12]

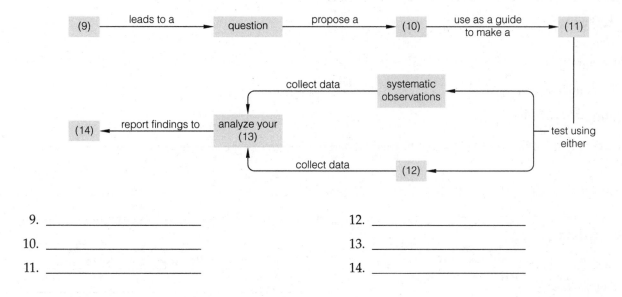

9. _____

10. _____

11. _____

12. _____

13. _____

14. _____

## 1.7. THE POWER OF EXPERIMENTAL TESTS [pp.14–15]

## 1.8. SAMPLING ERROR IN EXPERIMENTS [p.16]

### Selected Words

*If-then* process [pp.14–15], *quantitative* results [p.15]

### Boldfaced Terms

[p.16] sampling error _____

_____

### Matching

Review the experiment regarding peacock butterfly and birds [pp.14–15]. For each of the following, choose the appropriate component of the scientific method that is best described by the statement. Some answers may be used more than once.

1. _____ If the sounds deter predatory birds, then individuals who can't make the sound will be more likely to be eaten by predatory birds.

2. _____ All of the butterflies with unmodified wing spots survived.

3. _____ Some butterflies have the sound-making part of the wing removed.

4. _____ Why does the peacock butterfly flick its wings?

5. _____ The hissing and clicking sounds produced by the butterfly rubbing its wings deter predatory birds.

6. _____ If brilliant wing spots deter birds, then birds with no spots will be more likely to be eaten by predatory birds.

7. _____ Opening of the wings exposes spots that resemble owl eyes, thus scaring off predators.

8. _____ Some butterflies had both their spots painted black and the sound-producing portions of the wings removed.

9. _____ Most butterflies with neither spots nor sound structures survived.

a. question

b. prediction

c. hypothesis

d. experimental test

e. analysis of results

*TNOW:* To better understand the scientific method and the design of experiments, complete Chapter 1: Co-op Activities 1.1, 1.2, 1.3, & 1.4.

## Self-Quiz

1. All populations of all species occupying a specific area is the definition of a(n) _____. [pp.4–5]
   a. population
   b. community
   c. ecosystem
   d. biome
   e. domain

2. Which of the following is not a domain of life? [p.8]
   a. archaean
   b. eukarya
   c. protist
   d. bacteria
   e. all of the above are domains of life

3. In an ecosystem, energy _____ the ecosystem, while nutrients _____ the ecosystem. [p.6]
   a. flows through; cycles within
   b. flows through; flows through
   c. cycles within; cycles within
   d. cycles within; flows through

4. Which of the following terms describes the transformation of a single cell into an adult organism? [p.7]
   a. reproduction
   b. inheritance
   c. development
   d. homeostasis

5. Which of the following are involved with the initial capture of energy by an ecosystem? [p.6]
   a. decomposers
   b. consumers
   c. producers
   d. all of the above

6. In the scientific method, what step typically follows the creation of a hypothesis? [p.12]
   a. analysis of results
   b. report to the scientific community
   c. experimental testing
   d. forming a prediction

7. When breeders act as the selective agent, it is called _____. [p.10]
   a. supernatural
   b. artificial selection
   c. natural selection
   d. development

8. The ability of life to maintain an internal environment despite changes in the external environment is called _____. [p.7]
   a. inheritance
   b. homeostasis
   c. reproduction
   d. development
   e. evolution

9. The signature molecule of life is _____. [p.7]
   a. protein
   b. RNA
   c. DNA
   d. the cell

10. Changes in a line of descent over time are called _____. [p.10]
    a. evolution
    b. development
    c. inheritance
    d. homeostasis

## Chapter Objectives/Review Questions

This section lists general and detailed chapter objectives that can be used as review questions. You can make maximum use of these items by writing answers on a separate sheet of paper. Where blanks are provided, fill in the answers. Use the indicated page numbers or the glossary to check the accuracy of your answers.

1. Understand the levels of organization in nature. [pp.5–6]
2. Understand the concept of an emergent property. [p.6]
3. Define the role of producers, consumers, and decomposers in an ecosystem. [p.6]
4. Illustrate the difference between nutrient and energy flow in an ecosystem. [p.6]
5. Understand the importance of homeostasis. [p.7]
6. Explain the differences among inheritance, reproduction, and development. [p.7]
7. List the three domains of life. [p.8]
8. Distinguish among prokaryotes, protists, fungi, plants, and animals. [pp.8–9]
9. Explain the difference between artificial and natural selection. Give an example of each. [p.10]
10. Define the term *evolution*. [p.10]
11. Understand the importance of critical thinking to the study of science. [p.11]
12. Understand the basic steps of the scientific method. [pp.12–13]
13. Understand the concept of sampling error. [p.16]

## Chapter Summary

This section serves as a review of the Key Concepts presented at the beginning of the chapter. After completing the chapter you should be able to fill in the blanks without assistance from the textbook.

We study the world of life at different levels of (1) _____, which extend from atoms and molecules to the biosphere. The quality known as "life" emerges at the level of (2) _____.

The world of life shows (3) _____, because all organisms are alike in key aspects. They consist of one or more (4) _____, which stay alive through ongoing inputs of energy and raw materials. These sense and respond to changes in their external and internal (5) _____. Their cells contain (6) _____, a type of molecule that offspring inherit from parents and that encodes information necessary for growth, survival, and (7)_____.

The world of life also shows great diversity. Many millions of kinds of organisms, or species, have appeared and disappeared over time. Each species is unique in at least some trait—in some aspect of its body (8) _____ or behavior.

Theories of evolution, especially of theory of evolution by (9) _____, help explain why life shows both unity and (10) _____. Evolutionary theories guide research in all fields of biology.

Biologists make systematic (11) _____, predictions, and tests in the laboratory and in the field. They report their work so that others may repeat their work and check their reasoning.

---

## Integrating and Applying Key Concepts

1. What would happen if all of the decomposers were removed from an ecosystem?
2. What sorts of topics do scientists usually regard as untestable by applications of the scientific method?
3. Based upon your knowledge of the scientific method and critical thinking, what is the problem with calling religious beliefs "creationist theory"?

# 2

# LIFE'S CHEMICAL BASIS

## INTRODUCTION

All living organisms are based upon chemical reactions. This chapter introduces you to the principles of chemistry, namely atoms, elements, and chemical bonds. It is important that you develop an understanding of these basic chemical principles early in the course as many of the later topics will build on these concepts.

## FOCAL POINTS

- The basic characteristics of an atom. [p.22]
- Importance of ionic, covalent, and hydrogen bonds. [pp.26–27]
- The properties of water. [pp.28–29]
- Figure 2.14 [p.30] illustrates the pH scale.

## Interactive Exercises

*What Are You Worth?* [pp.20–21]

## 2.1. START WITH ATOMS [p.22]

## 2.2. PUTTING RADIOISOTOPES TO WORK [p.23]

*Selected Words*

Atomic number [p.22], mass number [p.22], radioactive decay [p.23], PET [p.23]

*Boldfaced Terms*

[p.22] atoms _____

_____

[p.22] protons _____

_____

[p.22] electrons _____

_____

[p.22] neutrons _____

_____

[p.22] elements _____

_____

[p.22] atomic number _____

_____

[p.22] isotopes _____

_____

[p.22] periodic table of the elements _____

_____

[p.23] radioisotopes _____

_____

[p.23] tracer _____     _____

_____

## Matching

Match each of the following terms to its correct description.

1. _____ atoms
2. _____ protons
3. _____ PET
4. _____ neutrons
5. _____ electrons
6. _____ atomic number
7. _____ radioactive decay
8. _____ elements
9. _____ isotopes
10. _____ radioisotopes
11. _____ tracers
12. _____ periodic table of the elements
13. _____ mass number

a. subatomic particles in the nucleus that carry no charge
b. particles that are the building blocks of all substances
c. an arrangement of the known elements based on their chemical properties
d. the number of protons and neutrons in the nucleus of a single atom
e. occurs when an element spontaneously emits energy or subatomic particles when its nucleus breaks down
f. an isotope that undergoes radioactive decay
g. the number of protons in an atom
h. atoms of a given element that differ in the number of neutrons
i. a procedure that is used to form images of tissues in the body
j. positively charged subatomic particles in the nucleus
k. negatively charged subatomic particles found outside the nucleus
l. a molecule with a detectable substance, such as a radioisotope, attached
m. pure substances containing atoms with the same number of protons

## 2.3. WHY ELECTRONS MATTER [pp.24–25]

## 2.4. WHAT HAPPENS WHEN ATOMS INTERACT? [pp.26–27]

### Selected Words

"Shells" [p.24], *single, double, triple* covalent bond [p.26], *polar* covalent bond [pp.26–27], *nonpolar* covalent bond [p.27]

## Boldfaced Terms

[p.24] shell model _____

_____

[p.25] ion _____

_____

[p.25] electronegativity _____

_____

[p.25] chemical bond _____

_____

[p.25] molecule _____

_____

[p.25] compounds _____

_____

[p.25] mixture _____

_____

[p.26] ionic bond _____

_____

[p.26] covalent bond _____

_____

[p.27] polarity _____

_____

[p.27] hydrogen bond _____

_____

## Matching

For each definition below, choose the most appropriate term from the list provided. [p.25]

1. _____ molecules that contain two or more elements in a constant proportion

2. _____ two or more substances that do not form chemical bonds and whose proportions may vary

3. _____ when two or more atoms of the same or different elements form a chemical bond

4. _____ an atom that carries an electrical charge

5. _____ an attractive force between two atoms

a. chemical bond
b. molecule
c. compound
d. ion
e. mixture

## Identification

Identify each of the following elements based on the number of electrons. Recall that the number of electrons equals the number of protons.

○ electron

6 _____     7 _____

8 _____     9 _____

10 _____

6. _____     9. _____

7. _____     10. _____

8. _____

*TNOW:* For an animation of this exercise, refer to Chapter 2: Art Labeling: Atomic Structure of Selected Elements and Co-op Activity 2.1: Atomic Structure.

## Choice

For each of the following, choose the most appropriate type of bond from the list provided. Some answers may be used more than once.

a. covalent bonds [pp.26–27]     b. hydrogen bonds [p.27]
c. ionic bond [p.26]             d. all of the above

11. _____ these are found in biological molecules

12. _____ occurs when one atom loses it electrons to another atom

13. _____ formed by the sharing of electrons

14. _____ atoms in this bond stay together due to a strong electrical attraction

15. _____ may be either polar or nonpolar

16. _____ does not form molecules from atoms

17. _____ may be single, double, or triple

18. _____ formed between the polar regions of two molecules

*TNOW:* For an interactive exercise of this material, refer to Chapter 2: Co-Op Activity 2.2: Atoms and Bonds.

## 2.5. WATER'S LIFE-GIVING PROPERTIES [pp.28–29]

## 2.6. ACIDS AND BASES [pp.30–31]

### Selected Words

Hydrogen bonds [p.28], hydrogen ions [p.30], *acidic* solutions [p.30], *basic* solutions [p.30], alkaline [p.30], *acid indigestion* [p.30], *acute respiratory acidosis* [p.31], *coma* [p.30], *alkalosis* [p.31], *tetany* [p.31]

### Boldfaced Terms

[p.28] hydrophilic _____

_____

[p.28] hydrophobic _____

_____

[p.28] temperature _____

_____

[p.29] evaporation _____

_____

[p.29] solvent _____

_____

[p.29] solutes _____

_____

[p.29] sphere of hydration _____

_____

[p.29] cohesion _____

_____

[p.30] pH scale _____

_____

[p.30] acids _____

_____

[p.30] bases _____

_____

[p.31] salt _____

_____

[p.31] buffer system _____

## Concept Map

The concept map below refers to the properties of water [pp.28–29]. Fill in the missing term or terms for each numbered space.

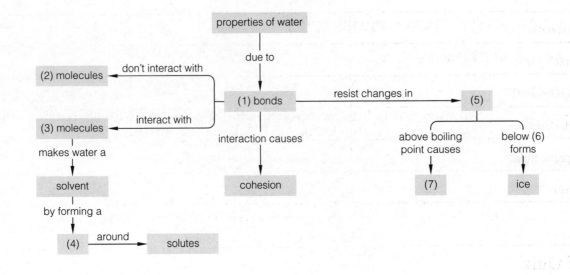

1. _____    5. _____

2. _____    6. _____

3. _____    7. _____

4. _____

*TNOW:* For an interactive exercise of this material, refer to Chapter 2: Co-op Activity 2.3: Properties of Water.

## Matching

Match each of the following terms to the correct description. [pp.30–31]

8. _____ salt

9. _____ pH scale

10. _____ acids

11. _____ buffer system

12. _____ bases

13. _____ acid indigestion

14. _____ tetany

15. _____ alkalosis

16. _____ acute respiratory acidosis

17. _____ coma

a. a substance that accepts hydrogen ions from a solution
b. an accumulation of carbon dioxide in the blood
c. a substance that releases ions other than $H^+$ and $OH^-$ in water
d. a dangerous level of unconsciousness
e. a measure of hydrogen ion concentration in a solution
f. a potentially lethal rise in blood pH
g. substances that donate hydrogen ions in a solution
h. excessive hydrogen ions in the stomach
i. a period of prolonged muscle spasms
j. chemicals that keep the pH of a solution stable

## Complete the Table

Use the information in Figure 2.14 [p.30] to complete the following table.

| Substance | pH value | Acid/Base |
|---|---|---|
| 18. seawater | | |
| 19. cola | | |
| 20. toothpaste | | |
| 21. black coffee | | |
| 22. gastric fluid | | |
| 23. beer | | |

# Self Quiz

1. Each element has a unique _____, which equals the number of protons in its atoms. [p.22]
   a. isotope
   b. atomic number
   c. mass number
   d. number of orbitals

2. A molecule is _____. [p.25]
   a. a substance that resists changes in pH
   b. a less stable form of an atom
   c. an element with a variable number of electrons
   d. a combination of two or more atoms

3. Lithium has an atomic number of 3 and an atomic mass of 7. How many neutrons are located in its nucleus? [p.22]
   a. 3
   b. 4
   c. 7
   d. can't be determined

4. A _____ substance is non-polar and repelled by water. [p.28]
   a. radioactive
   b. hydrophilic
   c. covalent
   d. hydrophobic

5. The properties of water are due to the _____ bonds in water. [pp.28–29]
   a. nonpolar
   b. hydrogen
   c. ionic
   d. covalent

6. A transfer of electrons from one atom to another forms a(n) _____. [p.26]
   a. hydrogen bond
   b. isotope
   c. acid
   d. ion

7. In a covalent bond, electrons are _____ between two atoms. [pp.26–27]
   a. shared
   b. transferred
   c. destroyed
   d. combined

8. The property of water that allows one water molecule to pull on another is called _____. [p.29]
   a. evaporation
   b. hydrophobic interactions
   c. buffering
   d. cohesion

9. A solution with a pH of 3.0 is _____ times more acidic than a pH of 5.0. [p.14]
   a. 2
   b. 20
   c. 100
   d. 200

10. In a solution, the substance that donates hydrogen ions is called the _____. [pp.30–31]
    a. buffer
    b. ion
    c. acid
    d. salt

---

## Chapter Objectives/Review Questions

1. Understand the location and charge of each of the subatomic particles. [p.22]
2. Understand how atomic number and mass number relate to an element. [p.22]
3. Define the terms isotope and radioisotope. [pp.22–23]
4. Understand the importance of electrons to an atom. [pp.24–25]
5. Understand the concept of a chemical bond. [p.25]
6. Distinguish between a molecule, compound, and mixture. [p.25]
7. Distinguish between an ionic, covalent, and hydrogen bond. [pp.26–27]
8. Understand polarity and be able to define the terms hydrophilic and hydrophobic. [p.28]
9. Understand the importance of hydrogen bonds to the properties of water. [pp.28–29]
10. Be able to list the properties of water. [pp.28–29]
11. Distinguish between a solvent and solute. [p.29]
12. Understand the importance of cohesion and how it occurs. [p.29]
13. Be able to interpret the pH scale. [p.30]
14. Distinguish between an acid and base. [p.30]
15. Understand the importance of a buffer. [p.31]

---

## Chapter Summary

Atoms are fundamental units of all (1) _____. Protons, electrons, and (2) _____ are their building blocks.

Elements are pure substances, each consisting entirely of atoms that have the same number of (3) _____.

(4) _____ are atoms of the same element that have different numbers of neutrons.

Whether one atom will bond with others depends on the number and arrangement of its (5) _____.

Atoms of many elements interact by acquiring, (6) _____, and giving up electrons. (7) _____, covalent, and (8) _____ bonds are the main interactions between atoms in biological molecules.

Life originated in (9) _____ and is adapted to its properties. Its temperature-stabilizing effects, (10) _____, and capacity to act as a solvent for so many other substances are the properties that make life possible on Earth.

Life is responsive to changes in the amounts of hydrogen (11) _____ and other substances dissolved in water.

# Integrating and Applying Key Concepts

1. Explain what would happen if water were a nonpolar molecule. In your explanation, include the implications for the properties of water including temperature stabilization, cohesion, action as a solvent, and evaporation.

2. Scientists have focused the search for life on other planets around the search for water. Why is liquid water important to life as we know it? Do you think that it is possible that life could be based on other molecules? Explain your answer.

3. Global climate change is based not only on carbon dioxide, but also water's temperature-stabilizing effects. How would this property of water serve to increase the world's temperature over time?

# 3

# MOLECULES OF LIFE

## INTRODUCTION

Students often are confused as to why introductory biology courses begin with a discussion of chemistry. As you will learn, the study of living organisms is the study of how they process chemicals to conduct the processes of life. In this chapter, you will be introduced to the principles of organic chemistry, which focuses on the element carbon. The chapter introduces the importance of organic molecules: carbohydrates, lipids, proteins, and nucleic acids.

## FOCAL POINTS

- Figure 3.3 [p.36] illustrates the major functional groups of organic molecules.
- Figure 3.5 [p.37] illustrates the difference between condensation and hydrolysis.
- Figure 3.16 [p.42] illustrates the formation of a polypeptide.
- Figure 3.17 [p.43] illustrates the levels of protein structure.
- Table 3.2 [p.48] summarizes the major types of organic molecules.

## Interactive Exercises

*Science or Supernatural?* [pp.34–35]

## 3.1. MOLECULES OF LIFE—FROM STRUCTURE TO FUNCTION [pp.36–37]

### Selected Words

Methane [p.34], methane hydrate [p.35], *hydroxyl* groups [p.36], *carbonyl* groups [p.36], *carboxyl* groups [pp.35–36], *phosphate* group [p.37]

### Boldfaced Terms

[p.36] organic compounds _____

_____

[p.36] functional groups _____

_____

[p.36] alcohols _____

_____

[p.37] metabolism _____

_____

[p.37] enzymes _____

_____

[p.37] condensation _____

_____

[p.37] hydrolysis _____

_____

[p.37] monomers _____

_____

[p.37] polymers _____

_____

## Labeling

Study the structure of the organic molecules below. Using the information presented in Figure 3.3 of the text [p.36], identify each of the circled functional groups.

1. _____        5. _____
2. _____        6. _____
3. _____        7. _____
4. _____

*TNOW:* For an animation of this exercise, refer to Chapter 3: Art Labeling; Common Functional Groups.

## Complete the Table

Complete the following table by providing either the name of the reaction category or a description of the reaction. [p.37]

| Reaction Category | Reaction Description |
|---|---|
| 8. | Juggling of internal bonds converts one type of organic molecule to another |
| Condensation | 9. |
| Electron transfer | 10. |
| 11. | One molecule splits into two smaller ones |
| 12. | One or more electrons taken from one molecule are donated to another molecule |

## Identification

13. Identify the metabolic reactions by writing *condensation* or *hydrolysis* in the space provided. [p.37]

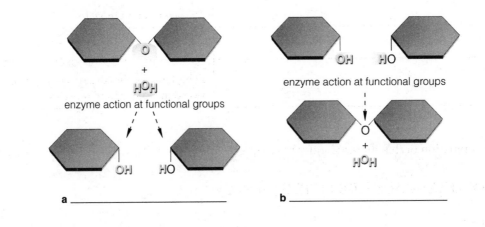

a. _____

b. _____

***TNOW:*** For an animation of this material, refer to Chapter 3: Co-op Activity 2.4: Synthesis and Hydrolysis.

## 3.2. CARBOHYDRATES—THE MOST ABUNDANT ONES [pp.38–39]

### Selected Words

*Mono*saccharides [p.38], "saccharide" [p.38], *oligo*saccharide [p.38], *oligo*- [p.38], *di*saccharides [p.38], "complex" carbohydrates [p.38], *poly*saccharides [p.38]

### Boldfaced Terms

[p.38] carbohydrates _____

_____

## Complete the Table

Complete the table below by entering the name and class of each carbohydrate described in terms of its function. For class, choose from *monosaccharide*, *oligosaccharide*, or *polysaccharide*. [pp.38–39]

| Carbohydrate | Class | Function |
|---|---|---|
| 1. | | Sugar present in milk |
| 2. | | Strengthens external skeletons and other hard body parts of some animals and fungi. |
| 3. | | Used by cells as an energy source |
| 4. | | The carbohydrate found in plant cell walls |
| 5. | | The plant storage carbohydrate for the products of photosynthesis |
| 6. | | Table sugar |
| 7. | | The sugar found in DNA |
| 8. | | The sugar found in RNA |
| 9. | | Long-term carbohydrates storage in animals; stored in the liver and muscles |

*TNOW:* For an animation of this material, refer to Chapter 3: Co-op Activity 2.7: Complex Carbohydrates.

## 3.3. GREASY, OILY—MUST BE LIPIDS [pp.40–41]

### Selected Words

"Essential fatty acids" [p.40], *neutral* fats [p.40], *saturated* fats [p.40], *unsaturated* fats [p.40], *cis* fatty acids [p.40], *trans* fatty acids [pp.40–41]

### Boldfaced Terms

[p.40] lipids _____

_____

[p.40] fatty acids _____

_____

[p.40] fats _____

_____

[p.40] triglycerides _____

_____

[p.41] phospholipids _____

_____

[p.41] waxes _____

_____

[p.41] sterols _____

_____

## Labeling

1. In the appropriate blanks, label the molecule shown as saturated or unsaturated. For the unsaturated molecules, circle the regions that make the unsaturated. [p.40]

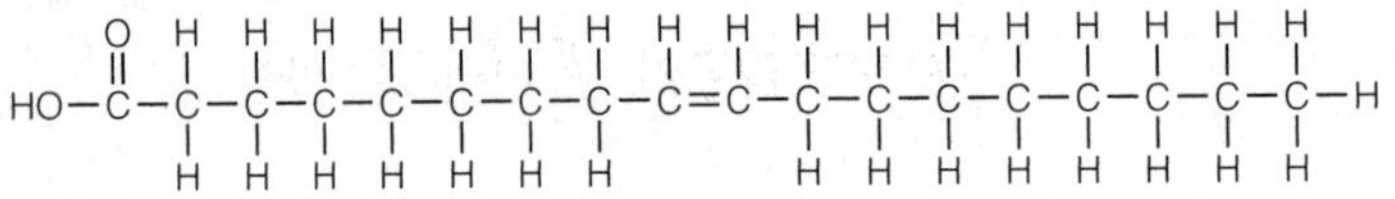

oleic acid

a _____

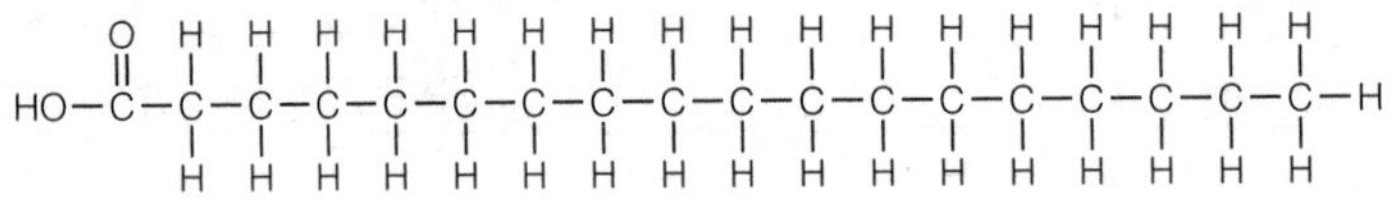

stearic acid

b _____

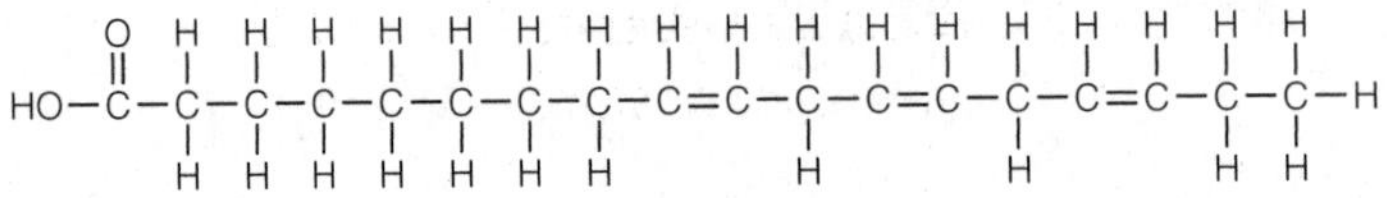

linolenic acid

c _____

a. _____

b. _____

c. _____

## Identification

2. Identify the lipid in each of the following diagrams. [pp.40–41]

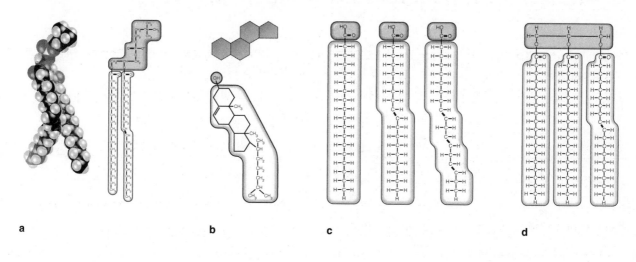

a                                        b                      c                                     d

a. _____

b. _____

c. _____

d. _____

## Choice

For each of the following statements, choose the most appropriate class of lipids from the list below.

        a. sterols [p.41]     b. phospholipids [p.41]     c. fats [pp.40–41]     d. waxes [p.41]

3. _____ Most abundant lipid in the cell membrane

4. _____ Firm, water-repellent lipids found in the plant cuticle

5. _____ Cholesterol belongs to this class

6. _____ May be either saturated or unsaturated

7. _____ Vitamin D is manufactured using this class

8. _____ The major component of adipose tissue

9. _____ Has a ring structure and no fatty acid tails

10. _____ Triglycerides are the most common type of these

11. _____ Contains both polar and nonpolar regions

12. _____ If saturated, these may be either cis or trans

13. _____ Steroid hormones are made from these

14. _____ The richest, most abundant source of energy in vertebrates

## 3.4. PROTEINS—DIVERSITY IN STRUCTURE AND FUNCTION [pp.42–43]

## 3.5. WHY IS PROTEIN STRUCTURE SO IMPORTANT? [pp.44–45]

### Selected Words

*Primary* structure [p.42], *secondary* structure [p.42], "domain" [p.42], *tertiary* structure [p.42], *quaternary* structure [p.42], *glycoprotein* [p.43], *fibrous* proteins [p.43], hemoglobin [p.44], globin [p.44]

### Boldfaced Terms

[p.42] protein _____

_____

[p.42] amino acid _____

_____

[p.42] peptide bond _____

_____

[p.42] polypeptide chain _____

_____

[p.44] denature _____

_____

### Identification

1. Label each of the following as being associated with the primary, secondary, tertiary, or quaternary structure of a polypeptide. Some answers may be used more than once.

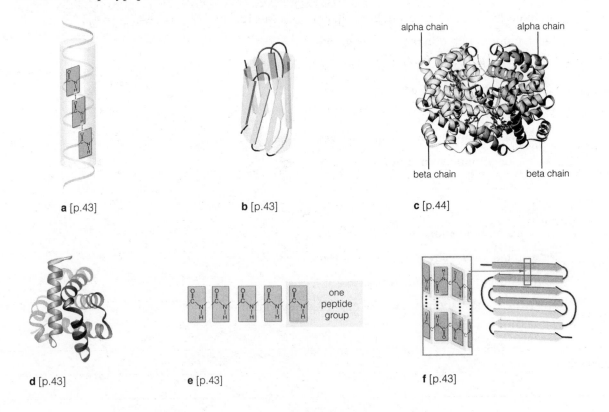

**a** [p.43]  **b** [p.43]  **c** [p.44]

one peptide group

**d** [p.43]  **e** [p.43]  **f** [p.43]

a. _____

b. _____

c. _____

d. _____

e. _____

f. _____

## Matching

Match each term with its description.

2. _____ fibrous protein [p.43]

3. _____ denature [p.44]

4. _____ globin [p.44]

5. _____ glycoproteins [p.43]

6. _____ quaternary structure [p.42]

7. _____ tertiary structure [p.42]

8. _____ domain [p.42]

9. _____ secondary structure [p.42]

10. _____ protein [p.42]

11. _____ primary structure [p.42]

12. _____ polypeptide chain [p.42]

13. _____ peptide bond [p.42]

14. _____ amino acid [p.42]

a. several amino acids linked together
b. two or more polypeptide chains bound together
c. an organic compound containing one or more chains of amino acids
d. a protein with an oligosaccharide attached
e. a type of protein associated with oxygen transport
f. the unraveling of a proteins secondary, tertiary and quaternary structure
g. an organic compound containing an amino and carboxyl group
h. contribute to the structure of cells and tissues
i. the structure of a protein that is based on hydrogen bonds
j. a linear, unique sequence of amino acids
k. the linkage between two amino acids
l. the level at which a protein becomes a working molecule
m. a part of a protein that is organized as a structurally stable unit

## 3.6. NUCLEOTIDES, DNA, AND THE RNAs [pp.46–47]

### Selected Words

"Base pairing" [p.46]

### Boldfaced Terms

[p.46] ATP _____

_____

[p.46] coenzymes _____

_____

[p.46] nucleic acids _____

_____

[p.46] DNA _____

_____

[p.47] RNA _____

_____

## Concept Map

Provide the missing term or terms in the numbered spaces of the concept map below. [pp.46–47]

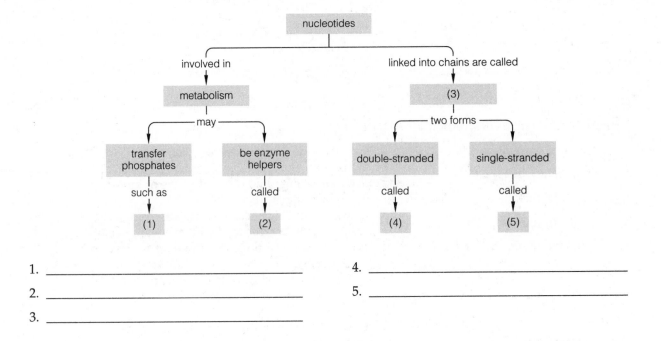

1. _____

2. _____

3. _____

4. _____

5. _____

*TNOW:* For a review of all of the organic compounds, refer to Chapter 3: Co-op Activity 2.8: Review of Organic Compounds.

---

# Self-Quiz

### Choice

For each of the molecules in the list below, choose the appropriate class of organic compounds from the list.

    a. carbohydrates [pp.38–39]
    b. proteins [pp.42–45]
    c. lipids [pp.40–41]
    d. nucleic acids [pp.46–47]

1. _____ glycoproteins

2. _____ enzymes

3. _____ glycogen

4. _____ ATP

5. _____ triglycerides

6. _____ cholesterol

7. _____ DNA and RNA

8. _____ oligosaccharides

9. _____ starch

10. _____ globins

11. Amino acids are linked by _____ to form the primary structure of a protein. [p.42]
    a. hydrogen bonds
    b. peptide bonds
    c. ionic bonds
    d. disulfide bonds

12. Primary, secondary, and tertiary structure are characteristic of the _____. [p.42]
    a. carbohydrates
    b. lipids
    c. proteins
    d. nucleic acids

13. Monosaccharides are joined by a process called _____ to form the oligosaccharides. [p.37]
    a. condensation
    b. hydrolysis
    c. functional group transfer
    d. rearrangement

14. Coenzymes that move hydrogen ions and electrons belong to the class of molecules called the _____. [p.46]
    a. nucleic acids
    b. proteins
    c. carbohydrates
    d. lipids

15. Which of the following may be either saturated or unsaturated? [p.40]
    a. complex carbohydrates
    b. polypeptides
    c. nucleotides
    d. triglycerides

---

## Chapter Objectives/Review Questions

1. Understand the definition of an organic compound. [p.36]
2. Be able to identify the major functional groups. [p.36]
3. Understand the process of condensation and hydrolysis. [p.37]
4. Understand the role of carbohydrates in a cell. [pp.38–39]
5. Be able to define the terms *monosaccharide*, *oligosaccharide*, and *polysaccharide* and give an example of each. [pp.38–39]
6. Understand the use of the major polysaccharides. [p.39]
7. Be able to define the role of the lipids in the cell. [pp.40–41]
8. Understand how the terms *saturated*, *unsaturated*, *cis*, and *trans* relate to fatty acids. [p.40]
9. Understand the role of phospholipids, cholesterol, and waxes in living organisms. [p.41]
10. Give examples of how proteins are used by a cell. [p.42]
11. Understand how amino acids are used as building blocks of proteins. [p.42]
12. Understand the four levels of protein structure. [p.42]
13. Recognize the influence of an incorrect amino acid on protein structure. [p.44]
14. Understand the role of nucleic acids and nucleotides in the cell. [pp.46–47]
15. Understand the importance of DNA and RNA. [pp.46–47]

---

## Chapter Summary

We define cells partly by their ability to build complex carbohydrates and lipids, (1) _____ and nucleic acids. The main building blocks are simple sugars, (2) _____, amino acids, and nucleotides. All of these organic compounds have a backbone of (3) _____ atoms with (4) _____ groups attached.

(5) _____ are the most abundant biological molecules. Simple sugars function as transportable forms of (6) _____ or as quick energy sources. The (7) _____ carbohydrates are structural materials or energy reservoirs.

Complex lipids function as energy (8) _____, structural materials of cell (9) _____, signaling molecules, and waterproofing or lubricating substances.

Structurally and functionally, (10) _____ are the most diverse molecules of life. They include (11) _____, structural materials, signaling molecules, and transporters.

Nucleotides have major metabolic roles and are building blocks of (12) _____ acids. Two kinds of nucleic acids, (13) _____ and RNA, interact as the cell's system of storing, retrieving, and translating information about building (14) _____.

## Integrating and Applying Key Concepts

1. All organic molecules can be used for energy. What is the common link between all organic compounds that makes this possible?
2. Humans can obtain energy from many different food sources. Do you think that this is an advantage or disadvantage in terms of long-term survival and human evolution?
3. We know that the ways in which atoms bond affect molecular shapes. Do you think that the shape of organelles and cells is influenced in the same way by organic molecules? Explain your answer.

# 4

# CELL STRUCTURE AND FUNCTION

## INTRODUCTION

The cell represents the fundamental unit of life, and in this chapter you will begin to explore some of the workings of this miniature factory. It is important that you develop an understanding of the major components of the cell early in the course, as you will keep returning to the material in this chapter in later units.

## FOCAL POINTS

- Figure 4.9 [p.57] illustrates the structure of the plasma membrane and how proteins interact with the membrane.
- Figure 4.11 [p.58] illustrates the general structure of a prokaryotic cell.
- Table 4.2 [p.60] provides the function of the major components of a eukaryotic cell.
- Figure 4.16 [pp.62–63] outlines the operation of the endomembrane system.
- Figure 4.19 [p.65] illustrates the structure of plant and animal eukaryotic cells.

## Interactive Exercises

*Animalcules and Cells Fill'd with Juices* [pp.50–51]

### 4.1. WHAT IS A CELL? [pp.52–53]

### 4.2. HOW DO WE SEE CELLS? [pp.54–55]

*Selected Words*

Hydrophobic [p.52], hydrophilic [p.52], *light microscopes* [p.54], micrographs [p.54], *fluorescence microscope* [p.54], *electron microscopes* [p.54], transmission electron microscopes [p.54], scanning electron microscopes [p.54]

*Boldfaced Terms*

[p.52] cell _____

_____

[p.52] eukaryotic cell _____

_____

[p.52] prokaryotic cells _____

_____

[p.52] plasma membrane _____

_____

[p.52] nucleus _____

_____

[p.52] nucleoid _____

_____

[p.52] cytoplasm _____

_____

[p.52] ribosomes _____

_____

[p.52] lipid bilayer _____

_____

[p.52] surface-to-volume ratio _____

_____

[p.54] cell theory _____

_____

[p.54] wavelength _____

_____

## Matching

Match each of the following terms with its correct description. [p.52]

1. _____ lipid bilayer
2. _____ cell
3. _____ ribosomes
4. _____ eukaryotic cell
5. _____ cytoplasm
6. _____ prokaryotic cell
7. _____ nucleoid
8. _____ nucleus
9. _____ plasma membrane

a. the smallest unit that has the properties of life
b. the cell's outer membrane
c. a region of cytoplasm that contains DNA in prokaryotic cells
d. a cell with internal functional compartments and a nucleus
e. semifluid mixture of water, ions, sugars, and proteins
f. smaller, simpler cells lacking a nucleus
g. location of the DNA in a eukaryotic cell
h. the structural foundation of the cell membranes
i. structures on which proteins are built

*Short Answer*

10. Explain how the concept of "surface-to-volume ratio" influences the size and shape of cells. [pp.52–53]

_____

_____

_____

_____

_____

11. List the three generalizations that constitute the cell theory. [p.54]

_____

_____

_____

_____

*Choice*

For each of the following statements, choose one of the following types of microscopes. Some answers may be used more than once. [p.54]

| | |
|---|---|
| a. light microscope | b. fluorescent microscope |
| c. scanning electron microscope | d. transmission electron microscope |

12. _____ visible light serves as the light source

13. _____ the specimen is coated first with a metal

14. _____ a cell or molecule emits light that is detected by the microscope

15. _____ the electrons detect surface structures

16. _____ electrons pass through the thin sample

17. _____ a light-emitting tracer is added to the object

## 4.3. MEMBRANE STRUCTURE AND FUNCTION [pp.56–57]

## 4.4. INTRODUCING PROKARYOTIC CELLS [pp.58–59]

## 4.5. MICROBIAL MOBS [p.59]

*Selected Words*

Phospholipid [p.56], *mosaic* [p.56], *fluid* [p.56], prokaryote [p.58], capsule [p.58], "sex" pilus [p.58]

*Boldfaced Terms*

[p.56] fluid mosaic model _____

_____

[p.56] transporters _____

_____

[p.56] receptors _____

_____

[p.56] recognition proteins _____

_____

[p.56] adhesion proteins _____

_____

[p.56] communication proteins _____

_____

[p.58] cell wall _____

_____

[p.58] flagella _____

_____

[p.58] pili _____

_____

[p.59] biofilms _____

_____

## Matching

Match each term with its description. [pp.56–57]

1. _____ fluid mosaic model
2. _____ phospholipid
3. _____ adhesion proteins
4. _____ communication proteins
5. _____ transporters
6. _____ receptors
7. _____ "fluid"
8. _____ "mosaic"

a. describes the mixed nature of the components of the cell membrane
b. allow the formation of cytoplasm-to-cytoplasm junctions
c. describes the movement of the phospholipids in the bilayer
d. trigger changes in cell activities
e. may either actively or passively move compounds
f. the description of the organization of cell membranes
g. helps cells stick to one another to form tissues
h. the major component of the cell membrane

## Labeling

Label each of the indicated parts of the prokaryotic cell. [p.58]

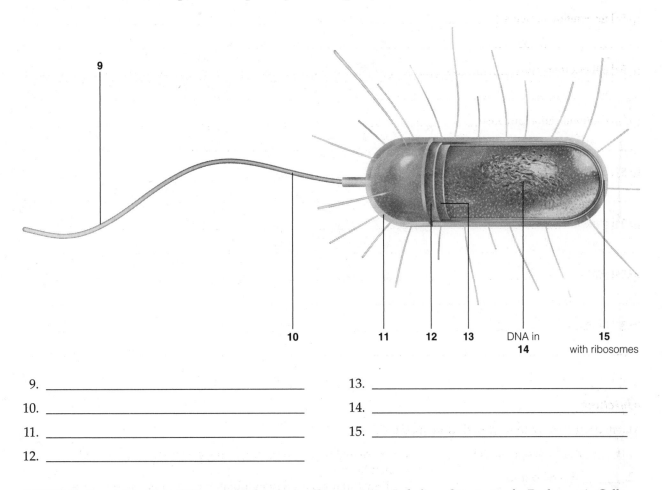

9. _____     13. _____

10. _____     14. _____

11. _____     15. _____

12. _____

*TNOW:* For an animation of this material, refer to Chapter 4: Art Labeling: Structure of a Prokaryotic Cell.

## 4.6. INTRODUCING EUKARYOTIC CELLS [p.60]

## 4.7. THE NUCLEUS [p.61]

## 4.8. THE ENDOMEMBRANE SYSTEM [pp.62-63]

### Selected Words

*Eu-* [p.60], *karyon* [p.60], *cyto-* [p.60], nucleus [p.61], *rough* ER [p.61], *smooth* ER [p.61], *endocytic* [p.63], *exocytic* [p.63]

### Boldfaced Terms

[p.60] organelle _____

_____

[p.61] nuclear envelope _____

_____

[p.61] nucleoplasm _____

_____

[p.61] nucleolus _____

_____

[p.61] chromosome _____

_____

[p.61] chromatin _____

_____

[p.62] endoplasmic reticulum (ER) _____

_____

[p.62] vesicles _____

_____

[p.62] Golgi body _____

_____

[p.63] lysosomes _____

_____

[p.63] central vacuole _____

_____

## Matching

Match each of the following cellular structures to its correct description. [p.60]

1. _____ cytoskeleton
2. _____ ribosomes
3. _____ chloroplast
4. _____ mitochondria
5. _____ vesicles
6. _____ Golgi body
7. _____ endoplasmic reticulum
8. _____ nucleus

a. protects and controls access to the DNA
b. contributes to cell shape, internal organization, and movement
c. routing and modifying new polypeptide chains, synthesizes lipids
d. assembles polypeptide chains
e. modifies new polypeptide chains, sorting, shipping proteins and lipids
f. makes sugars in plants and some protists
g. makes ATP by breaking down glucose
h. transporting, storing, or digesting substances in a cell

*TNOW:* For an animation of this exercise, refer to Chapter 4: Co-op Activity 3.1: Review of Cellular Organelles.

## Complete the Table

9. Complete the table about the eukaryotic nucleus by entering the name of each of the described nuclear components. [p.61]

| Nuclear Component | Description |
|---|---|
| a. | A construction site where large and small subunits of ribosomes are assembled |
| b. | Consists of two lipid bilayers studded with proteins |
| c. | The name for the collection of DNA molecules |
| d. | Semifluid interior of the nucleus |
| e. | One DNA molecule, along proteins attached to it |

## Concept Map

Complete the following concept map on the export of materials using the endomembrane system. [pp.62–63]

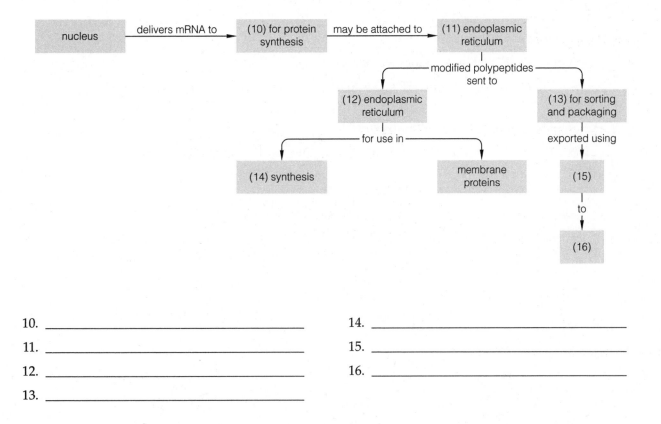

10. _____

11. _____

12. _____

13. _____

14. _____

15. _____

16. _____

## 4.9. MITOCHONDRIA AND CHLOROPLASTS [p.64]

## 4.10. VISUAL SUMMARY OF EUKARYOTIC CELL COMPONENTS [p.65]

### Selected Words

ATP [p.64], thylakoid membrane [p.64], chlorophyll [p.64]

### Boldfaced Terms

[p.64] mitochondrion _____

_____

[p.64] chloroplasts _____

_____

### Choice

For the following questions, choose from one of the following answers. [p.64]

    a. mitochondria    b. chloroplasts    c. both mitochondria and chloroplasts

1. _____ occur only in photosynthetic eukaryotic cells

2. _____ make far more ATP from the same compounds than prokaryotic cells

3. _____ hydrogen ions released from the breakdown of organic compounds accumulate in the inner compartment by operation of transport systems

4. _____ a muscle cell might have a thousand or more

5. _____ use sunlight energy to form ATP and NADPH, which in turn are used in sugar-producing reactions

6. _____ ATP-forming reactions require oxygen

7. _____ two outer membranes surround the stroma, a semifluid interior that bathes an inner membrane

8. _____ resemble bacteria in their size, structure, and biochemistry

9. _____ oxygen accepts the spent electrons and keeps reactions going

10. _____ contains the thylakoid membranes

## Identification

Identify each lettered organelle in the diagrams. Organelles that are found in both plants and animals may have two letters. [p.65]

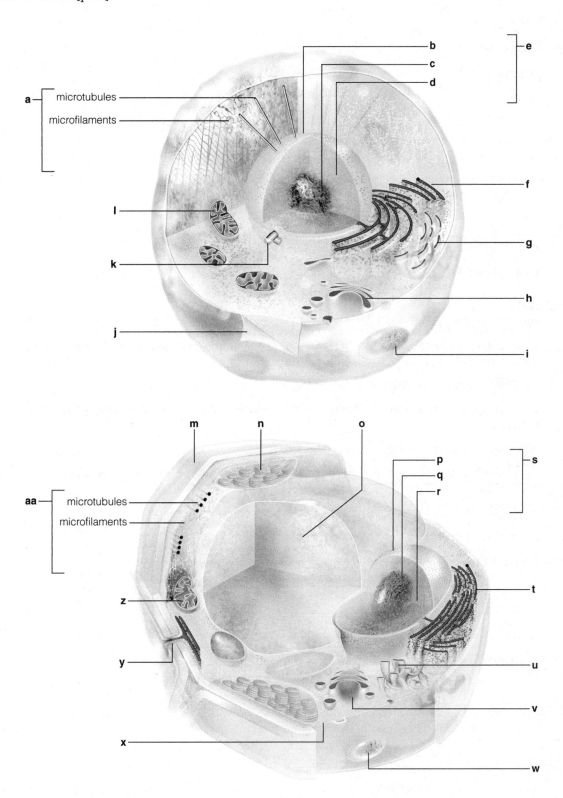

a — microtubules
microfilaments

b
c
d
e
f
g
h
i
j
k
l

m
n
o
p
q
r
s
t
u
v
w
x
y
z
aa — microtubules
microfilaments

11. _____ central vacuole

12. _____ lysosome

13. _____ nucleus

14. _____ Golgi body

15. _____ chloroplast

16. _____ plasma membrane

17. _____ nucleolus

18. _____ smooth ER

19. _____ mitochondrion

20. _____ rough ER

21. _____ plasmodesma

22. _____ cytoskeleton

23. _____ centrioles

24. _____ cell wall

25. _____ nuclear envelope

26. _____ DNA in nucleoplasm

*TNOW*: For an animation of this exercise, refer to Chapter 4: Art Labeling: Structures of a Cell and Chapter 4: Co-op Activity 3.2: Organelle Function.

## 4.11. CELL SURFACE SPECIALIZATIONS [pp.66–67]

## 4.12. THE DYNAMIC CYTOSKELETON [pp.68–69]

### Selected Words

Pectin [p.66], cellulose [p.66], chitin [p.67], collagen [p.67], plasmodesmata [p.67], *tight* junctions [p.67], *adhering* junctions [p.67], *gap* junctions [p.67], tubulin [p.68], actin [p.68], lamins [p.68], kinesins [p.68]

### Boldfaced Terms

[p.66] cell wall _____

_____

[p.66] primary wall _____

_____

[p.66] secondary wall _____

_____

[p.66] lignin _____

_____

[p.66] cuticle _____

_____

[p.66] extracellular matrix _____

_____

[p.67] cell junctions _____

_____

[p.68] cytoskeleton _____

_____

[p.68] microtubules _____

_____

[p.68] microfilaments _____

_____

[p.68] cell cortex _____

_____

[p.68] intermediate filaments _____

_____

[p.68] motor proteins _____

_____

[p.68] flagella _____

_____

[p.68] cilia _____

_____

[p.69] centriole _____

_____

[p.69] basal bodies _____

_____

[p.69] pseudopods _____

_____

## Matching

Match each of the following terms to its correct description. [pp.66–67]

1. _____ cuticle
2. _____ primary wall
3. _____ tight junctions
4. _____ adhering junctions
5. _____ secondary wall
6. _____ extracellular matrix
7. _____ gap junctions
8. _____ plasmodesmata

a. the cell wall made from lignin
b. a protective covering that prevents water loss in plants
c. seals cells together to prevent movement of water-soluble substances
d. a cell wall made of cellulose
e. connect the cytoplasm of adjacent cells
f. anchor cells and the extracellular matrix together
g. nonliving material secreted by cells for support and anchoring
h. connects the primary wall of two plant cells to allow quick movement of substances

*Choice*

For each of the following, choose the appropriate component of the cytoskeleton from the list below. [pp.68–69]

> a. microtubules    b. microfilaments    c. intermediate filaments

9. _____ made from the protein actin

10. _____ used to separate chromosomes prior to cell division

11. _____ lamins are an example

12. _____ made from the protein tubulin

13. _____ make of the cell cortex

14. _____ the major components of flagella and cilia

15. _____ responsible for the action of pseudopods

---

# Self-Quiz

1. As a cell increases in size, its volume increases with the _____ of its diameter, and the surface area increases with the _____ of the diameter. [pp.52–53]
   a. square ; square
   b. cube; cube
   c. cube; square
   d. square; cube

2. The major component of a cell membrane is _____. [p.56]
   a. protein
   b. triglycerides
   c. cellulose
   d. phospholipids

3. Which of the following would not be found in a prokaryotic cell? [p.58]
   a. ribosomes
   b. DNA
   c. cytoplasm
   d. cell membrane
   e. nucleus

4. Most cell membrane functions are carried out by _____ embedded in the bilayer or positioned at one of its surfaces. [p.56]
   a. carbohydrates
   b. proteins
   c. phospholipids
   d. cholesterol

5. Which of the following organelles sorts and ships molecules coming into and leaving the cell? [p.63]
   a. Golgi body
   b. plasma membrane
   c. rough endoplasmic reticulum
   d. ribosomes
   e. smooth endoplasmic reticulum

6. What is the site where the small and large subunits of the ribosome are assembled? [p.60]
   a. nucleus
   b. nucleoplasm
   c. Golgi body
   d. nucleolus

7. The _____ is free of ribosomes and makes the lipid molecules for the cell membranes. [p.62]
   a. smooth endoplasmic reticulum
   b. rough endoplasmic reticulum
   c. vesicles
   d. lysosomes

8. Which of the following part of the cytoskeleton is involved in moving chromosomes prior to cell division? [p.68]
   a. microfilaments
   b. microtubules
   c. intermediate filaments
   d. none of the above

9. Which of the following junctions forms a water-tight seal? [p.67]
   a. gap junctions
   b. plasmodesmata
   c. adhering junctions
   d. tight junctions

10. Which of the following is found in both plant and animal cells and produces the majority of ATP used by the cell? [p.64]
   a. chloroplasts
   b. mitochondria
   c. ribosomes
   d. peroxisomes

## Chapter Objectives/Review Questions

1. Understand the basic components of all cells. [p.52]
2. Understand the concept of the surface-to-volume ratio. [p.52]
3. Know the three principles of the cell theory. [p.54]
4. Recognize the major types of microscopes. [p.54]
5. Understand the principles of the fluid mosaic model. [pp.56–57]
6. Understand the structure of a prokaryotic cell. [p.58]
7. Define the term biofilm and explain its importance to the study of bacteria. [p.59]
8. Understand the function of the major eukaryotic organelles. [p.60]
9. Understand the structure of the nucleus and the role of the major components within the nucleus. [p.61]
10. Know the components of the endomembrane system. [pp.62–63]
11. Be able to trace the movement of a substance through the endomembrane system. [pp.62–63]
12. Distinguish between the mitochondria and chloroplast. [p.64]
13. Identify the major parts of a eukaryotic plant and animal cell. [p.65]
14. Understand the structure of a cell wall and the importance of the extracellular matrix to some organisms. [pp.66–67]
15. Be able to identify the major types of cell junctions. [p.67]
16. Give the function of microfilaments, microtubules, and intermediate filaments. [pp.68–69]

## Chapter Summary

Each cell has a (1) _____, a boundary between its interior and exterior environment. The interior consists of (2) _____ and an innermost region of DNA.

Microscopic analysis supports three generalizations of the cell theory. Each organism consists of one or more (3) _____ and their products, a cell has a capacity for independent life, and each new cell is descended from a living cell.

All cell membranes are mostly a lipid (4) _____— two layers of lipids—and a variety of (5) _____. The proteins have diverse tasks, including control over which (6) _____ substances cross the membrane at any given time.

Archaeans and (7) _____ are prokaryotic cells, which have few, if any, internal membrane-bound compartments. In general, they are the (8) _____ and structurally simplest cells.

Cells of protists, plants, fungi, and animals are (9)_____; they have a (10) _____ and other organelles. They differ in internal parts and surface specializations.

Diverse (11) _____ filaments reinforce a cell's shape and keep its parts organized. As some filaments lengthen and shorten, they move (12) _____ or other structures to new locations.

## Integrating and Applying Key Concepts

1. Cells are typically described as biological factories. Compare a typical eukaryotic cell to a modern factory by relating the function of the organelles to the functions you think are necessary for a factory to run efficiently (i.e., power, waste reduction, etc.).
2. How do you think that the process of evolution has made cells efficient?
3. In the past, some researchers believed that they had found evidence of "life" on the surface of Mars. Critics stated that since these "cells" were only 1/10th the size of a bacteria on Earth, that they were probably not alive. Using your knowledge of cells, explain the limitations that a cell this size would have.
4. Suppose that you wanted to develop a drug that inhibited the ability of a cell to export proteins. What organelles would you target and why?

# 5

# GROUND RULES OF METABOLISM

## INTRODUCTION

Having completed a tour of the cell in Chapter 4, it is now time to look at how cells distinguish themselves from their external environments. In this chapter you will begin to examine the role of enzymes in metabolic reactions and how cells move compounds needed for metabolic pathways across their membranes. Many of the principles introduced in this chapter will be used in later chapters on energy pathways.

## FOCAL POINTS

- Figure 5.6 [p.77] diagrams the effect of an enzyme on a chemical reaction.
- Figure 5.18 [p.87] diagrams the methods by which a cell moves compounds across the cell membrane.
- Figure 5.19 [p.84] illustrates the process of facilitated diffusion.
- Figure 5.20 [p.85] illustrates the process of active transport.
- Figure 5.22 [p.87] diagrams the concept of tonicity.
- Figure 5.24 [p.88] diagrams the relationship between endocytosis and exocytosis.

## Interactive Exercises

*Alcohol, Enzymes, and Your Liver* [pp.72–73]

## 5.1. ENERGY AND THE WORLD OF LIFE [pp.74–75]

## 5.2. ATP IN METABOLISM [p.76]

*Selected Words*

Potential energy [p.74], kinetic energy [p.74], thermal energy [p.74], endergonic reactions [pp.74–75], exergonic reactions [pp.74–75]

*Boldfaced Terms*

[p.74] energy_____

_____

[p.74] first law of thermodynamics_____

_____

[p.74] entropy_____

_____

[p.74] second law of thermodynamics _____

_____

[p.74] kilocalorie _____

_____

[p.75] reactants _____

_____

[p.75] products _____

_____

[p.76] ATP _____

_____

[p.76] phosphorylation _____

_____

[p.76] ATP/ADP cycle _____

_____

## Choice

For each of the following statements, choose the most appropriate law of thermodynamics to which the statement applies. [p.74]

      a. first law of thermodynamics      b. second law of thermodynamics

1. _____ States that entropy is increasing in the universe.

2. _____ Energy cannot be created or destroyed.

3. _____ Energy may be converted between forms.

4. _____ An example is the death and decay of an organism.

5. _____ An example is corn plants producing starch molecules.

6. _____ An example is oak logs burning in a fireplace.

7. _____ Energy disperses spontaneously.

## Matching

Match each of the following terms to the correct description.

8. _____ thermal energy [p.74]

9. _____ kinetic energy [p.74]

10. _____ kilocalorie [p.74]

11. _____ exergonic reaction [p.75]

12. _____ potential energy [p.74]

13. _____ endergonic reaction [p.75]

14. _____ energy [p.74]

a. a capacity to do work
b. the energy of motion
c. reactions that require an input of energy
d. capacity to cause change based on where an object is located or how its parts are arranged
e. reactions that release energy
f. the amount of energy required to heat 1000 grams of water by 1 degree Celsius at standard pressure
g. heat

## Concept Map

Provide the terms indicated in the following concept map of the ATP/ADP cycle. [p.76]

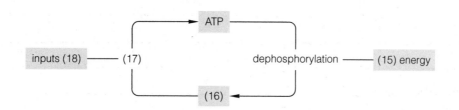

15. _____     17. _____

16. _____     18. _____

## 5.3. ENZYMES IN METABOLISM [pp.76–77]

## 5.4. ENZYMES DON'T WORK ALONE [pp.78–79]

### Selected Words

*Functional group transfers* [p.77], *electron transfers* [p.77], *rearrangements* [p.77], *condensation* [p.77], *cleavage* reactions [p.77], *allosteric* site [p.78], *allo-* [p.78], *steric* [p.78], temperature, [p.78], pH [p.78], salinity [p.78], *free radicals* [p.79]

### Boldfaced Terms

[p.76] activation energy _____

_____

[p.76] enzymes _____

_____

[p.76] substrates _____

_____

[p.76] active sites _____

_____

[p.77] transition state_____

_____

[p.77] induced-fit model_____

_____

[p.78] concentration_____

_____

[p.78] feedback inhibition_____

_____

[p.79] cofactors_____

_____

[p.79] coenzymes _____

_____

[p.79] antioxidant _____

_____

## Labeling

Choose the correct term for each numbered label in the diagram below. [p.77]

1. _____

2. _____

3. _____

4. _____

a. reactants

b. activation energy without enzyme

c. activation energy with enzyme

d. products

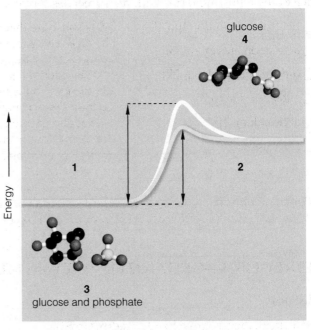

## Complete the Table

For each of the examples of enzyme-mediated reactions in the table below, provide the missing term or description. [p.77]

| Term | Description |
|---|---|
| 5. | Electrons are moved between molecules |
| 6. | Covalent binding of two or more molecules |
| 7. | Splitting of one molecule into two smaller molecules |
| 8. rearrangements | |
| 9. functional group transfer | |

## Matching

Match each of the following terms with its correct description.

10. _____ feedback inhibition [p.78]

11. _____ activation energy [p.76]

12. _____ induced-fit model [p.77]

13. _____ allosteric site [p.78]

14. _____ antioxidants [p.79]

15. _____ active sites [p.76]

16. _____ transition state [p.77]

a. a substrate that is practically complementary to the active site
b. an enzyme that reduces strong oxidizers such as free radicals
c. an activity that causes a change in conditions, which stops the reaction
d. the location where substrates bind in an enzyme
e. a site on an enzyme, other than the active site, that binds regulatory molecules
f. the minimum energy needed to start a reaction
g. when the substrate's bonds reach a breaking point, allowing the reaction to run spontaneously to product

*TNOW:* For animations of enzyme action, refer to Chapter 5: Art Labeling: Mechanism of Allosteric Regulation and Chapter 5: Co-op. Activity 5.1: Effect of pH on Enzyme Activity.

## 5.5. METABOLISM—ORGANIZED, ENZYME-MEDIATED REACTIONS [pp.80–81]

### Selected Words

Biosynthetic [p.80], degradative [p.80], "redox" [p.81]

### Boldfaced Terms

[p.80] metabolic pathways _____

_____

[p.80] autotrophs _____

_____

[p.80] heterotrophs _____

_____

[p.80] chemical equilibrium _____

_____

[p.81] oxidation-reduction reactions _____

_____

[p.81] electron transfer chains _____

_____

## Matching

Match each of the following terms to the correct definition. [pp.80–81]

1. _____ biosynthetic pathway
2. _____ heterotrophs
3. _____ redox reactions
4. _____ degradative pathway
5. _____ electron transfer chains
6. _____ chemical equilibrium
7. _____ autotrophs

a. membrane-bound arrays of enzymes that transfer electrons in a series of steps
b. build organic compounds from smaller molecules
c. obtain carbon from carbon dioxide in the environment
d. the point where the concentration of reactants and products does not change
e. the combination of oxidation and reduction in a single reaction
f. break down organic compounds, releasing energy
g. obtain carbon from organic compounds made by autotrophs

## Choice

For each of the following, choose whether the statement refers to *a* or *b*. [p.80]

a. biosynthetic     b. degradative

8. _____ anabolic reactions
9. _____ the process of aerobic respiration
10. _____ catabolic reactions
11. _____ the process of photosynthesis
12. _____ require an input of energy
13. _____ result in a release of energy

# 5.6. DIFFUSION, MEMBRANES, AND METABOLISM [pp.82–83]

# 5.7. WORKING WITH AND AGAINST GRADIENTS [pp.84–85]

## Selected Words

*Passive transport* [p.83], facilitated diffusion [p.83], *active transport* [p.83], *endocytosis* [p.83], *exocytosis* [p.83]

## Boldfaced Terms

[p.82] concentration gradient_____

_____

[p.82] diffusion_____

_____

[p.82] electric gradient _____

_____

[p.82] pressure gradient _____

_____

[pp.82–83] selective permeability _____

_____

[p.84] passive transport _____

_____

[p.84] active transport _____

_____

[p.84] calcium pumps _____

_____

[p.84] sodium-potassium pump _____

_____

## Matching

Match each of the following statements with the correct term. [pp.82–83]

1. _____ the net movement of like molecules or ions down
   a concentration gradient

2. _____ a difference in electric charge between two adjoining regions

3. _____ allows some substances, but not others, to move in specific ways

4. _____ the difference in the number per unit volume of molecules
   or ions between two adjacent areas

5. _____ a difference in pressure per unit volume between two adjoining
   areas

a. concentration gradient
b. selective permeability
c. pressure gradient
d. electric gradient
e. diffusion

## Short Answer

6. List three factors, besides electric and pressure gradients, that influence the rate of diffusion across a
   membrane. [p.82]

   _____

   _____

   _____

7. List two differences between passive and active transport. [p.83]

   _____

   _____

   _____

## Choice

For each statement, choose the form of transport that is described by the statement.

   a. passive transport [p.84]     b. active transport [pp.84–85]

8. _____ the movement of molecules down a concentration gradient

9. _____ requires an input of energy

10. _____ an example is facilitated diffusion

11. _____ small, non-polar molecules usually move by this mechanism

12. _____ assists the movement of polar molecules across the membrane

## Sequence

The diagrams below represent stages of the facilitated transport of glucose. Place these diagrams in order by writing the letter of the first step in #13, the second in #14, etc. [p.84]

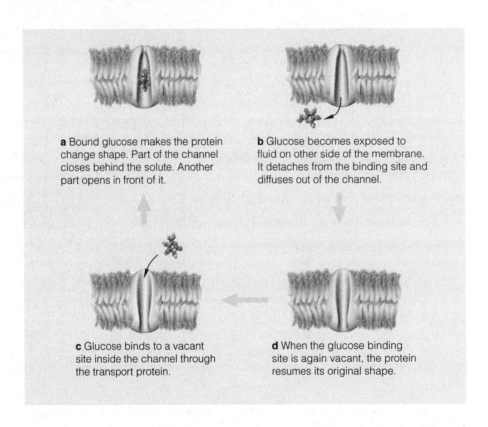

**a** Bound glucose makes the protein change shape. Part of the channel closes behind the solute. Another part opens in front of it.

**b** Glucose becomes exposed to fluid on other side of the membrane. It detaches from the binding site and diffuses out of the channel.

**c** Glucose binds to a vacant site inside the channel through the transport protein.

**d** When the glucose binding site is again vacant, the protein resumes its original shape.

13. _____     15. _____

14. _____     16. _____

## Sequence

The diagrams below represent stages of the active transport of calcium ions. Place these diagrams in order by writing the letter of the first step in #17, the second in #18, etc. [p.85]

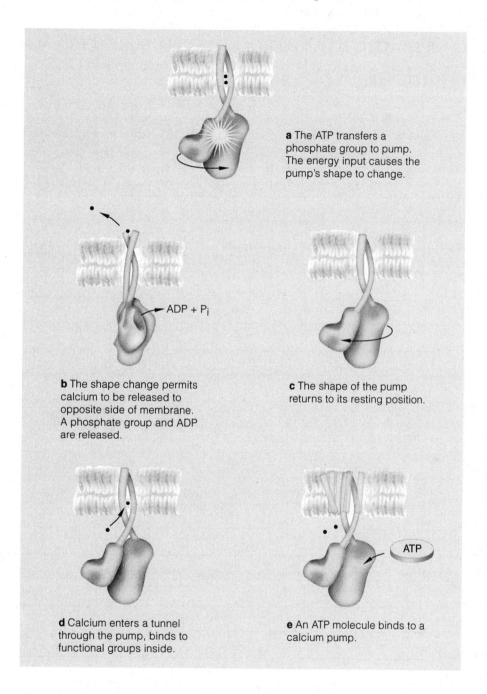

**a** The ATP transfers a phosphate group to pump. The energy input causes the pump's shape to change.

**b** The shape change permits calcium to be released to opposite side of membrane. A phosphate group and ADP are released.

**c** The shape of the pump returns to its resting position.

**d** Calcium enters a tunnel through the pump, binds to functional groups inside.

**e** An ATP molecule binds to a calcium pump.

17. _____

18. _____

19. _____

20. _____

21. _____

*TNOW:* For an exercise on this material, refer to Chapter 5: Art Labeling: Membrane Crossing Mechanisms and Chapter 5: Co-op. Activity 3.5: Experiments with Diffusion and Osmosis.

## 5.8.  WHICH WAY WILL WATER MOVE? [pp.86–87]

## 5.9.  MEMBRANE TRAFFIC TO AND FROM THE CELL SURFACE [pp.88–89]

## 5.10.  NIGHT LIGHTS [p.89]

### Selected Words

*Tonicity* [p.86], *receptor-mediated* endocytosis [p.88], *phagocytosis* [p.88], *bulk-phase* endocytosis [p.88]

### Boldfaced Terms

[p.86] osmosis_____

_____

[p.86] hypotonic fluid _____

_____

[p.86] hypertonic fluid _____

_____

[p.86] isotonic fluids _____

_____

[p.86] hydrostatic pressure_____

_____

[p.86] turgor _____

_____

[p.87] osmotic pressure_____

_____

[p.88] exocytosis_____

_____

[p.88] endocytosis _____

_____

[p.89] bioluminescence _____

_____

## Matching

Match each term with the correct definition.

1. _____ exocytosis [p.88]
2. _____ hydrostatic pressure [p.86]
3. _____ receptor-mediated endocytosis [p.88]
4. _____ bulk-phase endocytosis [p.88]
5. _____ osmotic pressure [p.86]
6. _____ phagocytosis [p.88]
7. _____ endocytosis [p.88]
8. _____ osmosis [p.86]
9. _____ tonicity [p.86]

a. a less selective form of endocytosis
b. the relative concentrations of solutes in two fluids that are separated by a semipermeable membrane
c. the general term for the movement of a vesicle to the cell membrane, releasing its contents into the external environment
d. this term means "cell eating"
e. receptors assist in the process of endocytosis
f. the general term for the taking up of substances near a cell's surface
g. the amount of hydrostatic pressure that can stop the diffusion of water
h. the diffusion of water across a membrane
i. the counterforce to osmosis

## Labeling

Indicate whether the fluid in each beaker is *hypotonic, hypertonic,* or *isotonic* in relation to the fluid in the bag. [p.87]

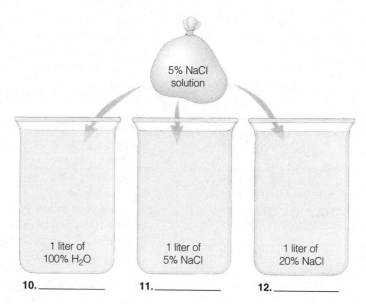

5% NaCl solution

1 liter of 100% H₂O

1 liter of 5% NaCl

1 liter of 20% NaCl

10. _____   11. _____   12. _____

## True–False

If the statement is true, write a T in the blank. If the statement is false, correct it by changing the underlined word(s) and writing the correct word(s) in the answer blank. [pp.86–87]

13. _____ Consider a membrane that has different concentrations of water molecules on each side; the side with fewer solute particles has a <u>higher</u> concentration of water molecules than the fluid on the other side of the membrane.

14. _____ Water moves from an area of <u>high</u> concentration to areas of <u>low</u> concentration.

15. _____ An animal cell placed in a <u>hypertonic</u> solution would swell and perhaps burst.

16. _____ Physiological saline is 0.9% NaCl; red blood cells placed in such a solution will not gain or lose water; therefore, one could correctly state that the fluid in red blood cells is <u>hypertonic</u> in relation to physiological saline.

17. _____ Red blood cells shrivel and shrink when placed in a <u>hypotonic</u> solution.

18. _____ Plant cells placed in a <u>hypotonic</u> solution will swell.

---

# Self-Quiz

1. _____ Essentially, the first law of thermodynamics states that _____. [p.74]
   a. one form of energy cannot be converted into
      another
   b. entropy is increasing in the universe
   c. energy cannot be created or destroyed
   d. energy cannot be converted into matter or
      matter into energy

2. _____ An important principle of the second law
   of thermodynamics states that _____. [p.74]
   a. energy can be transformed into matter, and
      because of this we can get something for
      nothing
   b. energy can be destroyed only during nuclear
      reactions, such as those that occur inside the
      sun
   c. if energy is gained by one region of the
      universe, another place in the universe also
      must gain energy in order to maintain the
      balance of nature
   d. matter tends to become increasingly more
      disorganized

3. _____ In _____ pathways, carbohydrates,
   lipids, and proteins are broken down in step-
   wise reactions that lead to products of lower
   energy. [pp.80–81]
   a. phosphorylation
   b. biosynthetic
   c. endergonic
   d. degradative

4. _____ Any substance (for example, NAD$^+$) that
   accepts electrons is _____. [p.81]
   a. oxidized
   b. a catalyst
   c. reduced
   d. a substrate

5. _____ With regard to major function, NAD$^+$ is
   classified as _____. [p.79]
   a. enzymes
   b. a phosphate carrier
   c. a coenzyme
   d. an end product of a metabolic pathway

6. _____ The majority of enzymes are _____.
   [p.78]
   a. acids
   b. proteins
   c. lipids
   d. nucleic acids

7. _____ Which is *not* true of enzyme behavior?
   [pp.78–79]
   a. Enzyme shape may change during
      catalysis.
   b. The active site of an enzyme orients its
      substrate molecules, thereby facilitating
      interaction of their reactive parts.
   c. All enzymes have an active site where
      substrates are temporarily bound.
   d. An individual enzyme can catalyze a wide
      variety of different reactions.

8. _____ An enzyme that has an allosteric
   site _____. [p.78]
   a. has an active site where substrate molecules
      bind and another site that binds with
      intermediate or end-product molecules
   b. is an important energy-carrying
      nucleotide
   c. carries out either oxidation reactions or
      reduction reactions but not both
   d. raises the activation energy of the chemical
      reaction it catalyzes

9. _____ Which of the following processes would
   require an expenditure of energy by the cell?
   [p.83]
   a. diffusion
   b. osmosis
   c. facilitated diffusion
   d. calcium pump

10. _____ Enzymes function by _____ the activa-
    tion energy of a reaction. [p.78]
    a. maintaining
    b. reversing
    c. raising
    d. lowering

# Chapter Objectives/Review Questions

1. Understand the two laws of thermodynamics and how they relate to biological systems. [p.74]
2. Define the terms entropy and kilocalorie. [p.74]
3. Be able to diagram the ATP-ADP cycle. [p.76]
4. Recognize the importance of enzymes to biological systems. [p.76]
5. Understand how enzymes act as catalysts. [p.76]
6. Understand the major types of enzyme-mediated reactions. [p.77]
7. Understand how the environmental and allosteric mechanisms control enzyme function. [pp.78–79]
8. Recognize the relationship between coenzymes and cofactors in enzyme function. [p.79]
9. Be able to distinguish between a biosynthetic and degradative metabolic pathway. [p.80]
10. Understand the difference between heterotrophs and autotrophs with regard to how they obtain their carbon. [p.80]
11. Understand the principles of a redox reaction. [p.81]
12. Understand the process of diffusion and the factors that influence the rate of diffusion across a membrane. [pp.82–83]
13. Distinguish between active and passive transport. [pp.84–85]
14. Define osmosis and tonicity. [p.86]
15. Be able to predict the tonicity relationship of two environments when given solute concentrations. [p.87]
16. Understand the difference between hydrostatic and osmotic pressure. [pp.86–87]
17. Be able to distinguish between endocytosis and exocytosis. [p.88]
18. Know the types of endocytosis. [pp.88–89]
19. Understand the process of membrane cycling. [pp.88–89]

# Chapter Summary

(1) _____ tends to disperse spontaneously. Each time energy is transferred, some of it disperses. Organisms maintain their complex (2) _____ only by continuously harvesting energy.

(3) _____ couples metabolic reactions that release usable energy with reactions that require energy input. On their own, those reactions proceed too slowly to sustain life. (4) _____ increase reaction rates. (5) _____ factors influence enzyme activity.

(6) _____ are energy-driven sequences of enzyme-mediated reactions. They concentrate, convert, or dispose of materials in cells. Controls over (7) _____ that govern key steps in these pathways can shift cell activities fast.

(8) _____ gradients drive the directional movements of ions and molecules into and out of cells. Transport (9) _____ raise and lower water and (10) _____ concentrations across the plasma membrane and internal cell membranes. Other mechanisms move larger cargo across the plasma membrane.

Knowledge about (11) _____, including how enzymes work, can help you interpret what you see in (12) _____.

# Integrating and Applying Key Concepts

1. A piece of dry ice left sitting on a table at room temperature vaporizes. As the dry ice vaporizes into $CO_2$ gas, does its entropy increase or decrease? Tell why you answered as you did.

# 6

# WHERE IT STARTS—PHOTOSYNTHESIS

## INTRODUCTION

The process of photosynthesis is responsible for capturing the energy necessary to sustain life on earth. While initially it may seem to be a complicated series of chemical processes, in reality it represents a series of logical steps in which the energy in sunlight is first captured and converted to chemical energy, and then this chemical energy is used to manufacture sugar. Pay close attention to the diagrams in this chapter, as they will assist you in understanding the basic flows of chemicals and ions in the photosynthetic pathway.

## FOCAL POINTS

- Figure 6.4 [p.96] illustrates the concept of an absorption spectrum.
- Figure 6.6c [p.97] provides an overview of the inputs and outputs of the stages of photosynthesis.
- Figure 6.8 [p.99] diagrams the noncyclic cycle of photosynthesis in relation to the location in the chloroplast.
- Figure 6.9 [p.100] diagrams the differences between the cyclic and noncyclic light-dependent reactions.
- Figure 6.10 [p.101] illustrates the Calvin-Benson cycle.
- Figure 6.11 [p.102] illustrates the variations to the Calvin-Benson cycle developed by C4 and CAM plants.

## Interactive Exercises

*Sunlight and Survival* [pp.92–93]

### 6.1. SUNLIGHT AS AN ENERGY SOURCE [pp.94–95]

### 6.2. EXPLORING THE RAINBOW [p.96]

*Selected Words*

*Hetero-* [p.92], *photo*autotroph [p.92], *chemo*autotroph [p.92]

*Boldfaced Terms*

[p.94] wavelength _____

_____

[p.94] photoautotrophs _____

_____

[p.94] pigment _____

_____

[p.95] chlorophyll a _____

_____

[p.95] carotenoids _____

_____

[p.96] absorption spectrum _____

_____

## Matching

Match each of the following terms with the correct definition.

1. _____ carotenoids [p.95]
2. _____ absorption spectrum [p.96]
3. _____ chemoautotroph [p.92]
4. _____ chlorophyll a [p.95]
5. _____ photoautotroph [p.94]
6. _____ pigment [p.94]
7. _____ wavelength [p.94]

a. extract energy and carbon from inorganic material in the environment
b. a unit that is used to identify light
c. the main photosynthetic pigment of plants
d. use light energy to build organic molecules from inorganic materials
e. first discovered by Engelmann, this is used to identify the wavelengths of light absorbed by a pigment
f. an accessory pigment that extends the range of photosynthesis
g. an organic compound that selectively absorbs light of specific wavelengths

## Short Answer

8. In an absorption spectrum, such as the one shown in Figure 6.4 [p.96], what do the peaks represent?

_____

_____

_____

## 6.3. OVERVIEW OF PHOTOSYNTHESIS [p.97]

## 6.4. LIGHT-DEPENDENT REACTIONS [pp.98–99]

### Boldfaced Terms

[p.97] chloroplast _____

_____

[p.97] stroma _____

_____

[p.97] thylakoid membrane _____

_____

[p.97] photosystems _____

_____

[p.97] light-dependent reactions _____

_____

[p.97] light-independent reactions _____

_____

[p.98] electron transport chains _____

_____

[p.98] photolysis _____

_____

## Labeling

Provide the missing term for each numbered item in the diagram below. [p.97]

1. _____

2. _____

3. _____

4. _____

5. _____

6. _____

7. _____

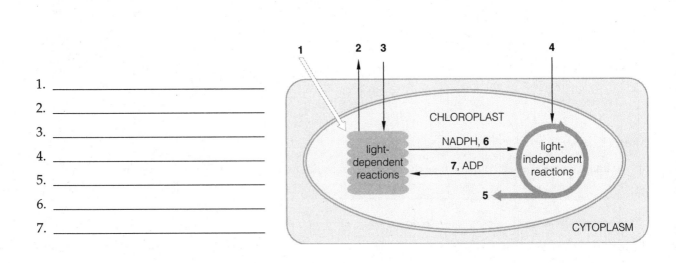

*TNOW:* For an interactive version of this exercise, refer to Chapter 6: Art Labeling: Overview of Photosynthesis.

## True–False

If the statement is true, place a T in the blank. If the statement is false, correct the underlined word to make the statement correct. [p.97]

8. _____ The chloroplast is the plant organelle that specializes in <u>photosynthesis</u>.

9. _____ During the <u>light-independent reactions</u>, light energy is converted to chemical energy.

10. _____ The light-harvesting pigments of photosynthesis are found in the <u>stroma</u> of a chloroplast.

11. _____ The sugars of photosynthesis are manufactured in the <u>thylakoids</u>.

12. _____ Glucose synthesis occurs in the <u>light-independent reactions</u> of photosynthesis.

## Labeling

Provide the missing term for each numbered item in the diagram below. [p.99]

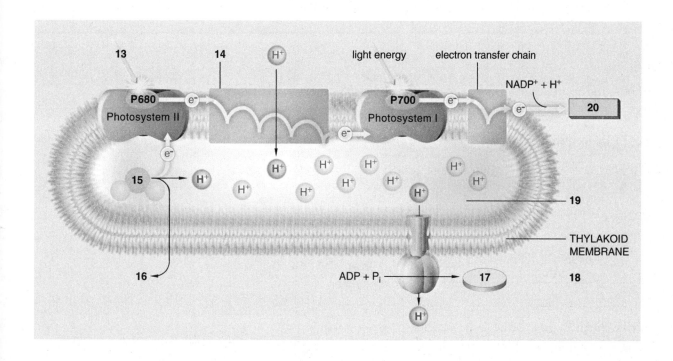

13. _____    17. _____

14. _____    18. _____

15. _____    19. _____

16. _____    20. _____

*TNOW:* For a comparison of photosynthesis and cellular respiration, refer to Chapter 6: Co-op. Activity 4.5: Comparing Photosynthesis with Cell Respiration.

## 6.5. ENERGY FLOW IN PHOTOSYNTHESIS [p.100]

## 6.6. LIGHT-INDEPENDENT REACTIONS: THE SUGAR FACTORY [p.101]

### Selected Words

Photophosphorylation [p.100]

## Boldfaced Terms

[p.101] Calvin-Benson cycle _____

_____

[p.101] carbon fixation _____

_____

[p.101] rubisco _____

_____

## Choice

Choose the appropriate form of electron flow for each of the following statements. [p.100]

   a. cyclic     b. noncyclic     c. both cyclic and noncyclic

1. _____ forms ATP

2. _____ forms NADPH

3. _____ lost electrons are replaced by photolysis

4. _____ electrons proceed through an electron transfer chain

5. _____ is used when NADPH supplies are plentiful in the cell

6. _____ oxygen is released as a byproduct.

7. _____ the first to evolve in photosynthetic organisms

## Identification

Identify each numbered part of the diagram, using abbreviations where appropriate. Complete the exercise by entering the letter of the correct statement in the parentheses following each label. [p.101]

8. _____ ( )

9. _____ ( )

10. _____ ( )

11. _____ ( )

12. _____ ( )

13. _____ ( )

14. _____ ( )

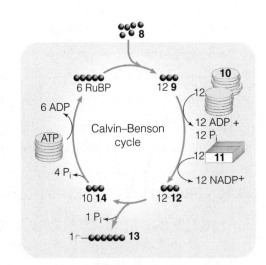

a. Donates hydrogens and electrons to PGA molecules
b. Gets phosphate groups from ATP, priming them for synthesis reactions that generates RuBP
c. Rubisco attaches the carbon atom of $CO_2$ to RuBP, which starts the Calvin-Benson cycle; this compound forms from splitting of an intermediate compound
d. Gets a phosphate group from ATP, plus hydrogen and electrons
e. In the air spaces inside a leaf; diffuses into photosynthetic cells
f. Formed by a combination of two of twelve PGAL molecules
g. Donates phosphate groups to PGA

## 6.7.  ADAPTATIONS: DIFFERENT CARBON-FIXING PATHWAYS [p.102]

## 6.8.  A BURNING CONCERN [p.103]

### Selected Words

*Mesophyll* cells [p.102], *bundle-sheath* cells [p.102]

### Boldfaced Terms

[p.102] stomata_____

_____

[p.102] C3 plants _____

_____

[p.102] photorespiration _____

_____

[p.102] C4 plants _____

_____

### Choice

For each of the following, choose the most appropriate form of adaptation from the list below. [p.102]

        a.  C3 plants     b.  C4 plants     c.  CAM plants

1. _____ Succulents such as cacti, which have juicy, water-storing tissues, and thick surface layers.

2. _____ The three-carbon PGA, the first stable intermediate of the Calvin-Benson cycle.

3. _____ Four-carbon oxaloacetate is the first to form.

4. _____ They open their stomata and fix carbon at night.

5. _____ Two types of photosynthetic cells occur in these plants.

6. _____ Evolved independently over millions of years in many lineages.

7. _____ With $CO_2$ levels rising, these plants may again have the edge.

8. _____ Do not grow well in hot, dry climates without steady irrigation.

9. _____ These plants survive prolonged droughts by closing stomata even at night.

10. _____ In the first cell type, the carbon-fixing enzyme ignores $O_2$, no matter how much there is; $CO_2$ is deposited in bundle sheath cells.

11. _____ $CO_2$ is fixed by repeated turns of a type of C4 cycle, then it enters the Calvin-Benson cycle the next day.

12. _____ An example is the basswood tree.

13. _____ These plants lose less water and make more sugar than the C3 plants can when days are dry.

14. _____ When oxygen level is high, rubisco uses oxygen instead of $CO_2$ in an alternate reaction that yields only one molecule of PGA.

15. _____ Common in grasses, corn, and other plants that evolved in the tropics.

## Short Answer

16. Briefly explain the primary cause of global warming. [p.103]

_____

_____

_____

_____

# Self-Quiz

___ 1. Plants need _____ and _____ as the raw materials to carry on photosynthesis. [p.97]
   a. oxygen; water
   b. oxygen; $CO_2$
   c. $CO_2$; $H_2O$
   d. sugar; water
   e. $CO_2$; NADPH

___ 2. Chlorophyll is found _____. [p.97]
   a. on the outer chloroplast membrane
   b. inside the mitochondria
   c. inside the stroma
   d. in the thylakoid membranes
   e. on the membranes of the stroma

___ 3. The cyclic route functions mainly to _____. [p.99]
   a. make NADPH
   b. make PGAL
   c. set up conditions for making ATP
   d. regenerate RuBP
   e. break down $CO_2$

___ 4. During the noncyclic pathway, the origin of the electrons passed to $NADP^+$ is _____. [pp.98–99]
   a. $CO_2$
   b. glucose
   c. sunlight
   d. water
   e. ATP

___ 5. Two products of the light-dependent reactions are required to drive the chemistry of the light-independent reactions. They are _____ and _____. [p.97]
   a. $O_2$; NADPH
   b. $CO_2$; $H_2O$
   c. $O_2$; inorganic phosphate
   d. ATP; NADPH
   e. RuBP; PGA

___ 6. ATP syntheses are involved in _____. [p.98]
   a. creating electron transport chains
   b. photosystems
   c. attaching inorganic phosphates to ADP during the cyclic route
   d. the noncyclic pathway
   e. allowing H+ flow to attach inorganic phosphate to ADP

___ 7. The Calvin-Benson cycle uses _____. [p.101]
   a. $CO_2$
   b. hydrogen and electrons from NADPH
   c. phosphate group transfers from ATP
   d. an enzyme known as rubisco
   e. all of the above

___ 8. In the Calvin-Benson cycle, RuBP is regenerated by _____. [p.101]
   a. rearrangement of PGAL molecules
   b. the combination of two PGAL molecules
   c. PGA receiving phosphate groups from ATPs
   d. glucose entering other reactions that form carbohydrates
   e. rubisco attaching carbon atoms to RuBP

___ 9. C4 plants have an advantage in hot, dry conditions because _____. [p.102]
   a. their leaves are covered with thicker wax layers than those of C3 plants
   b. their stomates open wider than those of C3 plants, thus cooling their surfaces

   c. special leaf cells possess a means of capturing $CO_2$ even in stress conditions
   d. they carry on normal photosynthesis at very high oxygen levels
   e. they can carry on carbon fixation at night

___10. The process of securing carbon from the environment by incorporating it into a stable organic compound is called _____ [p.102]
   a. fluorescence
   b. photolysis
   c. photorespiration
   d. the C4 cycle
   e. carbon fixation

---

## Chapter Objectives/Review Questions

1. Recognize the difference between photoautotrophs and chemoautotrophs. [p.92]
2. Understand how light is captured by photoautotrophs and the types of pigments involved in the process. [pp.94–95]
3. Recognize the importance of Englemann's experiments with absorption spectrums. [p.96]
4. Understand the basic flow of molecules and compounds in the process of photosynthesis. [p.97]
5. Understand the concept of a photosystem. [p.97]
6. Understand the difference between the light-dependent and light-independent reactions. [p.97]
7. Be able to identify the differences between the cyclic and noncyclic electron flows in photosynthesis. [pp.98–100]
8. Understand the principles of the Calvin-Benson cycle. [p.101]
9. Recognize the evolutionary adaptations presented by the C3, C4, and CAM plants. [p.102]
10. Recognize the relationship among photosynthesis, carbon concentrations, and global warming. [p.103]

---

## Chapter Summary

A great one-way flow of energy through the world of life starts after (1) _____ and other pigments absorb the energy of visible light from the sun's rays. In plants, some bacteria, and many protists, that energy ultimately drives the synthesis of (2) _____ and other carbohydrates.

Photosynthesis proceeds through two stages in the (3) _____ of plants and in many types of protists. First, pigments embedded in a membrane inside the chloroplast capture (4) _____ energy, which is then converted to (5) _____ energy. Next, that chemical energy drives the synthesis of (6) _____.

In the first stage of (7) _____, sunlight energy is converted to the chemical bond energy of (8) _____. The coenzyme (9) _____ forms in a pathway that also releases oxygen.

The second stage is the (10) _____ part of photosynthesis. Enzymes speed the assembly of sugars from carbon and oxygen atoms, both obtained from (11) _____. The reactions use ATP and NADPH that form in the first stage of (12) _____. ATP delivers energy, and NADPH delivers (13) _____ and hydrogens to the reaction sites. Details of the reactions vary among organisms.

Photosynthesis by (14) _____ removes carbon dioxide from the atmosphere; (15) _____ by all organisms puts it back in. Human activities have disrupted this balance, and so have contributed to (16) _____.

---

## Integrating and Applying Key Concepts

1. A scientist once proposed that human cells be injected with chloroplasts extracted from living plant cells. Speculate about that possibility in terms of changes in human anatomy, physiology, and behavior. Could this be successful? Why or why not?
2. Scientists are developing a spinach "battery" to regenerate cell phones. What principles of photosynthesis are these scientists using to generate the flow of electrons? What would the inputs to the device be?

# 7

# HOW CELLS RELEASE CHEMICAL ENERGY

## INTRODUCTION

In this chapter, you will explore the mechanisms by which the energy stored in glucose may be released to form ATP. The chapter examines the differences between aerobic and anaerobic pathways, as well as the use of alternative energy sources by the human body.

## FOCAL POINTS

- Figure 7.3 [p.109] provides an overview of aerobic respiration.
- Figure 7.6 [p.113] summarizes the reactions of acetyl-CoA formation and the Krebs cycle.
- Figure 7.8 [p.115] illustrates the relationship and inputs/outputs of the steps of aerobic respiration.
- Figure 7.12 [p.119] illustrates how the body can use non-glucose energy sources for energy.

## Interactive Exercises

*When Mitochondria Spin Their Wheels* [pp.106–107]

### 7.1. OVERVIEW OF CARBOHYDRATE BREAKDOWN PATHWAYS [pp.108–109]

*Selected Words*

Luft's syndrome [p.106], Friedreich's ataxia [p.106], *anaerobic* [p.108]

*Boldfaced Terms*

[p.108] fermentation pathways _____

_____

[p.108] aerobic respiration _____

_____

[p.108] pyruvate_____

_____

[p.109] glycolysis _____

_____

[p.109] Krebs cycle

[p.109] electron transfer phosphorylation

## Choice

For each statement, choose the most appropriate class of reactions from the list below. [pp.108–109]

       a. aerobic respiration    b. anaerobic reactions    c. both a and b

1. _____ Starts with a conversion of glucose to pyruvate.

2. _____ The first form of pathways to evolve on the earth.

3. _____ Requires oxygen.

4. _____ Fermentation pathways are an example.

5. _____ Electron transfer phosphorylation is an example.

6. _____ The Krebs cycle is an example.

7. _____ The process ends in the cytoplasm.

8. _____ The process ends in the mitochondria.

9. _____ Produces ATP.

## Labeling

Label each of the numbered components of the diagram below. [p.109]

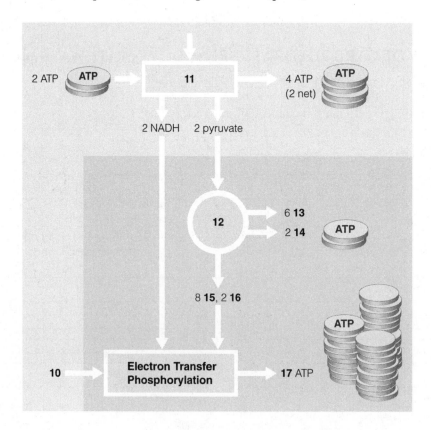

10. _____     14. _____

11. _____     15. _____

12. _____     16. _____

13. _____     17. _____

*TNOW:* For an animation of the material in this section, refer to Chapter 7: Art Labeling: Overview of Aerobic Respiration.

## 7.2. GLYCOLYSIS—GLUCOSE BREAKDOWN STARTS [pp.110–111]

## 7.3. SECOND STAGE OF AEROBIC RESPIRATION [pp.112–113]

### Selected Words

"Glycolysis" [p.110], *glyk-* [p.110], *-lysis* [p.110], Krebs cycle [p.112]

### Boldfaced Terms

[p.110] substrate-level phosphorylation _____

_____

### Concept Map

Complete the following concept map on the process of glycolysis. [pp.110–111]

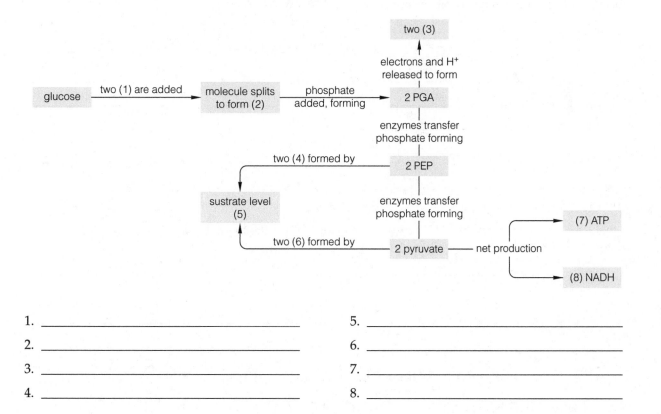

1. _____     5. _____

2. _____     6. _____

3. _____     7. _____

4. _____     8. _____

## Labeling

Provide the correct information for each numbered label in the diagram below.
*Note:* Some numbers are present at multiple locations in the diagram. [pp.112–113]

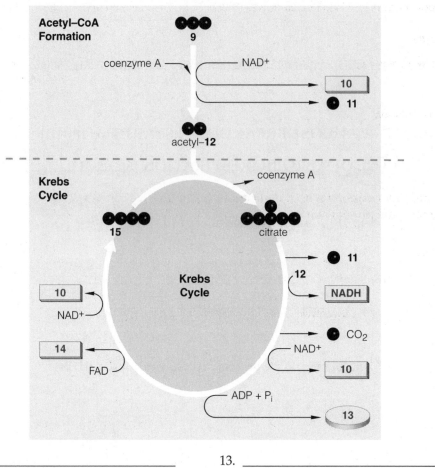

9. _____
10. _____
11. _____
12. _____

13. _____
14. _____
15. _____

## Short Answer

16. For each glucose that enters glycolysis, how many of each of the following are produced during acetyl-CoA formation and the Krebs cycle? [p.113]

a. $CO_2$ _____

b. NADH _____

c. $FADH_2$ _____

d. ATP _____

## 7.4. AEROBIC RESPIRATION'S BIG ENERGY PAYOFF [pp.114–115]

## 7.5. ANAEROBIC ENERGY-RELEASING PATHWAYS [pp.116–117]

# 7.6. THE TWITCHERS [p.117]

## Selected Words

Electron transport chains [p.114], *Saccharomyces cerevisiae* [p.116], *Lactobacillus acidophilus* [p.117]

## Boldfaced Terms

[p.116] alcoholic fermentation _____

_____

[p.117] lactate fermentation _____

_____

## Labeling

Provide the correct information for each numbered label in the diagram below.
*Note:* Some numbers are present at multiple locations in the diagram. [pp.114–115]

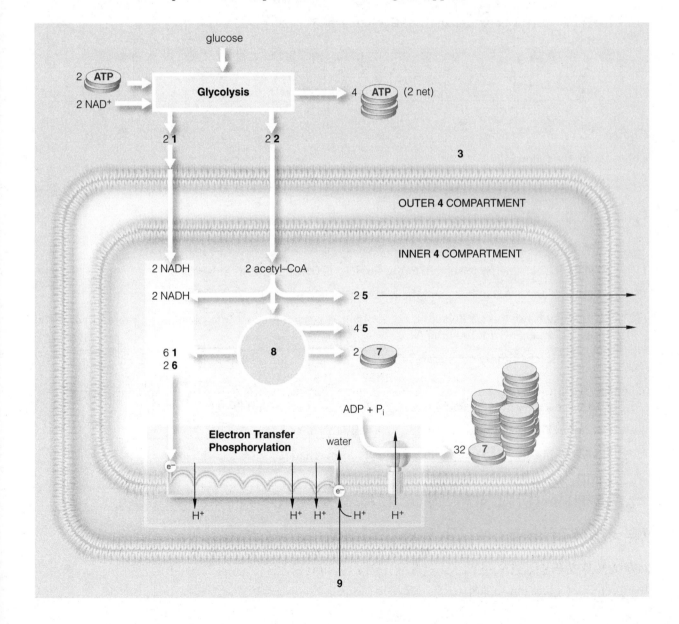

1. _____   6. _____

2. _____   7. _____

3. _____   8. _____

4. _____   9. _____

5. _____

## Short Answer

10. For each glucose that enters aerobic respiration, how many of each of the following are produced by the end of the pathway? [p.115]

    a. $CO_2$ _____

    b. NADH _____

    c. $FADH_2$ _____

    d. net ATP _____

11. What is the purpose of oxygen in the electron transport chain? [pp.114–115]

    _____

    _____

    _____

## Choice

Choose from the following:

       a. alcoholic fermentation [p.116]     b. lactate fermentation [p.117]
       c. applies to both types of fermentation

12. ___ Glycolysis is the first stage.

13. ___ Yeasts, such as *Saccharomyces cerevisea*, are famous for their use of this pathway.

14. ___ Yields enough energy to sustain many single-celled anaerobic organisms.

15. ___ Pyruvate and NADH form, and the net energy yield is 2 ATP.

16. ___ Muscle cells use this pathway but not for long; diverting glucose into this pathway would waste too much of its energy for too little ATP.

17. ___ The final steps simply regenerate $NAD^+$, the coenzyme that assists the breakdown reactions.

18. ___ Each pyruvate molecule that formed in glycolysis is converted to the intermediate acetaldehyde.

19. ___ These reactions do not completely degrade glucose to $CO_2$ and $H_2O$.

20. ___ *Lactobaccilus* and some other bacteria use only this anaerobic pathway.

21. ___ NADH transfers electrons and hydrogen to form ethanol.

## 7.7. ALTERNATIVE ENERGY SOURCES IN THE BODY [pp.118–119]

## 7.8. REFLECTIONS ON LIFE'S UNITY [p.120]

### Selected Words

Insulin [p.118], glucagon [p.118], triglycerides [p.118]

*Choice*

For each of the following, choose the appropriate class of molecules to which the statement corresponds. [pp.118–119]

        a. fats     b. proteins     c. glucose

1. _____ Enzymes covert these to glycerol and fatty acids.

2. _____ Carbon backbone may become acetyl-CoA, pyruvate or an intermediate of the Krebs cycle.

3. _____ The hormones insulin and glucagons are involved.

4. _____ Ammonia is an end-product of this nutrient's metabolism.

5. _____ Carbon backbone becomes acetyl-CoA.

6. _____ Glycerol is converted to PGAL by the liver.

7. _____ Pancreas and liver are involved in regulating blood levels.

8. _____ Too much glucose in the bloodstream ends up as this.

*Short Answer*

9. Explain how the process of photosynthesis and aerobic respiration are linked. [p.120]

_____

_____

_____

10. Explain why life on Earth would not be possible without the photosynthesizers. [p.120]

_____

_____

_____

*TNOW:* For additional material on this topic, refer to Chapter 7: Co-op Activity 4.5: Comparing Photosynthesis with Cell Respiration.

# Self-Quiz

___ 1. Anaerobic respiration pathways begin and end in the _____. [p.108]
  a. thylakoid of a chloroplast
  b. plasma membrane of the cell
  c. inner membrane of the mitochondrion
  d. cytoplasm
  e. outer compartment of the mitochondrion

___ 2. The first energy-releasing step in glycolysis splits activated glucose into two molecules of _____. [p.109]
  a. $NAD^+$
  b. PGAL
  c. ATP
  d. pyruvate
  e. PEP

___ 3. The *net* energy yield of glycolysis is _____ ATP molecules [pp.108–111]
  a. three
  b. four
  c. thirty-two
  d. two
  e. eight

___ 4. Pyruvate is regarded as the end-product of _____. [pp.108–111]
  a. glycolysis
  b. acetyl-CoA formation
  c. fermentation
  d. electron transfer phosphorylation
  e. the Krebs cycle

___ 5. During which of the following phases of the energy-releasing pathways is ATP produced directly by substrate-level phosphorylation? [p.111]
  a. glycolysis
  b. electron transfer chains
  c. the Krebs cycle
  d. formation of acetyl-Co A
  e. all of the above

___ 6. Select the process by which NADH and FADH$_2$ transfer electrons along a chain of acceptors to oxygen so as to form water and set up conditions for producing a large number of ATP molecules. [pp.114–115]
  a. glycolysis
  b. the Krebs cycle
  c. acetyl-CoA formation
  d. fermentation pathways
  e. electron transfer phosphorylation

___ 7. The total number of ATP molecules produced by the complete degradation of one glucose molecule is often 36. Which of the following is not correct? [p.115]
  a. glycolysis produces 2 ATP
  b. Krebs cycle produces 2 ATP
  c. electron transfer phosphorylation produces 32 ATP
  d. all of the above are correct

___ 8. Which of the following is incorrect regarding the fermentation reactions? [pp.116–117]
  a. the first step of the reaction is always glycolysis
  b. their purpose is to regenerate NAD+
  c. they are only used by prokaryotic organisms
  d. some produce alcohol as an end-product
  e. all of the above are correct

___ 9. Of the following, which one is not available as an alternative energy source for the human body? [pp.118–119]
  a. fats
  b. glycogen
  c. proteins
  d. carbohydrates
  e. all of the above are available

___ 10. The basic reason that we have an accumulation of oxygen in our atmosphere is _____. [p.120]
  a. that oxygen produces more oxygen
  b. cyclic photophosphorylation
  c. that oxygen is released in electron transfer phosphorylation
  d. the noncyclic pathway
  e. the production of ATP

## Chapter Objectives/Review Questions

1. Explain the similarities between Friedreich's ataxia and Luft's syndrome. [pp.106–107]
2. Understand the basic differences between aerobic and anaerobic pathways. [pp.108–109]
3. Know where the aerobic and anaerobic pathways start and stop in the cell. [p.108]
4. Know the basic formula of aerobic respiration. [p.109]
5. Understand the process of glycolysis, including the inputs and outputs of the process. [pp.110–111]
6. Understand the processes of acetyl-CoA formation and the Krebs cycle, including the inputs and outputs of the process. [pp.112–113]
7. Understand the process of electron transfer phosphorylation, including the inputs and outputs of the process. [pp.114–115]
8. Understand the role of oxygen in the Krebs cycle and electron transfer phosphorylation. [pp.112–115]
9. Understand the relationship between glycolysis, acetyl-CoA formation, the Krebs cycle, and electron transfer phosphorylation [p.115]
10. Understand the purpose and significance of lactose and alcoholic fermentation reactions. [pp.116–117]
11. Understand the alternative energy sources that a cell may use to generate ATP. [pp.118–119]
12. Understand the relationship between photosynthesis and aerobic respiration. [p.120]

# Chapter Summary

All organisms produce ATP by various (1) _____ pathways that extract (2) _____ energy from glucose and other organic compounds. (3) _____ respiration yields the most ATP from each glucose molecule. In eukaryotes, it is completed inside the (4) _____.

(5) _____ is the first stage of aerobic respiration and of anaerobic routes, such as fermentation pathways. As enzymes break down glucose to pyruvate, the coenzyme (6) _____ picks up (7) _____ and hydrogen atoms. The net energy yield is (8) _____ ATP.

Aerobic respiration has two more stages. In the (9) _____ and a few reactions before it, pyruvate is broken down to (10) _____, and many coenzymes pick up electrons and hydrogen atoms. In electron transfer (11) _____, coenzymes deliver electrons to transfer chains that set up conditions for ATP formation. (12) _____ accepts electrons at the end of the chains.

Fermentation pathways start with (13) _____. Substances other than oxygen are the final (14) _____ acceptor. Compared with aerobic respiration, the net yield of ATP is (15) _____.

Molecules other than (16) _____ are common energy sources. Different pathways convert (17) _____ and proteins to substances that may enter glycolysis or the Krebs cycle.

Life shows unity in its molecular and (18) _____ organization and in its dependence on a (19) _____ flow of energy.

# Integrating and Applying Key Concepts

1. What problems may humans and other organisms experience if their mitochondria were defective?
2. Evaluate this statement: Every atom in your body has first passed through the cell(s) of a photosynthetic organism.
3. Some diets propose that you eat 100% protein. Based on the information presented in this chapter, would that be a good idea?
4. What is the evolutionary benefit of anaerobic respiration in human muscles? Why not only use anaerobic respiration?

# 8

# HOW CELLS REPRODUCE

## INTRODUCTION

For single celled organisms to reproduce, or for the tissues of a multicellular organism to repair themselves, cell division must occur. This chapter discusses the mechanisms involved in cell division, or mitosis. Appropriate control of cellular division is critical to the health of the organism. The final portion of this chapter is concerned with the regulation of cell growth and the association of improperly regulated growth with the development of cancer in humans.

## FOCAL POINTS

- Figure 8.3 [p.126] depicts a eukaryotic chromosome before and after DNA replication.
- Figure 8.5 [p.128] illustrates and defines the parts of the eukaryotic cell cycle.
- Figure 8.7 [pp.130–131] defines the stages of mitosis and shows two chromosomes in a eukaryotic cell undergoing this process.
- Figure 8.8 [p.132] compares the process of cytokinesis in plant and animal cells.
- Figure 8.12 [p.135] compares benign and malignant tumors.

## Interactive Exercises

*Henrietta's Immortal Cells* [pp.124–125]

### 8.1. OVERVIEW OF CELL DIVISION MECHANISMS [pp.126–127]

### 8.2. INTRODUCING THE CELL CYCLE [pp.128–129]

*Selected Words*

HeLa cells [p.124]; *somatic cells* [p.126]; histone proteins [p.127]; G1, S, and G2 [p.128]; *2n* [p.129]; XY [p.129]; prophase, metaphase, anaphase, and telophase [p.129]; "bipolar" spindle [p.129]

*Boldfaced Terms*

[p.126] mitosis _____

_____

[p.126] meiosis _____

_____

[p.126] sister chromatids _____

_____

[p.127] nucleosome _____

_____

[p.127] centromere _____

_____

[p.128] cell cycle _____

_____

[p.128] interphase _____

_____

[p.129] chromosome number _____

_____

[p.129] diploid _____

_____

[p.129] bipolar spindle _____

_____

## Short Answer

1. Describe the origin and the significance of HeLa cells. [pp.124–125]

_____

_____

_____

_____

## Matching

Match each term with its description.

2. _____ centromere [p.127]

3. _____ chromosome [p.126]

4. _____ somatic cells [p.126]

5. _____ sister chromatids [p.126]

6. _____ mitosis and meiosis [p.126]

7. _____ nucleosome [p.127]

8. _____ gametes [p.126]

9. _____ mitosis [p.126]

10. _____ meiosis [p.126]

11. _____ histones [p.127]

12. _____ prokaryotic fission [p.126]

a. sex cells, such as sperm and eggs that function in sexual reproduction
b. composed of a chromosome and its copy attached to each other until late in the division process
c. cell divisions that give rise to gametes or spores; functions in sexual reproduction
d. an organizational unit composed of a histone-DNA spool
e. chromosomal proteins; appear like beads on a string
f. in bacterial cells only, the basis of asexual reproduction
g. body cells
h. a pronounced constricted region on a chromosome; docking site for certain microtubules that take part in nuclear division
i. cell divisions by which an organism grows, replaces worn-out cells, and repairs tissues; used by single-celled organisms for asexual reproduction
j. composed of a molecule of DNA and its proteins
k. the nuclear division mechanisms

## Identification

Identify the stage in the cell cycle indicated by the numbers in each diagram. [p.128]

13. _____

14. _____

15. _____

16. _____

17. _____

18. _____

19. _____

20. _____

21. _____

22. _____

23. _____

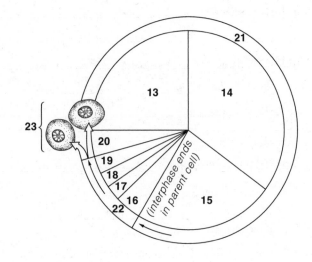

## Matching

Link each time span identified below with the most appropriate number in the preceding diagram. [p.128]

24. _____ interval following DNA replication; cell prepares to divide

25. _____ the complete period of nuclear division, followed by cytoplasmic division (a separate event)

26. _____ interval of cell growth, when DNA replication is completed (chromosomes duplicated)

27. _____ interval of cell growth, before DNA duplication (chromosomes unduplicated)

28. _____ usually the longest part of a cell cycle

29. _____ interval of cytoplasmic division

30. _____ period that includes G1, S, G2

## 8.3. A CLOSER LOOK AT MITOSIS [pp.130–131]

## 8.4. DIVISION OF THE CYTOPLASM [pp.132–133]

### Selected Words

Threadlike form [p.130], thick, compact, rod-shaped forms [p.130], two barrel-shaped centrioles [p.130], aligned midway between the poles [p.130], *pull* it to a spindle pole [p.131], spindle itself lengthens [p.131], decondense [p.131], two nuclei form [p.131]

### Boldfaced Terms

[p.130] prophase _____

_____

[p.130] centrosome _____

_____

[p.131] metaphase _____

_____

[p.131] anaphase _____

_____

[p.131] telophase _____

_____

[p.132] contractile ring _____

_____

[p.133] cell plate formation _____

_____

## Identification

Identify each mitotic stage shown. Select from *late prophase, transition to metaphase (prometaphase), cell at interphase, metaphase, early prophase, telophase, interphase-daughter cells,* and *anaphase*. Complete the exercise by entering the letter of the correct phase in the parentheses.

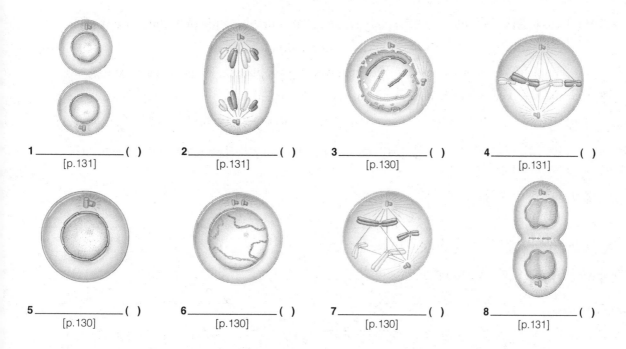

1 _____ ( )
[p.131]

2 _____ ( )
[p.131]

3 _____ ( )
[p.130]

4 _____ ( )
[p.131]

5 _____ ( )
[p.130]

6 _____ ( )
[p.130]

7 _____ ( )
[p.130]

8 _____ ( )
[p.131]

a. Attachments between two sister chromatids of each chromosome break; the two are now separate chromosomes, which microtubules move to opposite spindle poles.
b. Microtubules penetrate the nuclear region and collectively form the bipolar spindle apparatus; microtubules become attached to the two sister chromatids of each chromosome.
c. The DNA and its associated proteins have started to condense.
d. All the chromosomes are now fully condensed and lined up at the equator of the fully formed microtubule spindle; chromosomes are now in their most tightly condensed form.
e. The cell duplicates its DNA and prepares for nuclear division.

f. Two daughter cells have formed; each is diploid with two of each type of chromosome, just like the parent cell's nucleus.

g. Chromosomes continue to condense; new microtubules become assembled; they move one of two centriole pairs toward the opposite end of the cell; the nuclear envelope begins to break up.

h. There are two clusters of chromosomes, which decondense; patches of new membrane fuse to form a new nuclear envelope; mitosis is complete.

## Choice

Choose from the following two choices to fill in the blanks below:

a. plant cells    b. animal cells

_____ 9. formation of a cell plate [p.133]

_____10. as the microfilament ring shrinks in diameter, it pulls the cell surface inward. [p.132]

_____11. cellulose is deposited inside the sandwich; in time, these deposits will form two cell walls. [p.133]

_____12. the primary wall of the growing cell is still thin; new material is deposited on it. [p.133]

_____13. a cleavage furrow [p.132]

_____14. a ring of actin and myosin filaments attached to the plasma membrane contracts. [p.132]

_____15. contractions continue; the cell is pinched in two [p.132]

_____16. as mitosis ends, vesicles cluster at the spindle equator. [p.132]

_____17. cellulose deposits build up at the plate; in time, they are thick enough to form a cross-wall through the cell. [p.133]

## 8.5. WHEN CONTROL IS LOST [pp.134–135]

### Selected Words

*Inhibit* mitosis [p.134], *stimulate* mitosis [p.134], oncogene products [p.134], *benign* [p.135], *malignant* neoplasm [p.135], *metastasis* [p.135]

### Boldfaced Terms

[p.134] kinases _____

_____

[p.134] growth factors _____

_____

[p.135] tumor _____

_____

[p.135] neoplasms _____

_____

## Short Answer

1. List the four characteristics displayed by *all* cancer cells. [p.135]

_____

_____

_____

# Self-Quiz

___ 1. The replication of DNA occurs _____.
   [p.128]
   a. between the growth phases of interphase
   b. immediately before prophase of mitosis
   c. during prophase of mitosis
   d. during anaphase of mitosis
   e. in daughter cells during telophase

___ 2. Each histone-DNA spool is a single structural unit called a _____. [p.127]
   a. kinetochore
   b. motor protein
   c. centromere
   d. nucleosome
   e. gene

___ 3. In the cell life cycle of a particular cell,
   _____. [p.128]
   a. mitosis occurs immediately after G1
   b. G2 precedes S
   c. G1 precedes S
   d. mitosis and S precede G1
   e. S occurs immediately prior to mitosis

___ 4. The correct order of the stages of mitosis is
   _____. [pp.130–131]
   a. prophase, metaphase, telophase, anaphase
   b. telophase, anaphase, metaphase, prophase
   c. telophase, prophase, metaphase, anaphase
   d. anaphase, prophase, telophase, metaphase
   e. prophase, metaphase, anaphase, telophase

___ 5. During _____, sister chromatids of each chromosome are separated from each other, and those former partners, now chromosomes, are moved toward opposite spindle poles. [pp.130–131]
   a. prophase
   b. metaphase
   c. anaphase
   d. telophase
   e. transition to metaphase

___ 6. In the process of cytokinesis, cleavage furrows are associated with _____ cell division, and cell plate formation is associated with _____ cell division. [pp.132–133]
   a. animal; animal
   b. plant; animal
   c. plant; plant
   d. animal; plant

___ 7. In eukaryotic cells, which of the following can occur during mitosis? [p.130]
   a. Two mitotic divisions to maintain the parental chromosome number
   b. The replication of DNA
   c. A long growth period
   d. The disappearance of the nuclear envelope and nucleolus
   e. A G2 event

___ 8. *Diploid* refers to _____. [p.129]
   a. having two chromosomes of each type in somatic cells
   b. twice the parental chromosome number
   c. half the parental chromosome number
   d. having one chromosome of each type in somatic cells
   e. The chromosome conditions in a sex cell

___ 9. Somatic cells are _____ cells; germ cells are _____ cells. [p.126]
   a. meiotic; body
   b. body; body
   c. meiotic; meiotic
   d. body; meiotic

___10. If a parent cell has sixteen chromosomes and undergoes mitosis, the resulting cells will have _____ chromosomes. [p.131]
a. 64
b. 32
c. 16
d. 8
e. 4

___11. "Microtubules penetrate the nuclear region. Collectively, they form a bipolar spindle apparatus." These sentences describe the _____ of mitosis. [p.130]
a. prophase
b. metaphase
c. prometaphase
d. anaphase
e. telophase

___12. During _____, sister chromatids of each chromosome are separated from each other, and those former partners, now chromosomes, are moved toward opposite poles. [p.133]
a. prophase
b. metaphase
c. anaphase
d. telophase
e. prometaphase

## Chapter Objectives/Review Questions

1. Define *HeLa cells*; explain their origin. [pp.124–125]
2. Mitosis and meiosis refer to the division of the cell's _____. [p.126]
3. Define *somatic cells, gametes, prokaryotic fission, chromosome, nucleosome, sister chromatids*, and *centromere*. [pp.126–127]
4. List and describe, in order, the various activities occurring in the eukaryotic cell life cycle. [p.128]
5. Interphase of the cell cycle consists of G1, _____, and G2. [p.128]
6. S is the time in the cell cycle when _____ replication occurs. [p.128]
7. Each species has a characteristic _____ number, the sum of all chromosomes in cells of a given type. [p.129]
8. The _____ number of chromosomes in your body cells is 46. [p.129]
9. Describe, in detail, the cellular events occurring in the prophase, metaphase, anaphase, and telophase of mitosis. [pp.130–131]
10. Describe the number and movements of centrioles in the cell division of some cells. [pp.130–131]
11. Compare and contrast cytokinesis as it occurs in plant and animal cell division; use the following concepts: *cleavage furrow, microfilaments at the cell's midsection*, and *cell plate formation*. [pp.132–133]
12. Be able to define *growth factors, kinases, oncogenes, neoplasms, benign, malignant neoplasm, basal cell carcinoma, squamous cell carcinoma, malignant melanoma, metastasis*. [pp.134–135]
13. List the four characteristics displayed by *all* cancer cells. [p.135]

## Chapter Summary

Individuals of a (1) _____ have a characteristic number of (2) _____ in each of their cells. The chromosomes differ in length and (3) _____, and they carry different portions of the cell's (4) _____ information. Division mechanisms parcel out the information to each (5) _____ cell, along with enough (6) _____ for that cell to start up its own operation.

A cell (7) _____ starts when a daughter cell forms and ends when that cell completes its own division. A typical cell cycle goes through interphase, (8) _____, and cytoplasmic division. In interphase, a cell increases its (9) _____ and number of components, and copies its (10) _____.

Mitosis divides the (11) _____, not the cytoplasm. It has four sequential stages: prophase, metaphase, (12) _____, and telophase. A (13) _____ spindle forms. It moves the cell's duplicated chromosomes into two parcels, which end up in two genetically identical (14) _____.

After (15) _____ division, the (16) _____ divides and typically puts a nucleus in each daughter cell. The cytoplasm of an (17) _____ cell is simply pinched into two. In (18) _____ cells, a cross-wall forms in the cytoplasm and divides it.

Built-in mechanisms monitor and control the (19) _____ and (20) _____ of cell division. On rare occasions, the surveillance mechanisms fail, and cell division becomes uncontrollable. (21) _____ formation and (22) _____ are the outcome.

## Integrating and Applying Key Concepts

1. Runaway cell division is characteristic of cancer. Imagine the various points of the mitotic process that might be sabotaged in cancerous cells in order to halt their multiplication. Then, try to imagine how one might discriminate between cancerous and normal cells to guide those methods of sabotage most effective in combating cancer.
2. The side effects of cancer chemotherapy in humans are very significant and most often result in nausea and hair loss. Think about cell division, and more importantly, the regulation of cell division, and hypothesize why these two particular side effects are so common.

# 9

# MEIOSIS AND SEXUAL REPRODUCTION

## INTRODUCTION

Survival and success of a species is certainly possible without sexual reproduction. The bacteria as a group are probably the most successful on the planet and reproduce strictly by asexual means. Many species, however, such as humans, are dependent upon sexual reproduction for the continuation of the species. This chapter details the cellular mechanisms that allow for sexual reproduction and the subsequent variation that accompanies this process. It is this variation that accounts for the success of sexually reproducing species, such as humans.

## FOCAL POINTS

- Figure 9.5 [pp.142–143] animates the stages of meiosis in an animal cell.
- Figure 9.6 [p.144] illustrates how crossing over may occur in the first stage of meiosis.
- Figure 9.7 [p.145] shows how random alignment during metaphase 1 contributes to diversity.
- Figure 9.8 [p.146] diagrams the general life cycle of both plants and animals.
- Figure 9.11 [pp.148–149] compares the processes of mitosis and meiosis.

## Interactive Exercises

*Why Sex?* [pp.138–139]

### 9.1. INTRODUCING ALLELES [p.140]

### 9.2. WHAT MEIOSIS DOES [pp.140–141]

### 9.3. VISUAL TOUR OF MEIOSIS [pp.142–143]

*Boldfaced Terms*

[p.140] asexual reproduction _____

_____

[p.140] genes _____

_____

[p.140] clones _____

_____

[p.140] sexual reproduction _____

_____

[p.140] allele _____

_____

[p.140] gametes _____

_____

[p.140] germ cells _____

_____

[p.140] chromosome number _____

_____

[p.140] diploid number (2n) _____

_____

[p.141] homologous chromosomes _____

_____

[p.141] haploid number (*n*) _____

_____

[p.141] sister chromatids _____

_____

## Choice

Choose from the following two options to fill in the blanks below:

a. asexual reproduction      b. sexual reproduction

_____ 1. Offspring inherit different combinations of alleles. [p.140]

_____ 2. One parent alone produces offspring. [p.140]

_____ 3. Each offspring inherits the same number and kinds of genes as its parent. [p.140]

_____ 4. Commonly involves two parents [p.140]

_____ 5. The production of "clones" [p.140]

_____ 6. Involves meiosis, formation of gametes, and fertilization [p.140]

_____ 7. Offspring are genetically identical copies of the parent. [p.140]

_____ 8. Produces the variation in traits that is the foundation of evolutionary change [p.140]

_____ 9. Change can occur only by rare mutations. [p.140]

_____ 10. The first cell of a new individual has pairs of genes on pairs of chromosomes; usually, one of each pair is maternal and the other paternal in origin. [p.140]

## Dichotomous Choice

Circle one of two possible answers given in parentheses in each statement.

11. (Meiosis/Mitosis) divides chromosomes into separate parcels not once but twice prior to cell division. [p.141]

12. Sperm cells and eggs are also known as (germ/gamete) cells. [p.140]

13. (Haploid/Diploid) germ cells produce haploid gametes. [p.141]

14. (Haploid/Diploid) cells possess pairs of homologous chromosomes. [p.140]

15. (Meiosis/Mitosis) produces haploid cells. [p.141]

16. Identical alleles are found on (homologous chromosomes/sister chromatids). [p.141]

17. Two attached DNA molecules are known as (sister chromatids/homologous chromosomes). [p.141]

18. Two attached sister chromatids represent (one/two) chromosome(s). [p.141]

19. One pair of duplicated chromosomes would be composed of (two/four) chromatids. [p.141]

20. With meiosis, chromosomes proceed through (one/two) divisions to yield four haploid nuclei. [p.143]

21. Germ cell DNA is replicated during (prophase I of meiosis I/interphase preceding meiosis I). [p.141]

22. During meiosis I, each duplicated (chromosome/chromatid) lines up with its partner, homologue to homologue, and then the partners are moved apart from one another. [p.141]

23. Cytoplasmic division following meiosis I results in two (diploid/haploid) daughter cells. [p.141]

24. During (meiosis I/meiosis II), the two sister chromatids of each chromosome are separated from each other, and each sister chromatid is now a separate chromosome. [p.141]

25. If human body cell nuclei contain 23 pairs of homologous chromosomes, each resulting gamete will contain (23/46) chromosomes. [p.141]

## Identification

Identify each of the meiotic stages shown below by entering the correct stage of either meiosis I or meiosis II in the blank. Choose from *prophase I, metaphase I, anaphase I, telophase I, prophase II, metaphase II, anaphase II,* and *telophase II*. Complete the exercise by matching and entering the letter of the correct stage description in the parentheses.

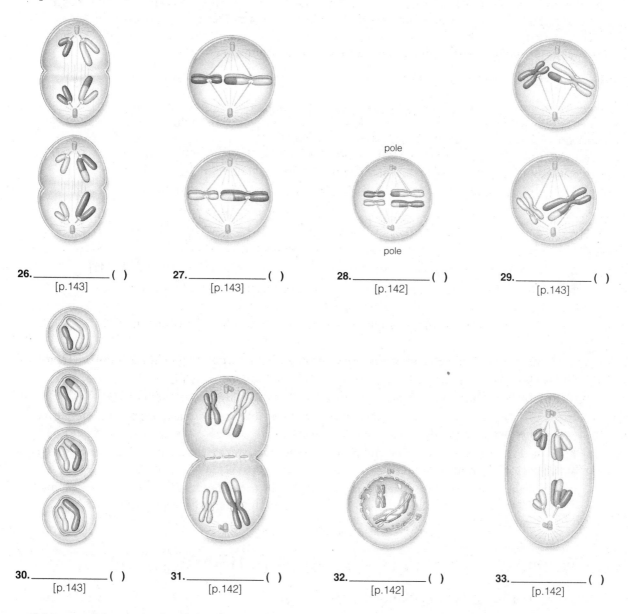

26. _____ (  )
    [p.143]

27. _____ (  )
    [p.143]

28. _____ (  )
    [p.142]

29. _____ (  )
    [p.143]

30. _____ (  )
    [p.143]

31. _____ (  )
    [p.142]

32. _____ (  )
    [p.142]

33. _____ (  )
    [p.142]

a. Motor proteins projecting from the microtubules move the chromosomes and spindle poles apart; chromosomes are tugged into position midway between the spindle poles; the spindle becomes fully formed.

b. Microtubules have moved one member of the centriole pair to the opposite spindle pole in each of two daughter cells; microtubules attach to the chromosomes, which motor proteins slide toward the spindle's equator.

c. Four daughter nuclei form; when cytoplasmic division is over, each daughter is haploid (n); all chromosomes are in the unduplicated state.

d. Some microtubules extend from the spindle poles and overlap at its equator; these lengthen and push the poles apart; other microtubules extending from the poles shorten and pull each chromosome away from its homologous partner; these motions move the homologous partners to opposite poles.

e. At the end of interphase, chromosomes are duplicated and in threadlike form; now they start to condense; each pairs with its homologue, and the two usually cross over and swap segments; newly forming spindle microtubules become attached to each chromosome.

f. In each of two cells, the microtubules, motor proteins, and duplicated chromosomes interact, which positions all of the duplicated chromosomes midway between the two spindle poles.

g. The cytoplasm of the cell divides at some point; there are now two haploid (n) cells with one of each type of chromosome that was present in the parent (2n) cell; all chromosomes are still in the duplicated state.

h. Attachment between the sister chromatids of each chromosome breaks, and the two are moved to opposite spindle poles; each former "sister" is now a chromosome on its own.

## Matching

Following careful study of the major stages of meiosis shown in Figure 9.5 of the main text [pp.142–143], apply what you have learned by matching the following written descriptions with the appropriate sketch. Assume that the cell in this model initially has one pair of homologous chromosomes (one from a paternal source and one from a maternal source) and crossing over does not occur. Complete the exercise by indicating the diploid (2n = 2) or haploid (n = 1) chromosome number of the cell in the parentheses.

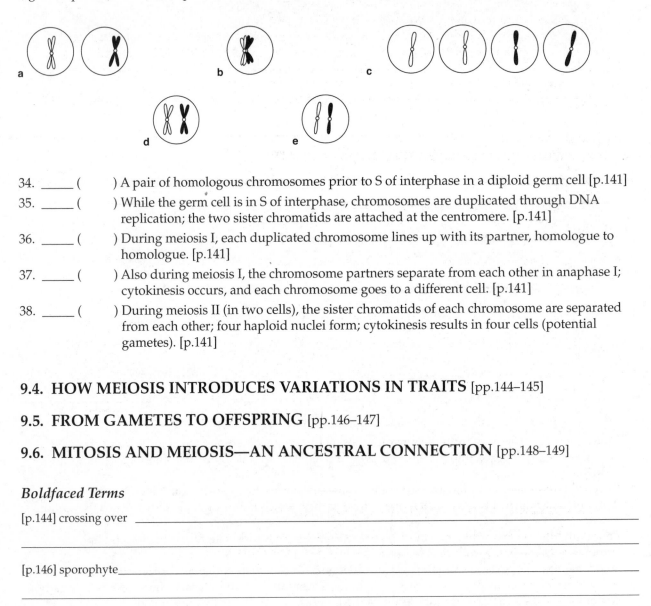

34. _____ (          ) A pair of homologous chromosomes prior to S of interphase in a diploid germ cell [p.141]

35. _____ (          ) While the germ cell is in S of interphase, chromosomes are duplicated through DNA replication; the two sister chromatids are attached at the centromere. [p.141]

36. _____ (          ) During meiosis I, each duplicated chromosome lines up with its partner, homologue to homologue. [p.141]

37. _____ (          ) Also during meiosis I, the chromosome partners separate from each other in anaphase I; cytokinesis occurs, and each chromosome goes to a different cell. [p.141]

38. _____ (          ) During meiosis II (in two cells), the sister chromatids of each chromosome are separated from each other; four haploid nuclei form; cytokinesis results in four cells (potential gametes). [p.141]

## 9.4. HOW MEIOSIS INTRODUCES VARIATIONS IN TRAITS [pp.144–145]

## 9.5. FROM GAMETES TO OFFSPRING [pp.146–147]

## 9.6. MITOSIS AND MEIOSIS—AN ANCESTRAL CONNECTION [pp.148–149]

### Boldfaced Terms

[p.144] crossing over _____

_____

[p.146] sporophyte_____

_____

[p.146] gametophyte _____

_____

[p.146] sperm _____

_____

[p.146] egg _____

_____

[p.146] fertilization _____

_____

## Matching

Match each term with its description.

1. _____ $2^{23}$ [p.145]
2. _____ paternal chromosomes [p.144]
3. _____ prophase I [p.142]
4. _____ nonsister chromatids [p.141]
5. _____ crossing over [p.144]
6. _____ function of meiosis [p.140]
7. _____ maternal chromosomes [p.144]
8. _____ metaphase I [p.142]

a. random positioning of maternal and paternal chromosomes at the spindle equator
b. a time of much gene shuffling
c. reduction of the parental chromosomal number by half
d. break at the same places along their length and then exchange corresponding segments—including genes
e. 23 chromosomes inherited from your mother
f. combinations of maternal and paternal chromosomes possible in gametes from one germ cell
g. 23 chromosomes inherited from your father
h. breaks up old combinations of alleles and puts new ones together in pairs of homologous chromosomes

## Choice

Choose from the following three options to fill in the blanks below: [p.146]

a. animals     b. plants     c. animals and plants

_____ 9. Meiosis results in the production of haploid spores.

_____ 10. Germ cells give rise to gametes.

_____ 11. Meiosis results in the production of haploid gametes.

_____ 12. A diploid germ cell becomes an oocyte, or immature egg.

_____ 13. A spore develops into a haploid gametophyte.

_____ 14. A haploid gametophyte produces haploid gametes by mitosis.

_____ 15. Three polar bodies are formed that do not function as gametes.

_____ 16. A haploid spore divides by mitosis to produce a gametophyte.

_____ 17. A gamete-producing body and a spore-producing body develop during the life cycle.

_____ 18. The egg receives most of the cytoplasm.

## Sequence

Arrange the following entities in correct order of development, entering "1" by the stage that appears first and "4" by the stage that completes the process of spermatogenesis. Complete the exercise by indicating in parentheses whether each cell is *n* or *2n* in the parentheses.

19. _____ ( ) primary spermatocyte [p.147]

20. _____ ( ) sperm [p.147]

21. _____ ( ) spermatid [p.147]

22. _____ ( ) secondary spermatocyte [p.147]

## Matching

Choose the most appropriate answer to match with each oogenesis concept.

23. ___ primary oocyte [p.147]

24. ___ oogonium [p.147]

25. ___ secondary oocyte [p.147]

26. ___ ovum and three polar bodies [p.147]

27. ___ first polar body [p.147]

a. the cell in which synapsis, crossing over, and recombination occur
b. a cell that is equivalent to a diploid germ cell
c. a haploid cell formed after division of the primary oocyte that does not form an ovum at second division
d. haploid cells, but only one of which functions as an egg
e. a haploid cell formed after division of the primary oocyte, the division of which forms a functional ovum

## Short Answer

29. List the various mechanisms that contribute to the huge number of new gene combinations that may result from fertilization. [pp.146–147]

_____

_____

_____

_____

# Self-Quiz

___ 1. Which of the following is not a characteristic of sexual reproduction? [p.140]
  a. sperms and eggs
  b. zygotes
  c. meiosis
  d. clones
  e. germ cells

___ 2. Which of the following does *not* occur in prophase I of meiosis? [p.142]
  a. chromosome duplication
  b. a cluster of four chromatids
  c. homologues pairing tightly
  d. crossing over
  e. chromosomes condensing

___ 3. "The two identical DNA molecules and their proteins stay attached at the centromere." This is a description of _____. [p.144]
  a. homologous chromosomes
  b. a diploid chromosome
  c. sister chromatids
  d. two homologous chromosomes
  e. chromosome duplication

___ 4. Two sister chromatids of each chromosome are separated from each other during _____. [p.143]
   a. anaphase I
   b. metaphase II
   c. metaphase I
   d. telophase II
   e. anaphase II

___ 5. The first haploid cells to appear during meiosis are seen at _____. [pp.142–143]
   a. telophase II
   b. telophase I
   c. anaphase II
   d. anaphase I
   e. prophase II

___ 6. Crossing over is one of the most important events in meiosis because _____. [p.144]
   a. it leads to genetic recombination
   b. homologous chromosomes must be separated into different daughter cells
   c. the number of chromosomes allotted to each daughter cell must be halved
   d. homologous chromatids must be separated into different daughter cells
   e. nonsister chromatids do not break during the process

___ 7. Crossing over _____. [p.144]
   a. generally results in pairing of homologues and binary fission
   b. is accompanied by gene-copying events
   c. involves breakages and exchanges between sister chromatids
   d. is a molecular interaction between two of the nonsister chromatids of a pair of homologous chromosomes
   e. results in the recombination of genes located on sister chromatids

___ 8. In plants, gametes are produced by _____. [p.146]
   a. meiosis in gametophyte plants
   b. mitosis in sporophyte plants
   c. germ cells
   d. mitosis in gametophyte plants
   e. meiosis in sporophyte plants

___ 9. In female animals, a diploid germ cell develops into a(n) _____. [pp.146–147]
   a. primary spermatocyte
   b. primary oocyte
   c. egg
   d. secondary oocyte
   e. first polar body

___10. Which of the following does *not* increase genetic variation? [p.148]
   a. crossing over
   b. random fertilization
   c. prophase II of meiosis
   d. random homologue alignments at metaphase I
   e. prophase I of meiosis

___11. The cell in the diagram is a diploid with three pairs of chromosomes. From the number and pattern of chromosomes, the cell _____. [pp.142–143]
   a. could be in the first division of meiosis
   b. could be in the second division of meiosis
   c. could be in prophase of mitosis
   d. could be in neither mitosis nor meiosis, because this stage is not possible in a cell with three pairs of chromosomes
   e. could be in anaphase of mitosis

## Chapter Objectives/Review Questions

1. List advantages of sexual reproduction in a species. [pp.138–140]
2. List the characteristics of asexual reproduction in a species. [p.140]
3. Define *germ cells, meiosis, gametes, zygote, homologous chromosomes,* and *alleles.* [p.140]
4. Name the sites of meiosis in human males and human females. [p.140]
5. If a cell has a _____ chromosome number, it has a pair of each type of chromosome, often from two parents. [p.141]
6. Gametes have a _____ chromosome number, or half the parental number. [p.141]
7. At what point in the cell cycle does a germ cell duplicate its DNA? [p.141]

8. The two nuclear divisions of meiosis are called meiosis _____ and meiosis _____. [pp.142–143]
9. Be able to tell the story of the stages of meiosis; include proper terminology and chromosome movement descriptions. [pp.142–143]
10. Describe the ways that meiosis adds variation in traits. [pp.144–145]
11. List and define the terms that describe gamete formation in plants and animals. [pp.146–147]
12. _____ also adds to variation among offspring. [p.147]
13. List the events that provide millions of chromosome combinations in gametes. [pp.146–147]
14. How does fertilization provide additional staggering numbers of different gene combinations in possible offspring? [p.147]

## Chapter Summary

By (1) _____ reproduction, one parent alone transmits its (2) _____ information to its offspring. By (3) _____ reproduction, offspring typically inherit information from two parents that differ in their (4) _____. Alleles are different forms of the same (5) _____; they specify different versions of a (6) _____.

(7) _____ cells have a pair of each type of chromosome, one maternal and one paternal. (8) _____, a nuclear division mechanism, reduces the chromosome number. It occurs only in cells set aside for (9) _____ reproduction. Meiosis sorts out a reproductive cell's chromosomes into four (10) _____ nuclei, which are distributed to daughter cells by way of (11) _____ division.

During meiosis, each pair of maternal and paternal (12) _____ swaps segments and exchanges (13) _____. The pairs get randomly (14) _____, so forthcoming gametes end up with different mixes of maternal and paternal chromosomes. (15) _____ also governs which gametes combine during fertilization. All three events contribute to (16) _____ in traits among offspring.

In animals (17) _____ form by different mechanisms in males and females. In most plants, (18) _____ formation and other events intervene between meiosis and gamete formation.

Recent (19) _____ evidence suggests that meiosis originated through mechanisms that already existed for (20) _____ and, before that, for repairing damaged (21) _____.

## Integrating and Applying Key Concepts

During the year 2001, several scientists from various countries, including the United States, claimed to have the knowledge and technology to clone human beings. Other scientists suspect that human cloning may have already been achieved in secret. Those who criticize human cloning experiments suggest that each successful human clone would be accompanied by repeated failures in the form of severely deformed and dead human embryos. If, indeed, human cloning is currently possible, speculate about the public acceptance of such experimentation and the effects of reproduction without sex on human populations.

# 10

# OBSERVING PATTERNS
# IN INHERITED TRAITS

## INTRODUCTION

Among species that undergo sexual reproduction, it has long been observed that certain traits may be predictably passed from parent to offspring. The mechanism responsible for this, however, was not understood until the 1860s, when Gregor Mendel conducted his famous and enlightening experiments on peas. This chapter outlines the influential work of Mendel, and the impact this work had on modern genetics and our understanding of inheritance. The chapter then goes on to examine the effect that nature can have on our inherited traits.

## FOCAL POINTS

- Figure 10.4 [p.155] depicts a pair of homologous chromosomes and introduces crucial vocabulary.
- Figure 10.5 [p.156] illustrates the segregation of alleles.
- Figure 10.7 [p.157] shows the step-by-step construction of a Punnett square.
- Figure 10.8 [p.158] illustrates the process of independent assortment.
- Figure 10.9 [p.159] animates a dihybrid cross.

## Interactive Exercises

*In Pursuit of a Better Rose* [pp.152–153]

## 10.1. MENDEL, PEA PLANTS, AND INHERITANCE PATTERNS [pp.154–155]

## 10.2. MENDEL'S THEORY OF SEGREGATION [pp.156–157]

## 10.3. MENDEL'S THEORY OF INDEPENDENT ASSORTMENT [pp.158–159]

### Selected Words

*Observable* evidence [p.154], *Pisum sativum* [p.155], *identical* alleles [p.155], *nonidentical* alleles [p.155], *homozygous* condition [p.155], *heterozygous* condition [p.155], *dominant* allele [p.155], *recessive* allele [p.155]

### Boldfaced Terms

[p.155] genes_____

_____

[p.155] diploid _____

_____

[p.155] mutation _____

_____

[p.155] alleles _____

_____

[p.153] hybrids _____

_____

[p.155] homozygous dominant _____

_____

[p.153] homozygous recessive _____

_____

[p.153] gene expression _____

_____

[p.155] genotype _____

_____

[p.155] phenotype _____

_____

[p.156] monohybrid experiment _____

_____

[p.156] probability _____

_____

[p.157] Punnett square _____

_____

[p.157] testcrosses _____

_____

[p.157] segregation _____

_____

[p.158] dihybrid experiment _____

_____

[p.158] independent assortment _____

_____

(*Hint:* As you study each section of this chapter, make sure you understand all the selected words and bold-faced terms. Try the genetics problems where they are provided. The best way to understand genetics is by solving problems.)

## Matching

Please designate the following parts of the following figure: [p.155]

1. Indicate the homologous pair of chromosomes
2. Label a specific gene locus.
3. Designate a pair of alleles.
4. How many genes are on these chromosomes?

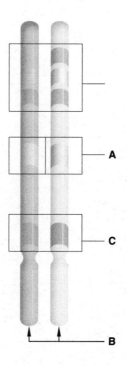

## Genetics Problems

5. In garden pea plants, tall (*T*) is dominant over dwarf (*t*). Use the Punnett-square method (refer to the main text, Figure 10.7a, p.157) to determine the genotype and phenotype probabilities of offspring from the above cross, Tt × tt:

_____

_____

_____

_____

6. Using the gene symbols (tall and dwarf pea plants) in exercise 19, apply the six Mendelian ratios listed above to complete the following table of single-factor crosses by inspection. State results as phenotype and genotype ratios. [pp.154–155]

| Cross | Phenotype Ratio | Genotype Ratio |
|---|---|---|
| a. Tt × tt | | |
| b. TT × Tt | | |
| c. tt × tt | | |
| d. Tt × Tt | | |
| e. tt × Tt | | |
| f. TT × tt | | |
| g. TT × TT | | |
| h. Tt × TT | | |

7. Assume that in humans, pigmented eyes (*B*) (an eye color other than blue) are dominant over blue (*b*), and right-handedness (*R*) is dominant over left-handedness (*r*). To solve this problem, you will cross the parents: *BbRr* × *BbRr*. A sixteen-block Punnett square is required with gametes from each parent arrayed on two sides of the Punnett square (refer to text Figure 10.9, p.159).

   (*Hint:* When working genetic problems that deal with two gene pairs, use a fork-line device to visualize the independent assortment of gene pairs located on nonhomologous chromosomes as gametes.)

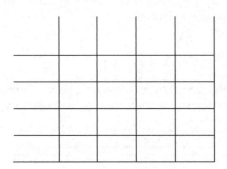

8. Array the gametes at the right on two sides of the Punnett square; combine these haploid gametes to form diploid zygotes in the squares. In the blank spaces below the Punnett square, enter the probability ratios derived in the Punnett square for the phenotypes listed: [pp.156–157]

_____

_____

_____

_____

_____

_____

_____ a. pigmented eyes, right-handed

_____ b. pigmented eyes, left-handed

_____ c. blue-eyed, right-handed

_____ d. blue-eyed, left-handed

9. In horses, black coat color is influenced by the dominant allele (*B*), and chestnut coat color is influenced by the recessive allele (*b*). Trotting gait is due to a dominant gene (*T*), pacing gait to the recessive allele (*t*). A homozygous black trotter is crossed to a chestnut pacer. [pp.158–159]

a. What will be the appearance of the $F_1$ and $F_2$ generations? _____

_____

b. Which phenotype will be most common? _____

c. Which genotype will be most common? _____

d. Which of the potential offspring will be certain to breed true? _____

## 10.4. BEYOND SIMPLE DOMINANCE [pp.160–161]

## 10.5. LINKAGE GROUPS [p.162]

## 10.6. GENES AND THE ENVIRONMENT [pp.162–163]

## 10.7. COMPLEX VARIATION IN TRAITS [pp.164–165]

### Selected Words

*ABO blood typing* [p.160], *A, B, O* [p.160], *incomplete* dominance [p.160], fibrillin [p.161], *Marfan Syndrome* [p.161], *Daphnia pulex* [p.163], *camptodactyly* [p.165]

### Boldfaced Terms

[p.160] codominance _____

_____

[p.160] multiple allele system _____

_____

[p.160] incomplete dominance _____

_____

[p.161] epistasis _____

_____

[p.161] pleiotropy _____

_____

[p.162] linked genes _____

_____

[p.164] continuous variation _____

_____

[p.164] polygenic inheritance _____

_____

[p.164] bell curves _____

_____

## Genetics Problems

Study the example of ABO blood typing in the text. Note that it is a multiple allele system with codominance and then solve the following problems. [p.160]

1. $AO \times AB =$ _____

2. $BO \times AO =$ _____

3. $AB \times AB =$ _____

4. Study the examples of snapdragons given in the text, which demonstrate incomplete dominance; then determine the phenotypes and genotypes of the offspring of the following crosses. $R$ = red flower color, $R'$ = white flower color, and $RR'$ is pink. [p.160]

| Cross | Phenotypes | Genotypes |
|---|---|---|
| a. $RR \times R'R' =$ | | |
| b. $RR' \times RR' =$ | | |

5. In poultry, an interaction occurs in which two genes produce a phenotype that neither gene can produce alone. The two interacting genes ($R$ and $P$) produce comb shape in chickens. Here are the possible genotypes and phenotypes [p.161]:

| Genotypes | Phenotypes |
|---|---|
| R__P__ | Walnut comb |
| R__pp | Rose comb |
| rrP__ | Pea comb |
| rrpp | Single comb |

What is this process called?

(*Hint:* Where a blank appears in the genotypes above, either the dominant or the recessive symbol in that blank yields the same phenotype.)

What are the genotype and phenotype ratios of the offspring of a heterozygous walnut-combed male and a single-combed female? [p.161] _____

_____

_____

6. In the inheritance of the coat (fur) color of Labrador retrievers, allele *B* specifies black that is dominant to brown (chocolate), *b*. Allele *E* permits full deposition of color pigment but the presence of two recessive alleles, *ee*, reduces deposition, and a yellow coat results. [p.161]

Predict the phenotypes of the coat color and their proportions resulting from the following cross:

*BbEe* × *Bbee* = _____

_____

## Complete the Table

7. Complete the following table by supplying the type of inheritance illustrated by each example. Choose from these gene interactions: pleiotropy, multiple allele system, incomplete dominance, codominance, and epistasis.

| Type of Inheritance | Example |
| --- | --- |
| [p.160] a. | Pink-flowered snapdragons produced from red- and white-flowered parents |
| [p.160] b. | AB type blood from a gene system of three alleles, *A*, *B*, and *O* |
| [p.160] c. | A gene with three or more alleles such as the *ABO* blood typing alleles |
| [p.161] d. | Black, brown, or yellow fur of Labrador retrievers and comb shape in poultry |
| [p.161] e. | The multiple phenotypic effects of the gene causing human sickle-cell anemia |

## Choice

Choose from the following two options to fill in the blanks below:

      a. environment as a primary effect     b. a number of genes affect a trait

_____ 8. height of human beings [p.164]

_____ 9. continuous variation in a trait [p.164]

_____ 10. flower color in *Hydrangea macrophylla* [p.163]

_____ 11. the range of eye colors in the human population [p.164]

_____12. heat-sensitive version of one of the enzymes required for melanin production in Himalayan rabbits [p.163]

_____13. three yarrow cuttings planted at three elevations [p.163]

_____14. coat color in Siamese cats [p.163]

## Short Answer

14. What is the take-home lesson when dealing with questions of heritable and environmental factors that give rise to variations in traits? [p.163]

_____

_____

# Self-Quiz

___ 1. The best statement of Mendel's principle of independent assortment is that _____. [p.157]
   a. one allele is always independently dominant to another
   b. independent hereditary units from the male and female parents are blended in the offspring
   c. the two hereditary units that influence certain independent traits to separate during gamete formation
   d. genes on pairs of homologous chromosomes are distributed into one gamete or another independently of genes on pairs of other chromosomes
   e. the two genes of each pair are separated from each other during meiosis, so they end up in different gametes

___ 2. All the different molecular forms of the same gene are called _____. [p.153]
   a. chiasmata
   b. alleles
   c. autosomes
   d. genes
   e. mutations

___ 3. In the $F_2$ generation of a monohybrid cross involving complete dominance, the expected phenotypic ratio is _____. [pp.154–155]
   a. 3:1
   b. 1:1:1:1
   c. 1:2:1
   d. 1:1
   e. 1:3:1

___ 4. In the $F_2$ generation of a cross between a red-flowered snapdragon (homozygous) and a white-flowered snapdragon, the expected phenotypic ratio of the offspring is _____. [p.158]
   a. 3/4 red, 1/4 white
   b. 100 percent red
   c. 1/4 red, 1/2 pink, 1/4 white
   d. 100 percent pink
   e. 1/2 red, 1/2 white

___ 5. In a testcross, $F_1$ hybrids are crossed to an individual known to be _____ for the trait. [p.155]
   a. heterozygous
   b. homozygous dominant
   c. homozygous
   d. homozygous recessive
   e. heterozygous dominant

___ 6. The tendency for dogs to bark while trailing is determined by a dominant gene, S, while silent trailing is due to the recessive gene, s. In addition, erect ears, D, are dominant over drooping ears, d. What combination of offspring would be expected from a cross between two erect-eared barkers who are heterozygous for both genes? [pp.156–157]
   a. 1/4 erect barkers, 1/4 drooping barkers, 1/4 erect silent, 1/4 drooping silent
   b. 9/16 erect barkers, 3/16 drooping barkers, 3/16 erect silent, 1/16 drooping silent
   c. 1/2 erect barkers, 1/2 drooping barkers
   d. 9/16 drooping barkers, 3/16 erect barkers, 3/16 drooping silent, 1/16 erect silent
   e. 3/4 erect barkers, 1/4 drooping barkers

___ 7. A man with type A blood could be the father of _____. [p.158]
   a. a child with type A blood
   b. a child with type B blood
   c. a child with type O blood
   d. a child with type AB blood
   e. all of the above

___ 8. Alleles at one chromosome locus may affect two or more traits in good or bad ways. This outcome of the activity of one gene's product is _____. [p.159]
   a. pleiotropy
   b. epistasis
   c. a mosaic effect
   d. continuous variation
   e. gene interaction

___ 9. Suppose two individuals, each heterozygous for the same characteristic, are crossed. The characteristic involves complete dominance. The expected genotypic ratio of their progeny is _____. [pp.154–155]
   a. 1:2:1
   b. 1:1
   c. 100 percent of one genotype
   d. 3:1
   e. 1:1:1:1

___ 10. If the two homozygous classes in the $F_1$ generation of the cross in exercise 9 are allowed to mate, the observed genotypic ratio of the offspring will be _____. [pp.154–155]
   a. 1:1
   b. 1:2:1
   c. 100 percent of one genotype
   d. 3:1
   e. 1:1:1:1

___ 11. Applying the types of inheritance learned in this chapter in the text, the skin color trait in humans exhibits _____. [pp.164–165]
   a. pleiotropy
   b. epistasis
   c. environmental effects
   d. continuous variation
   e. gene interactions

## Chapter Objectives/Review Questions

1. How is genetic mapping currently being used in the rose industry? [pp.152–153]
2. What was the prevailing method of explaining the inheritance of traits before Mendel's work with pea plants? [p.154]
3. Garden pea plants are naturally _____-fertilizing, but Mendel took steps to _____-fertilize them for his experiments. [p.155]
4. _____ are units of information about specific traits; they are passed from parents to offspring. [p.155]
5. What is the general term applied to the location of a gene on a chromosome? [p.155]
6. Define *allele*; how many types of alleles are present in the genotypes *Tt? tt? TT?* [p.155]
7. Explain the meaning of a true-breeding lineage. [p.155]
8. When two alleles of a pair are identical, it is a _____ condition; if the two alleles are different, this is a _____ condition. [p.155]
9. Distinguish a dominant allele from a recessive allele. [p.155]
10. _____ refers to the genes present in an individual; _____ refers to an individual's observable traits. [p.155]
11. Offspring of _____ crosses are heterozygous for the one trait being studied. [p.156]
12. Explain why probability is useful to genetics. [pp.156–157]
13. Be able to use the Punnett-square method of solving genetics problems. [p.157]
14. Define the *testcross* and cite an example. [p.157]
15. Mendel's theory of _____ states that during meiosis, the two genes of each pair separate from each other and end up in different gametes. [p.158]
16. Be able to solve dihybrid genetic crosses. [p.159]
17. Mendel's theory of _____ _____ states that gene pairs on homologous chromosomes tend to be sorted into one gamete or another independently of how gene pairs on other chromosomes are sorted out. [p.157]

18. Define *multiple allele system* and cite an example. [p.160]
19. Distinguish among incomplete dominance, epistasis, pleiotropy, and codominance. [p.160–161]
20. Explain why Marfan Syndrome is a good example of pleiotropy. [p.161]
21. List possible explanations for less predictable trait variations that are observed. [p.163]
22. List two human traits that are explained by continuous variation. [p.164]
23. Himalayan rabbits, yarrow plants, and garden hydrangeas are good examples of environmental effects on _____ expression. [p.163]

## Chapter Summary

Gregor Mendel gathered the first indirect (1) _____ evidence of the genetic basis of (2) _____. His meticulous work tracking traits in many generations of pea plants gave him clues that (3) _____ traits are specified in (4) _____. The units, which are distributed into gametes in predictable patterns, were later identified as (5) _____.

Some experimenters yielded evidence of gene (6) _____. When one chromosome separates from its (7) _____ partner during meiosis, the pairs of alleles on those chromosomes also separate and end up in different (8)_____.

Other experiments yielded evidence of (9) _____ assortment. During meiosis, the members of a pair of homologous chromosomes are (10) _____ into gametes independently of how all other pairs are distributed.

Not all traits have clearly dominant or (11) _____ forms. One allele of a pair may be fully or partially dominant over its nonidentical partner, or (12) _____ with it. Two or more gene pairs often influence the same trait, and some single genes influence many (13) _____. The (14) _____ also influences variation in gene expression.

## Integrating and Applying Key Concepts

Solve the following genetics problems:

1. In garden peas, one pair of alleles controls the height of the plant and a second pair of alleles controls flower color. The allele for tall (*D*) is dominant to the allele for dwarf (*d*), and the allele for purple (*P*) is dominant to the allele for white (*p*). A tall plant with purple flowers crossed with a tall plant with white flowers produces 3/8 tall purple, 3/8 tall white, 1/8 dwarf purple, and 1/8 dwarf white. What are the genotypes of the parents?
2. A child is born with type B blood. The mother also has type B blood. Two men claim to be the father: James has type A blood and Mark is Type AB. Could either of these men be the father of this child? If so, which one could possibly be the father? Give all possible phenotypes of the individuals involved.

# 11

---

# CHROMOSOMES AND HUMAN INHERITANCE

## INTRODUCTION

Human health and human disease have been topics of interest and discussion since the advent of man. These issues still pervade society today; for example, certain genetic diseases are currently the subject of popular television shows (*Small People, Big World*). Research has shown us that many diseases are inherited on chromosomes. This chapter discusses genetic diseases with respect to how they are passed on and how certain genetic characteristics actually cause disease. The final part of the chapter describes the technology involved in detecting genetic abnormalities and the ethical consequences of this knowledge.

## FOCAL POINTS

- Figure 11.3 [p.171] depicts an animated look at a human karyotype, or chromosome map.
- Figure 11.4 [p.172] shows the inheritance patterns of both autosomal dominant and autosomal recessive diseases.
- Figure 11.7 [p.174] shows the inheritance pattern of a sex-linked disease.
- Figure 11.17 [p.180] illustrates a typical pedigree used to study disease patterns in families.
- Table 11.1 [p.181] summarizes some of the more common chromosomally determined and inherited diseases.

---

## Interactive Exercises

*Strange Genes, Tortured Minds* [pp.168–169]

## 11.1. HUMAN CHROMOSOMES [pp.170–171]

## 11.2. EXAMPLES OF AUTOSOMAL INHERITANCE PATTERNS [pp.172–173]

### Selected Words

*Schizophrenia* [p.169], neurobiological disorders (NBDs), [p.169], *bipolar disorder* [p.169], *gene* [p.169], *achondroplasia* [p.172], *Huntington disease* [p.172], *galactosemia* [p.172]

## Boldfaced Terms

[p.170] sex chromosomes _____

_____

[p.170] autosomes _____

_____

[p.171] karyotype _____

_____

## Short Answer

1. Describe the characteristics of two NBDs and generally state the cause of these disorders. [pp.168–169]

   _____

   _____

   _____

   _____

2. Describe differences in the X and Y chromosome. [p.170]

   _____

   _____

   _____

   _____

3. Compare autosomal dominant with autosomal recessive traits. [p.172]

   _____

   _____

   _____

   _____

4. Huntington disorder is a rare form of autosomal dominant inheritance, $H$; the normal gene is $h$. The disease causes progressive degeneration of the nervous system with onset exhibited near middle age. An apparently normal man in his early twenties learns that his father has recently been diagnosed as having Huntington disorder. What are the chances that the son will develop this disorder? [p.172]

   _____

5. The autosomal allele that causes galactosemia (g) is recessive to the allele for normal lactose metabolism (G). A normal woman whose father is a galactosemic marries a man with galactosemia whose parents are normal. They have three children, two of which are normal and one of which is affected with galactosemia. List the genotypes for each person involved. [pp.172–173]

   _____

   _____

## Dichotomous Choice

Circle the word in parentheses that makes each statement correct. Refer to the Punnett-square table shown in the text in Figure 11.3. [p.171]

6. Male humans transmit their Y chromosome only to their (sons/daughters). [p.172]
7. Male humans receive their X chromosome only from their (mothers/fathers). [p.172]
8. Human mothers and fathers each provide an X chromosome for their (sons/daughters). [p.172]

## Short Answer

9. State the significance of the *SRY* gene on the Y chromosome. [p.171]

_____

_____


## Choice

10. A preparation of an individual's metaphase chromosomes, sorted by their defining features is called a _____. [p.171]

    a. blood type     b. DNA sample     c. karyotype     d. chromosome linkage chart

11. Technicians add a microtubule poison, colchicine, that blocks spindle formation at the _____ stage of mitosis. Here, the chromosomes are seen at their most condensed state. [p.171]

    a. anaphase     b. prophase     c. telophase     d. prometaphase     e. metaphase

12. Following fixing and staining the chromosomes on a slide, a technician takes a photograph of the chromosomes. The purpose of cutting the photograph into pieces is to _____. [p.171]

    a. remove chromosome abnormalities     b. separate homologs     c. apply colored fluorescent dyes
    d. sort individual chromosomes by size and shape     e. use a technique called spectral analysis


# 11.3.  TOO YOUNG, TOO OLD [p.173]

# 11.4.  EXAMPLES OF X-LINKED INHERITANCE PATTERNS [pp.174–175]

## Selected Words

*Progeriac* [p.173], *hemophilia A* [p.174], *color blindness* [p.175], *Duchenne muscular dystrophy* [p.175]

## Genetics Problems

1. A color-blind man and a woman with normal vision whose father was color-blind have a son. Color blindness, in this case, is caused by an X-linked recessive gene. If only the male offspring are considered, what is the probability that their son is color-blind? [p.175] _____

2. Hemophilia A is caused by an X-linked recessive gene. A woman who is seemingly normal but whose father was a hemophiliac marries a normal man. What proportion of their sons will have hemophilia? What proportion of their daughters will have hemophilia? What proportion of their daughters will be carriers? [p.175]

_____

3. The pedigree below shows the pattern of inheritance of color blindness in a family (persons with the trait are indicated by black circles). What is the chance that the third-generation female indicated by the arrow (below) will have a color-blind son if she marries a normal male? A color-blind male? [p.175]

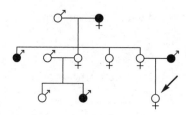

## Short Answer

4. Describe the frequency and cause of progeria; list the characteristics of progeriacs. [p.176]

## 11.5. HERITABLE CHANGES IN CHROMOSOME STRUCTURE [pp.176–177]

## 11.6. HERITABLE CHANGES IN THE CHROMOSOME NUMBER [pp.178–179]

### Selected Words

*Cri-du-chat* [p.176], *Down syndrome* [p.179], *Turner syndrome* [p.179], XO [p.179], *XXX* condition [p.179], *XXY mosaic condition* [p.179], *Klinefelter syndrome* [p.179], *XYY condition* [p.179]

### Boldfaced Terms

[p.176] duplications _____

[p.176] deletion _____

[p.176] inversion _____

[p.176] translocation _____

[p.178] nondisjunction _____

[p.178] aneuploidy _____

[p.178] polyploid _____

## Labeling

On rare occasions, chromosome structure becomes abnormally rearranged. Such changes can have profound effects on the phenotype of an organism. Label these diagrams of abnormal chromosome structure as a *deletion*, a *duplication*, an *inversion*, or a *translocation*.

1. _____ (  )
   [p.176]

2. _____ (  )
   [p.176]

3. _____ (  )
   [p.176]

4. _____ (  )
   [p.176]

## Short Answer

5. Which of the abnormal chromosomal rearrangements in the preceding diagrams results in the cri-du-chat syndrome? [p.177]

   _____

6. Provide an example showing that changes in chromosome structure have become adaptive? [p.177]

   _____

   _____

   _____

   _____

## Complete the Table

7. Complete the table by describing the mechanisms of chromosome number change in organisms for each category. [p.178]

| Category of Change | Description |
| --- | --- |
| a. Aneuploidy | |
| b. Polyploidy | |
| c. Nondisjunction | |

## Short Answer

8. If a nondisjunction occurs at anaphase I of the first meiotic division, what will be the proportion of abnormal gametes (for the chromosomes involved in the nondisjunction)? [p.178]

   _____

9. If a nondisjunction occurs at anaphase II of the second meiotic division, what will be the proportion of abnormal gametes (for the chromosomes involved in the nondisjunction)? [p.178]

   _____

10. List generally known polyploid organisms. [p.178]

_____

_____

11. Define the terms *trisomic* and *monosomic*. [p.178]

_____

_____

_____

_____

## Choice

Choose from the following:

       a. Down syndrome    b. Turner syndrome    c. Klinefelter syndrome    d. XXY condition
       e. XYY condition      f. XXX syndrome

12. _____ About one of every 500 to 1,000 males inherits one Y and two or more X chromosomes, mainly by nondisjunction. [p.179]

13. _____ Most don't have functional ovaries and can't produce enough sex hormones. [p.179]

14. _____ The testes and the prostate gland usually are smaller than average; hair is sparse, the voice is pitched high, and the breasts are a bit enlarged. [p.179]

15. _____ Tend to be taller than average, with mild mental impairment, but most are phenotypically normal. [p.179]

16. _____ Most affected individuals show moderate to severe mental impairment and heart defects; as a group, they tend to be cheerful and sociable people. [p.178]

17. _____ Occurs at a frequency of about 1 in 1,000 live births; adults are fertile; no physical or medical problems. [p.179]

18. _____ Which diagram represents a chromosome that has undergone crossover and recombination? (Assume that the organism involved is heterozygous with the genotype.) [p.176]

$$\text{a.} \quad \left|\begin{array}{c} A \\ B \end{array}\right| \quad\quad \text{b.} \quad \left|\begin{array}{c} a \\ b \end{array}\right| \quad\quad \text{c.} \quad \left|\begin{array}{c} B \\ A \end{array}\right| \quad\quad \text{d.} \quad \left|\begin{array}{c} a \\ B \end{array}\right|$$

## Genetics Problems

19. If genes A and B are twice as far apart on a chromosome as genes C and D, how often would you expect that crossing over occurs between genes A and B as compared to genes C and D? [p.176]

_____

_____

_____

20. Explain how a child can have the genotype *Ab/Ab* when one parent has a chromosome with genes A and b linked (*Ab/aB*) and the other does not (*AB/ab*). [p.176]

_____

_____

_____

## 11.7.  HUMAN GENETIC ANALYSIS [pp.180–181]

## 11.8.  PROSPECTS IN HUMAN GENETICS [pp.182–183]

### Selected Words

Genetic abnormality [p.180], genetic disorder [p.180], genetic disease [p.180], *prenatal diagnosis* [p.182], *embryo* [p.182], *fetus* [p.182], amniocentesis [p.182], *chorionic villi sampling* (CVS) [p.182], *fetoscopy* [p.182], *in vitro fertilization* [p.183], *phenylketonuria* or PKU [p.183], *abortion* [p.183]

### Boldfaced Terms

[p.180] pedigrees _____

_____

[p.180] syndrome _____

_____

[p.180] disease _____

_____

### Matching

The following standardized symbols are used in the construction of pedigree charts. Choose the appropriate description for each. [p.180]

1. _____ ■ ●

2. _____ ●

3. _____ ◆

4. _____ ■

5. _____ I, II, III, IV…

6. _____ ■—●

7. _____ ●■■●

a. individual showing the trait being studied
b. male
c. generation
d. offspring in order of birth, from left to right
e. female
f. sex not specified
g. marriage/mating

### Short Answer

8. Distinguish among these terms: *genetic abnormality, genetic disorder, syndrome,* and *disease.* [p.180]

_____

_____

_____

_____

_____

_____

## Complete the Table

9. Complete the table below by indicating whether the genetic disorder listed is due to inheritance that is autosomal recessive, autosomal dominant, X-linked recessive, or due to changes in chromosome number or changes in chromosome structure.

*Genetic Disorder/Abnormality*          *Inheritance Pattern or Cause*

| Genetic Disorder/Abnormality | Inheritance Pattern or Cause |
|---|---|
| a. galactosemia | |
| b. achondroplasia | |
| c. hemophilia | |
| d. Huntington disorder | |
| e. Turner and Down syndromes | |
| f. cri-du-chat syndrome | |
| g. XYY condition | |
| h. color blindness | |
| i. XXX syndrome | |
| j. fragile X syndrome | |
| k. progeria | |
| l. Klinefelter syndrome | |
| m. phenylketonuria | |

## Complete the Table

10. To complete the table, which summarizes methods of dealing with the problems of human genetics, choose from *phenotypic treatments, abortion, preimplantation diagnosis, genetic counseling,* and *prenatal diagnosis*.

| Method | Description |
|---|---|
| [p.182] a. | Detects genetic disorders before birth; may use biochemical tests, amniocentesis, CVS, and fetoscopy; an example is a pregnancy at risk in a mother 45 years old. |
| [p.183] b. | Expulsion of a pre-term embryo or fetus from the uterus. |
| [p.182] c. | Compares the risks of diagnostic procedures against the risk that a child will be affected by a severe genetic disorder; may involve diagnosis of parental genotypes, pedigrees, and genetic testing for known disorders. |
| [p.183] d. | Relies on in-vitro fertilization; a fertilized egg mitotically divides into a ball of eight cells; genes of one of these undifferentiated cells can be analyzed for genetic defects. |
| [p.183] e. | The symptoms of a number of genetic disorders can be minimized or alleviated by surgery, drugs, hormone replacement therapy, or in some cases by controlling diet; PKU is one example. |

# Self-Quiz

___ 1. All the genes located on a given chromosome compose a _____. [p.171]
   a. karyotype
   b. bridging cross
   c. wild-type allele
   d. linkage group
   e. homologous chromosome

___ 2. Chromosomes other than those involved in sex determination are known as _____. [p.170]
   a. nucleosomes
   b. heterosomes
   c. alleles
   d. autosomes
   e. heterozygous chromosomes

___ 3. The farther apart two genes are on a chromosome, _____. [p.176]
   a. the less likely that crossing over and recombination will occur between them.
   b. the greater will be the frequency of crossing over and recombination between them.
   c. the more likely they are to be in two different linkage groups.
   d. the more likely they are to be segregated into different gametes when meiosis occurs.
   e. the more likely one of the them will mutate.

___ 4. Karyotype analysis is _____. [p.178]
   a. a means of detecting and reducing mutagenic agents.
   b. a surgical technique that separates chromosomes that have failed to segregate properly during meiosis II.
   c. a means of detecting abnormalities in the structure or number of chromosomes.
   d. a process that substitutes defective alleles with normal ones.
   e. a means of detecting Mendelian ratios.

___ 5. The Cri-du-chat disorder is known to be caused by _____. [p.176]
   a. sex-linked inheritance
   b. autosomal recessive inheritance
   c. an inversion
   d. a deletion
   e. a change in chromosome number

___ 6. Red-green color blindness is a sex-linked recessive trait in humans. A color-blind woman and a man with normal vision have a son. What are the chances that the son is color blind? If the parents ever have a daughter, what is the chance for each birth that the daughter will be color blind? (Consider only the female offspring.) [p.175]
   a. 100 percent, 0 percent
   b. 50 percent, 0 percent
   c. 100 percent, 100 percent
   d. 50 percent, 100 percent
   e. none of the above

___ 7. Suppose that a hemophilic male (X-linked recessive allele) and a female carrier for the hemophilic trait have a nonhemophilic daughter with Turner syndrome. Nondisjunction could have occurred in _____. [pp.175, 179]
   a. both parents
   b. neither parent
   c. the father only
   d. the mother only
   e. the nonhemophilic daughter

___ 8. Nondisjunction involving the X chromosome occurs during oogenesis and produces two kinds of eggs, XX and O (no X chromosome). If normal Y sperm fertilize the two types, which genotypes are possible? [p.179]
   a. XX and XY
   b. XXY and YO
   c. XYY and XO
   d. XYY and YO
   e. YY and XO

___ 9. Of all phenotypically normal males in prisons, the type once thought to be genetically predisposed to becoming criminals was the group with _____. [p.179]
   a. XXY disorder
   b. XYY disorder
   c. Turner syndrome
   d. Down syndrome
   e. Klinefelter syndrome

___10. Amniocentesis is _____. [p.182]
   a. a surgical means of repairing deformities.
   b. a form of chemotherapy that modifies or inhibits gene expression or the function of gene products.
   c. used in prenatal diagnosis; a tiny sample of amniotic fluid is drawn; cell samples are analyzed for many severe genetic disorders.
   d. a form of gene-replacement therapy.
   e. a diagnostic procedure; cells for analysis are withdrawn from the chorion.

## Chapter Objectives/Review Questions

1. NBDs such as bipolar disorder and schizophrenia are associated with abnormal _____ biochemistry. [p.168]
2. Distinguish between human sex chromosomes and autosomes. [p.170]
3. Explain how sex determination occurs in humans. [p.170]
4. _____, one of 330 genes in a human Y chromosome, is the master gene for male sex determination. [p.171]
5. A _____ is a preparation of an individual's metaphase chromosomes, sorted by their defining visual features. [p.171]
6. Carefully characterize patterns of autosomal recessive inheritance, autosomal dominant inheritance, and X-linked recessive inheritance. [pp.172–174]
7. Describe the characteristics of Hutchinson-Gilford progeria syndrome. [p.173]
8. A(n) _____ is a loss of a chromosome segment; a(n) _____ is a gene sequence separated from a chromosome but then was inserted at the same place, but in reverse; a(n) _____ is a repeat of several gene sequences on the same chromosome; a(n) _____ is the transfer of part of one chromosome to a nonhomologous chromosome. [p.176]
9. Cite evidence that tends to support the idea that chromosome structure does evolve. [p.177]
10. Crossing over between homologous chromosomes disrupts gene _____ and results in nonparental combinations of genes in chromosomes. [p.171]
11. _____ is the failure of the chromosomes to separate in either meiosis or mitosis. [p.178]
12. When gametes or cells of an affected individual end up with one extra or one less than the parental number of chromosomes, it is known as _____; relate this concept to monosomy and trisomy. [p.178]
13. Trisomy 21 is known as _____ syndrome; Turner syndrome has the chromosome constitution _____; XXY chromosome constitution is _____ syndrome; taller than average males with sometimes mild mental impairment have the _____ condition. [pp.178–179]
14. A _____ is a chart showing genetic connections among individuals; be familiar with the standardized symbols used in such charts. [p.180]
15. A genetic _____ is a rare, uncommon version of a trait whereas an inherited genetic _____ is an inherited condition that sooner or later will cause mild to severe medical problems. [p.180]
16. A _____ is a recognized set of symptoms that characterize an abnormality or disorder. [p.180]
17. Describe what is meant by a genetic disease. [p.180]
18. Explain the procedures used in three types of prenatal diagnosis: amniocentesis, chorionic villi analysis, and fetoscopy; compare their risks. [p.182]
19. A procedure known as preimplantation diagnosis relies on _____ _____ fertilization. [p.182]
20. A _____ _____ will work with expectant parents by emulating pedigrees and the results of genetic tests to predict risks for future children. [p.182]

# Chapter Summary

All animals have a pair of (1) _____-chromosomes that are identical in length, shape, and which (2) _____ they carry. Sexually reproducing species also have a pair of (3) _____ chromosomes. The members of this pair differ between females and males. A (4) _____ on one of the human sex chromosomes dictates the male sex. (5) _____, a diagnostic tool, reveals changes in the structure or number of an individual's (6) _____.

Many genes on autosomes are expressed in (7) _____ patterns of simple (8) _____.

Some (9) _____ are affected by genes on the X chromosome. Inheritance patterns of such traits differ in (10) _____ and (11) _____.

On rare occasions, a chromosome may undergo permanent change in its (12) _____, as when a segment is deleted, (13) _____, inverted, or (14) _____.

On rare occasions, the (15) _____ of autosomes or sex chromosomes changes. In humans, the change usually results in a (16) _____ disorder.

Various analytical and (17) _____ procedures often reveal genetic disorders. (18) _____ questions are associated with what individuals as well as society at large do with the information.

# Integrating and Applying Key Concepts

1. The parents of a young boy bring him to their doctor. They explain that the boy does not seem to be going through the same vocal developmental stages as his older brother. The doctor orders a common cytogenetics test to be done, and it reveals that the young boy's cells contain two X chromosomes and one Y chromosome. Describe the test that the doctor ordered and explain how and when such a genetic result, XXY, most logically occurred.
2. Solve the following genetics problem. Show rationale, genotypes, and phenotypes. A husband sues his wife for divorce, arguing that she has been unfaithful. His wife gave birth to a girl with a fissure in the iris of her eye, an X-linked recessive trait. Both parents have normal eye structure. Can the genetic facts be used to argue for the husband's suit? Explain your answer.

# 12

# DNA STRUCTURE AND FUNCTION

## INTRODUCTION

The DNA molecule has a history full of scientific intrigue, experimentation, and even scandal. While the molecule itself was discovered in the 1800s, it was not until the 1940s that researchers became certain that DNA was the heritable genetic material. Then, the race was on to discover the structure and replication scheme of this all-important molecule. This chapter describes the experiments that led to the elucidation of the function of DNA and the scandal which resulted in the discovery of the structure of this molecule. The chapter concludes with a look at DNA technology with respect not to how far we have come in our understanding, but to where this understanding may take us.

## FOCAL POINTS

- Figure 12.3 [p.188] animates the experiment by Fredrick Griffith in which he discovered "the transforming principle."
- Figure 12.5 [p.190] illustrates the four nucleotide bases in the DNA molecule.
- Figure 12.6 [p.191] shows an animation of the double helix molecule of DNA.
- Figure 12.9 [p.193] depicts the formation of a newly synthesized strand during replication.
- Table 12.1 [p.194] compares various cloning methods.

## Interactive Exercises

*Here Kitty, Kitty, Kitty, Kitty, Kitty* [pp.186–187]

### 12.1. THE HUNT FOR FAME, FORTUNE, AND DNA [pp.188–189]

*Selected Words*

Telomeres [p.186], cloning mammals [p.186], Johann Miescher [p.188], Frederick Griffith [p.188], *Streptococcus pneumoniae* [p.188], Oswald Avery [p.188], Linus Pauling [p.189], Hershey and Chase [p.189], James Watson [p.189], Francis Crick [p.189]

## Boldfaced Terms

[p.188] deoxyribonucleic acid, or DNA _____

_____

[p.188] bacteriophages _____

_____

## Short Answer

1. The cloned sheep, Dolly, developed health problems and died after five years of life. List some of the health problems typically exhibited by surviving cloned mammals. [p.185]

   _____

   _____

   _____

2. Explain how Fredrick Griffith's experiments with *Streptococcus pneumoniae* bacteria showed that a cell could be changed forever by materials present in another cell. [p.188]

   _____

   _____

   _____

## Complete the Table

3. Complete the table by noting the name of the investigator or the contribution made in the discovery of the structure of DNA.

| Investigators | Contribution |
|---|---|
| a. Miescher [p.188] | |
| b. _____ [p.188] | In 1928, discovered the transforming principle in *Streptococcus pneumoniae*; live harmless cells were mixed with dead S cells, R cells became S cells |
| c. _____ [p.188] | Reported that the transforming substance in Griffith's bacteria experiments was probably DNA |
| d. Pauling [p.189] | |
| e. _____ [p.189] | Worked with radioactve sulfur (found in protein) and phosphorus (found in DNA) labels; T4 bacteriophage and *E. coli* cells demonstrated that labeled phosphorus was in bacteriophage DNA and contained hereditary instructions for new bacteriophages |
| f. Watson and Crick [p.189] | |

## 12.2. THE DISCOVERY OF DNA'S STRUCTURE [pp.190–191]

### Selected Words

Erwin Chargaff [p.190], pyrimidines [p.190], purines [p.190], Maurice Wilkins [p.190], Rosalind Franklin [p.190], DNA double helix [p.190]

### Boldfaced Terms

[p.190] nucleotide _____

_____

[p.190] adenine _____

_____

[p.190] guanine _____

_____

[p.190] thymine _____

_____

[p.190] cytosine _____

_____

[p.190] x-ray diffraction images _____

_____

### Short Answer

1. List the three parts of every nucleotide. [p.188] _____

_____

### Labeling

For the four nucleotides shown in questions 2–5, label each nitrogen-containing base as *guanine, thymine, cytosine,* or *adenine*. Then, indicate in the parentheses whether that nucleotide base is a purine (pu) or a pyrimidine (py). [p.190]

2. _____ (　)　　3. _____ (　)　　4. _____ (　)　　5. _____ (　)

For questions 6–12, label each numbered part of the DNA model as *phosphate, purine, pyrimidine, nucleotide,* or *deoxyribose.* [p.191]

6. _____

7. _____

8. _____

9. _____

10. _____

11. _____

12. _____

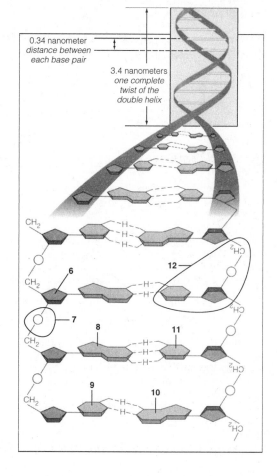

*True–False*

If the statement is true, write a "T" in the answer blank. If the statement is false, correct it by writing the correct word(s) for the underlined word(s) in the answer blank.

_____ 13. <u>Maurice Wilkins</u> shared two crucial insights into the composition of DNA with the scientific community. First, the amount of A relative to G differs from one species to the next. Second, the amounts of T and A in a DNA molecule are exactly the same, and so are the amounts of C and G. [pp.189–190]

_____ 14. Rosalind Franklin, a colleague of Wilkins, obtained especially good <u>electron microscope</u> images of DNA fibers. [p.190]

_____ 15. <u>Linus Pauling</u> made a big chemical mistake. His model had the negatively charged hydrogen bonds holding DNA together at phosphate groups. [p.191]

_____ 16. As <u>Wilkins</u> and <u>Pauling</u> had perceived, DNA consists of two strands of nucleotides, held together at their bases by hydrogen bonds. [p.191]

_____ 17. Much of the actual research data that elucidated the chemical structure of DNA came through the work of <u>Rosalind Franklin</u>. [p.190]

## 12.3. ROSALIND'S STORY [p.192]

## 12.4. DNA REPLICATION AND REPAIR [pp.192–193]

*Selected Words*

*Semiconservative* replication [p.190]

## Boldfaced Terms

[p.192] DNA replication _____

_____

[p.192] DNA polymerase _____

_____

[p.192] DNA ligase _____

_____

[p.193] DNA repair mechanisms _____

_____

## Choice

1. ___ Before her work at Cambridge University, Rosalind Franklin had refined an x-ray diffraction method while studying the structure of _____. [p.192]

    a. wood      b. stainless steel      c. quartz      d. coal

2. ___ At Cambridge University, Franklin was given the assignment to investigate the structure of DNA in the laboratory of _____. [p.192]

    a. Francis Crick      b. Linus Pauling      c. Maurice Wilkins      d. James Watson

3. ___ When Watson and Crick finally focused on Franklin's x-ray diffraction image of wet DNA fibers, it gave them which bit of information? [p.192]

    a. DNA had two pairs of chains      b. two helically twisted chains running in opposing directions
    c. the names of the four nucleotide bases      d. nucleotide bases are held together by weak hydrogen bonds

## Fill in the Blanks

Until Watson and Crick presented their model, no one could explain DNA (4) _____ [p.192], or how the molecule of inheritance is duplicated before a cell divides. (5) _____ [p.192] easily break the hydrogen bonds between the two-nucleotide strands of a DNA molecule. When enzymes and other proteins act on DNA, one (6) _____ [p.192] unwinds from the other and exposes stretches of its nucleotide bases. Cells contain stockpiles of free (7) _____ [p.192] that can pair with the exposed bases.

Each parent strand stays intact, and a companion strand is assembled on each one according to this base-pairing rule: A to (8) _____ [p.192], and G to (9) _____ [p.192]. As soon as a stretch of a new, partner strand forms on a stretch of the parent strand, the two twist up together into a (10) _____

_____ [p.192]. Because the parent DNA strand is (11) _____ [p.190] during the replication process, half of every double-stranded DNA molecule is "old" and half is "new." The process is called (12)

_____ [p.192] replication. DNA replication uses a team of molecular workers. (13) _____ [p.192] enzymes become active along the length of the DNA molecule. Along with other proteins, some enzymes unwind the strands in both directions and prevent them from (14) _____. [p.192] (15) _____ [p.191] action jump-starts the unwinding process but isn't necessary to unzip hydrogen bonds between the strands; (16) _____ [p.192] bonds are individually weak.

Enzymes called DNA (17) _____ [p.192] attach short stretches of free nucleotides to unwound parts of the parent template. Free nucleotides themselves drive the strand assembly. Each has three (18) _____ [p.192] groups. DNA polymerase splits off two, and it is this release of (19) _____ [p.192] that drives the attachments.

DNA (20) _____ [p.191] fill in the tiny gaps between the new short stretches to form one continuous strand. Then enzymes wind the template strand and complementary strand together to form a DNA (21) _____ _____. [p.192]

Sometimes a molecule of DNA breaks. (22) _____ [p.193] can fix breaks, and specialized DNA (23) _____ [p.193] can fix mismatched base pairs or replace mutated ones. These repair processes confer a (24) _____ [p.193] advantage on cells. This maintains the integrity of genetic information. When proofreading and repair mechanisms fail, a mistake becomes a (25) _____ [p.193], which is a permanent change in the DNA.

## Labeling

26. The term *semiconservative replication* refers to the fact that each new DNA molecule resulting from the replication process is "half-old, half-new." In the diagram, complete the replication required in the middle of the molecule by adding the required letters representing the missing nucleotide bases. Recall that ATP energy and the appropriate enzymes are actually required to complete this process. [p.193]

T - _____        _____ - A
G - _____        _____ - C
A - _____        _____ - T
C - _____        _____ - G
C - _____        _____ - G
C - _____        _____ - G

old  new        new  old

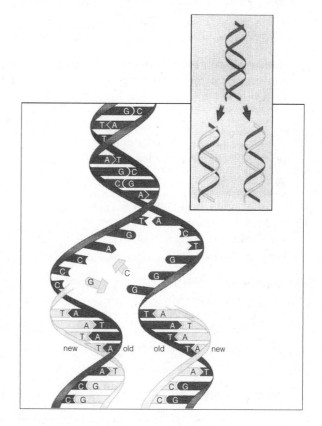

# 12.5. USING DNA TO DUPLICATE EXISTING MAMMALS [p.194]

## Selected Words

*Embryo cloning* [p.194], therapeutic cloning [p.194], adult cloning [p.194], *artificial twinning* [p.194]

# Self-Quiz

___ 1. _____ was the first scientist to isolate DNA. [p.188]
a. Oswald Avery
b. Linus Pauling
c. James Watson
d. Johann Miescher
e. Frederick Griffith

___ 2. The scientist who demonstrated that harmless pneumonia-causing bacteria cells had become permanently transformed into pathogens through a change in the bacterial hereditary material was _____. [p.188]
a. Oswald Avery
b. Linus Pauling
c. James Watson
d. Johann Miescher
e. Frederick Griffith

___ 3. _____ were the scientists who demonstrated that radioactively labeled bacteriophages transfer their DNA but not their protein coats to their host bacteria. [p.189]
a. Watson and Crick
b. Hershey and Chase
c. Hershey and Griffith
d. Watson and Pauling
e. Wilkins and Chargaff

___ 4. In 1953, _____ built a model of DNA that fit all the pertinent biochemical rules and insights they had gleaned from other sources. [p.189]
a. Watson and Crick
b. Hershey and Chase
c. Wilkins and Franklin
d. Watson and Pauling
e. Wilkins and Chargaff

___ 5. Adenine and guanine are _____. [p.190]
a. double-ringed purines
b. single-ringed purines
c. double-ringed pyrimidines
d. single-ringed pyrimidines
e. neither purines nor pyrimidines

___ 6. _____ discovered that in DNA, the amounts of A = T and the amounts of G = C. [p.190]
a. Maurice Wilkins
b. Rosalind Franklin
c. Erwin Chargaff
d. Hershey and Chase
e. Watson and Crick

___ 7. A single strand of DNA with the base-pairing sequence C-G-A-T-T-G is compatible only with the sequence _____. [p.191]
a. C-G-A-T-T-G
b. G-C-T-A-A-G
c. T-A-G-C-C-T
d. G-C-T-A-A-C
e. G-C-T-A-T-C

___ 8. Rosalind Franklin's x-ray diffraction research on DNA established _____. [p.192]
a. phosphate groups project outward from the molecule
b. one pair of chains
c. one pair of chains oriented in opposite directions
d. a helical structure
e. all of the above

___ 9. Enzymes known as _____ attach short stretches of free nucleotides to unwound parts of the parent template. [p.192 ]
a. semiconservative enzymes
b. DNA ligases
c. conservative enzymes
d. DNA polymerases
e. replication enzymes

___10. Embryo splitting or artificial twinning was used to produce _____. [p.194]
a. Dolly, a sheep
b. CC, a cat
c. a clone of an adult with a rewound developmental clock
d. Tetra, a monkey
e. cloned pigs with heart problems

# Chapter Objectives/Review Questions

1. Describe the cloning technique used to clone Dolly the sheep, and list the problems that have arisen as a result of cloning mammals with this technique. [p.187]
2. Summarize the research carried out by Miescher, Griffith, Avery, and colleagues; Pauling; and Hershey and Chase; state the specific advances made by each in the understanding of DNA structure. [pp.188–189]
3. Summarize the specific research that demonstrated that DNA, not protein, governed inheritance. [p.188]
4. The two scientists who assembled the clues to DNA structure and produced the first model were _____ and _____. [p.189]
5. Name the four nucleotides in DNA and indicate which are purines and which are pyrimidines. [p.190]
6. Sketch a DNA molecule, using letters to represent the nitrogenous bases, phosphates, sugars, and hydrogen bonds. [p.191]
7. Describe the relationship between Rosalind Franklin and Maurice Wilkins. [p.192]
8. List the pieces of information about DNA structure that Rosalind Franklin discovered through her x-ray diffraction research. [p.192]
9. Assume that the two parent strands of DNA have been separated and that the base sequence on one parent strand is A-T-T-C-G-C; the base sequence that will complement that parent strand is _____. [p.193]
10. Describe how double-stranded DNA replicates from stockpiles of nucleotides. [p.193]
11. Define *semiconservative replication*. [p.192]
12. During DNA replication, enzymes called DNA _____ attach short stretches of free nucleotides to unwound parts of the parent template. [p.192]
13. DNA _____ fill in the tiny gaps between the new short stretches of DNA to form one continuous strand; then enzymes wind the template strand and complementary strand together to form a DNA _____ _____. [p.192]
14. DNA _____ can fix breaks in DNA, and specialized DNA _____ can fix mismatched base pairs or replace mutated ones. [p.193]
15. Define and give examples of *embryo-splitting* and *nuclear transfers* as cloning methods. [p.194]

# Chapter Summary

In all living cells, (1) _____ molecules store information that governs heritable traits.

A DNA molecule consists of two chains of (2) _____, (3)_____-bonded together along their length and coiled into a (4) _____ helix. Four kinds of nucleotides make up the chains: adenine, thymine, guanine, and (5) _____.

The (6) _____ in which one kind of nucleotide base follows the next along a DNA strand encodes heritable information. The order of some regions of DNA is unique for each (7) _____.

Like any race, the one that led to the discovery of DNA's (8) _____ had its winners—and its losers.

Before a cell divides, enzymes and other proteins (9) _____ its DNA. Newly formed DNA strands are monitored for (10) _____, most of which are corrected. Uncorrected errors are (11) _____.

Knowledge about the structure and (12) _____ of DNA is the basis of several methods of (13) _____.

# Integrating and Applying Key Concepts

1. Review the stages of mitosis and meiosis, as well as the process of fertilization. Include what has now been learned about DNA replication and the relationship of DNA to a chromosome. As you cover the stages, be sure each cell receives the proper number of DNA threads.
2. What would happen to the amount of DNA in a eukaryotic cell if that cell did not duplicate its DNA prior to cell division? (Refer to the cell cycle, p.128.)
3. AT HOME: Using materials that you can find around your home, build a three dimensional model of DNA. Pay particular attention to base pairing and to the formation of the double helix.

# 13

# FROM DNA TO PROTEIN

## INTRODUCTION

Ultimately, most of the structure of a cell and most of the work done by a cell is due to proteins that the cell produces. A defective protein can be deadly to a cell and perhaps to an entire organism. But where do these proteins originate? How are they made? This chapter explains protein synthesis by describing the two cellular processes that are used to make proteins—transcription and translation. Molecular information flows from DNA to RNA and then to protein. The integrity of the DNA molecule is vital to correct protein formation, and this chapter explains why. Changes in the DNA, mutations, may be detrimental because they may cause a protein to be incorrectly made. These same changes, however, are the mechanisms by which cells, and therefore organisms, change and undergo evolution.

## FOCAL POINTS

- Figure 13.2 [p.198] illustrates the difference between DNA and RNA, and between DNA replication and transcription—the cellular process of RNA synthesis.
- Table 13.1 [p.199] compares DNA and RNA.
- Figure 13.4 [p.199] animates RNA processing.
- Figure 13.5 [p.200] illustrates the genetic code and the process of translation.
- Figure 13.8 [p.202] shows the stages of translation.
- Figure 13.9 [p.204] depicts the different types of mutations.
- Figure 13.12 [p.206] summarizes eukaryotic translation.

## Interactive Exercises

*Ricin and Your Ribosomes* [pp.196–197]

## 13.1. TRANSCRIPTION [pp.196–197]

### Selected Words

*Protein-building* instructions [p.196], "pre-mRNA" [p.197], guanine "cap" [p.197], ribosomes [p.197], translation [p.197]

### Boldfaced Terms

[p.198] transcription _____

[p.198] messenger RNA or mRNA _____

_____

[p.198] ribosomal RNA or rRNA _____

_____

[p.198] transfer RNA or tRNA_____

_____

[p.198] uracil _____

_____

[p.198] RNA polymerase _____

_____

[p.198] promoter _____

_____

[p.199] introns _____

_____

[p.199] exons _____

_____

[p.199] alternative splicing _____

_____

## Choice

1. Ricin is a product of the _____. [p.196]

    a. fava bean    b. deadly nightshade    c. *Clostridium* bacterium    d. castor oil plant

2. Traces of Ricin have shown up _____. [p.196]

    a. in Afghanistan caves    b. on a Chechen fighter killed in Moscow    c. in a U.S. Senate mailroom
    d. all of the preceding

3. Ricin itself is a _____. [p.196]

    a. carbohydrate    b. protein    c. nucleic acid    d. fat

4. A dose of Ricin the size of _____ can kill you. [p.196]

    a. one teaspoon    b. one tablespoon    c. a grain of salt    d. 500 grams

5. Ricin has two polypeptide chains; one helps Ricin insert itself into cells, the other is an enzyme that
    _____. [p.196]

    a. destroys mitochondria    b. wrecks part of the ribosome    c. inactivates the nucleus
    d. breaks down the cytoskeleton

## Complete the Table

6. Three types of RNA are transcribed from DNA in the nucleus. Provide information about these
    molecules. [p.198]

| RNA Molecules | Abbreviation | Description/Function |
|---|---|---|
| a. messenger RNA | | |
| b. ribosomal RNA | | |
| c. transfer RNA | | |

## Short Answer

7. List three ways in which a molecule of RNA is structurally different from a molecule of DNA. [p.199]

_____

_____

_____

## Fill in the Blanks

Transcription differs from DNA replication in three respects. Part of a (8) _____ [p.198] strand, not the whole molecule is the template. The enzyme (9) _____ _____ [p.198], not DNA (10) _____ [p.198], adds ribonucleotides to the end of a growing (11) _____ [p.198] strand, in the 5′ to 3′ direction. Also, transcription produces (12) _____ [p.198] free strand of RNA, not a hydrogen-bonded double helix.

A (13) _____ [p.198] region is a particular base sequence coded in the DNA that serves as the "start" signal for (14) _____ [p.198]. RNA synthesis begins when several proteins, including (15) _____ _____ [p.198], attach to a promoter. The (16) _____ [p.198] then swiftly moves down the DNA strand, joining one (17) _____ [p.197] after another, using the DNA as a template to make a new (18) _____ [p.199] as it goes. Eventually, it arrives at another particular (19) _____ [p.199] sequence in the DNA that signals "the end," and the new RNA is released as a free (20) _____ [p.199].

## Labeling

Refering to Figure 13.4 [p.199], select the letters on the sketch that best fit the numbered descriptions.

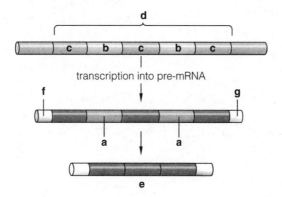

transcription into pre-mRNA

21. ___ Unit of transcription in a DNA strand.

22. ___ Enzymes attach a modified guanine "cap" that will bind the mRNA to a ribosome.

23. ___ Different enzymes attach a "poly-A tail" of about 100-300 adenine ribonucleotides.

24. ___ Introns, or gene sequences that are removed before an mRNA is used for protein synthesis.

25. ___ Exons, coding parts of a gene sequence; exons alternate with introns.

26. ___ Introns are snipped out before the mRNA leaves the nucleus in mature form; introns remain in the nucleus, where they are recycled.

27. ___ Mature mRNA transcript.

## 13.2. THE GENETIC CODE [p.200]

## 13.3. tRNA AND mRNA [p.201]

### Selected Words

Base triplets [p.200], molecular "hook" [p.201], "wobble effect" [p.201]

### Boldfaced Terms

[p.200] codons _____

_____

[p.200] genetic code _____

_____

[p.201] anticodon _____

_____

### Matching

Match each term with its description.

1. ___ codon [p.200]

2. ___ three bases at a time [p.200]

3. ___ 61 of the base triplets [p.200]

4. ___ the genetic code [p.200]

5. ___ ribosome subunits [p.201]

6. ___ anticodon [p.200]

7. ___ the "stop" codons [p.200]

8. ___ Niremberg, Leder, Ochoa, and Korana [p.200]

9. ___ mitochondria with their own genetic code [p.200]

10. ___ "Wobble effect"; isoleucine is an example [p.200]

a. large and small are assembled in the nucleus from rRNA and proteins; shipped separately to the cytoplasm; intact and functional when translation is to occur

b. reading frame of the nucleotide bases in mRNA

c. freedom in third-position codon-anticodon pairing

d. UAA, UAG, UGA

e. deduced the correlation between genes and proteins

f. a sequence of three tRNA nucleotide bases that can pair with a specific mRNA codon

g. probably evolved from ingested foreign bacteria that managed to survive inside host cells

h. name for each base triplet in mRNA

i. the number of mRNA base triplets that actually specify amino acids

j. the set of 64 different codons

### Completion

11. Given the following DNA sequence, deduce the composition of the mRNA transcript: [p.200]

TAC   AAG   ATA   ACA   TTA   TTT   CCT   ACC   GTC   ATC

___   ___   ___   ___   ___   ___   ___   ___   ___   ___

(mRNA transcript)

12. Deduce the composition of the tRNA anticodons that would pair with the above specific mRNA codons as these tRNAs deliver the amino acids (to be identified below) to the P and A binding sites of the small ribosomal subunit. [p.198]

———  ———  ———    ———  ———  ———  ———  ———  ———  ———
(tRNA anticodons)

13. From the mRNA transcript in question 12, use Figure 13.6 (the genetic code) of the text to deduce the composition of the amino acids of the polypeptide sequence. [p.198]

———  ———  ———    ———  ———  ———  ———  ———  ———  ———
(amino acids)

14. Write the RNA sequence for a start codon _____. Now, write the DNA sequence for this start codon _____. What amino acid is coded for by this sequence? [p.200] _____

15. Write one of the codons that would signal the end of translation. [p.200] _____

## 13.4. THE THREE STAGES OF TRANSLATION [pp.202–203]

### Selected Words

*Initiation* [p.202], *elongation* [p.202], *termination* [p.202]

### Boldfaced Terms

[p.202] translation _____

_____

### Complete the Table

1. Complete the following table, which distinguishes the stages of translation.

| Translation Stage | Description |
|---|---|
| [p.202] a. | The ribosome encounters the mRNA's STOP codon; proteins called release factors find to the ribosome and trigger enzyme activity that detaches the mRNA and the polypeptide chain from the ribosome. |
| [p.202] b. | A polypeptide chain is assembled as the mRNA passes between the two ribosomal subunits; molecules of tRNA bring amino acids to the ribosome, then bind to the mRNA in the order specified by its codon; peptide bond formation between the amino acids is catalyzed by part of an rRNA molecule functioning as an enzyme at the center of the large ribosomal subunit. |
| [p.202] c. | Initiator tRNA binds with a small ribosomal subunit; then mRNA's START codon, AUG, joins up with the initiator tRNA's anticodon; a complex is formed by the ribosome, mRNA, and initiator tRNA. |

## 13.5. MUTATED GENES AND THEIR PROTEINS PRODUCTS [pp.204–205]

*Selected Word*

*Frameshift* mutations [p.204]

*Boldfaced Terms*

[p.204] mutations _____

_____

[p.204] base-pair substitution _____

_____

[p.204] insertions _____

_____

[p.204] deletions _____

_____

[p.204] transposable elements _____

_____

[p.204] ionizing radiation _____

_____

[p.205] nonionizing radiation _____

_____

*Choice*

1. ___ In genetic mutations called _____, one base is copied incorrectly during DNA replication. [p.204]

    a. transposons     b. base substitutions     c. insertions     d. deletions

2. ___ Insertions and deletions result in _____. [p.204]

    a. transposon mutations     b. base substitutions     c. frameshift mutations
    d. a break in RNA molecules

3. ___ When _____ land in a gene, they often block or alter the timing or extent of its activity. [p.204]

    a. base substitutions     b. frameshift mutations     c. insertions     d. transposonable elements

4. ___ _____ breaks chromosomes into pieces. [pp.204–205]

    a. a base substitution     b. ionizing radiation     c. nonionizing radiation
    d. a frameshift mutation

5. ___ _____ boosts electrons to a higher energy level; when DNA absorbs UV light, cytosine and thymine are susceptible to changing their base-pairing properties (T – T, not A – T). [p.203]

    a. a base substitution     b. ionizing radiation     c. nonionizing radiation
    d. a frameshift mutation

# Self-Quiz

___ 1. rRNA _____. [p.201]
  a. serves as the "start" signal for transcription
  b. is a ribosomal component
  c. carries protein-building instructions
  d. delivers amino acids, one at a time to a ribosome
  e. receives RNA polymerase

___ 2. _____ carry(ies) amino acids to ribosomes, where amino acids are linked into the primary structure of a polypeptide. [p.201]
  a. mRNA
  b. tRNA
  c. introns
  d. rRNA
  e. RNA polymerase

___ 3. Of the following, which *one* is not a part of a mature mRNA transcript. [p.199]
  a. exons
  b. a poly-A tail
  c. a modified guanine "cap"
  d. introns

___ 4. Transfer RNA differs from other types of RNA because it _____. [p.201]
  a. transfers genetic instructions from cell nucleus to cytoplasm
  b. specifies the amino acid sequence of a particular protein
  c. carries an amino acid at one end
  d. contains codons
  e. is a promoter

___ 5. The codons UAA, UAG, and UGA _____. [p.200]
  a. are all code for glutamine
  b. are all code for methionine
  c. are all start signals
  d. are all stop signals
  e. serve as universal codons all amino acids

___ 6. The genetic code is a set of _____ codons, or sets of three ribonucleotide bases. [p.200]
  a. 32
  b. 16
  c. 46
  d. 61
  e. 64

___ 7. In _____, a polypeptide chain is assembled as the mRNA passes through the two ribosomal subunits. [p.202]
  a. elongation
  b. transcription
  c. initiation
  d. mRNA transcript processing
  e. termination

___ 8. Many newly formed polypeptide chains have a _____, a special sequence of amino acids that allows them to enter the ribosome-studded, flattened sacs of rough ER. [p.202]
  a. poly-A cap
  b. "shipping label"
  c. polysome
  d. guanine tail
  e. stop codon

___ 9. Frameshift mutations involve _____. [p.204]
  a. base substitutions
  b. insertions
  c. transposon
  d. deletions
  e. both b and d

___10. T – T pairings in DNA are induced by _____. [p.205]
  a. ionizing radiation
  b. transposons
  c. frameshift mutations
  d. nonionizing radiation
  e. base substitutions

---

# Chapter Objectives/Review Questions

1. Only plutonium and botulism toxin are more deadly than _____. [p.196]
2. _____ RNA is a component of ribosomes; _____ RNA carries protein-building instructions; _____ RNA delivers amino acids one at a time to a ribosome. [p.198]
3. State how RNA differs from DNA in structure and function, and indicate what features RNA has in common with DNA. [p.199]

4. Generally describe the process of transcription, and indicate three ways in which it differs from replication. [p.198]
5. What is the function of RNA polymerase? A promoter region? [p.198]
6. Describe the alterations of pre-mRNA to produce a mature mRNA transcript. [p.199]
7. Distinguish introns from exons and describe the fate of each. [p.199]
8. What RNA code would be formed from the following DNA code: TAC-CTC-GTT-CCC-GAA? [p.198]
9. Each base triplet in mRNA is called a(n) _____. [p.199]
10. The set of 64 different codons is the _____ _____. [p.200]
11. Explain how the DNA message TAC-CTC-GTT-CCC-GAA would be used to code for a segment of protein, and state what its amino acid sequence would be. [p.200]
12. Describe the "wobble effect." [p.201]
13. Describe events occurring in the following stages of translation: initiation, elongation, and termination. [p.201]
14. What is the fate of the new polypeptides produced by protein synthesis? [p.202]
15. Briefly describe the spontaneous DNA mutations known as base substitutions, frameshift mutation, and transposons. [p.204]
16. Distinguish between mutations caused by ionizing radiation and nonionizing radiation. [pp.204–205]
17. Using a diagram, summarize the steps involved in the transformation of genetic messages into proteins. [Figure 13.9, p.204]

## Chapter Summary

Life depends on (1) _____ and other proteins. All proteins consist of (2) _____ chains. The chains are sequences of amino acids that correspond to sequences of (3) _____ bases in DNA called (4) _____. The path leading from genes to proteins has two steps: (5) _____ and (6) _____.

During transcription, the two strands of the (7) _____ double helix are unwound in a gene region. Exposed bases of one strand become the (8) _____ for assembling a single strand of (9) _____ (a transcript). (10) _____ RNA is the only type of RNA that carries DNA's protein-building instructions.

The nucleotide sequence of RNA is read (11) _____ bases at a time. Sixty-four base triplets that correspond to specific amino acids represent the genetic (12) _____, which has been highly (13) _____ over time.

During (14) _____, amino acids become bonded together into a polypeptide chain in a sequence specified by base triplets in (15) _____ RNA. (16) _____ RNAs deliver amino acids one at a time to ribosomes. (17) _____ RNA catalyzes the formation of (18) _____ bonds between the amino acids.

(19) _____ in genes may result in changes in protein structure, protein function, or both. The changes may lead to (20) _____ in traits among individuals.

## Integrating and Applying Key Concepts

1. Genes code for specific polypeptide sequences. Not every substance in living cells is a polypeptide. Explain how genes might be involved in the production of a storage starch (such as glycogen) that is constructed from simple sugars.
2. Explain why a frameshift mutation would be considered much more drastic to the resulting protein than a simple point mutation. Give an example with a hypothetical DNA sequence containing at least four codons.

# 14

# CONTROLS OVER GENES

## INTRODUCTION

This chapter introduces the various mechanisms involved in the regulation of gene expression, or more simply put, when certain proteins are made. It begins to answer the question, if all cells have the same DNA, then why are some cells different from others? A liver cell, for example, certainly looks and functions differently from a skin cell, yet both have the same DNA. The regulation of gene expression allows cells to be specialized in both structure and function. Examples of gene regulation in both prokaryotic and eukaryotic cells are illustrated in this chapter. Finally, the importance of gene regulation in the processes involved in cancer is explained and emphasized.

## FOCAL POINTS

- Figure 14.2 [p.210] illustrates various levels at which gene expression can be controlled in the eukaryote.
- Figure 14.6 [p.213] shows how gene expression is involved in flower formation.
- Figure 14.9 [p.216] depicts the control of the lactose operon in bacteria, a model for gene regulation in prokaryotes.
- Table 14.1 [p.217] compares genetic regulation in prokaryotes and eukaryotes.

## Interactive Exercises

*Between You and Eternity* [pp.208–209]

## 14.1. GENE EXPRESSION IN EUKARYOTIC CELLS [pp.210–211]

## 14.2. A FEW OUTCOMES OF GENE CONTROLS [pp.212–213]

### Selected Terms

*Polytene chromosomes* [p.211], *Arabidopsis thaliana* [p.213]

### Boldfaced Terms

[p.210] cell differentiation _____

_____

[p.210] gene expression _____

_____

[p.210] enhancers _____

_____

[p.210] repressor _____

_____

[p.211] transcription factors _____

_____

[p.211] RNA interference _____

_____

[p.212] X-chromosome inactivation _____

_____

[p.213] dosage compensation _____

_____

[p.213] ABC Model _____

_____

[p.213] master genes _____

_____


## 14.3.  THERE'S A FLY IN MY RESEARCH [pp.214–215]

## 14.4.  PROKARYOTIC GENE CONTROL [pp.212–213]

### Selected Terms

*Drosophila melanogaster* [p.214], *antennapedia* [p.214], *eyeless* [p.214], *dunce* [p.214], *wingless* [p.214], *wrinkless* [p.214], *minibrain* [p.214], *tinman* [p.214], *groucho* [p.214], *Pax-6* [p.214], *Escherichia coli* [p.216]

### Boldfaced Terms

[p.214] homeotic genes _____

_____

[p.214] knockout experiments _____

_____

[p.215] pattern formation _____

_____

[p.217] operon _____

_____

# Labeling

For each of the following statements, choose the correct letter from the diagram below. [p.210]

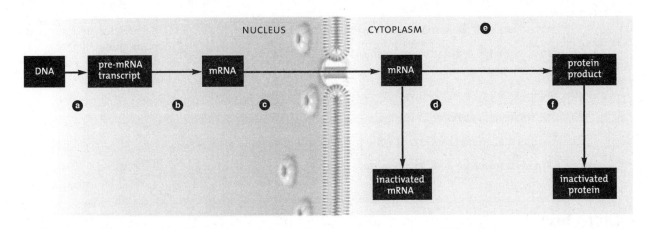

_____ 1. Newly formed peptide chains are activated or disabled.

_____ 2. Chemical modifications to the DNA restrict access to genes.

_____ 3. Transport proteins deliver mRNA to the correct region of the cytoplasm.

_____ 4. The length of the poly-A tail determines how long the mRNA is available for translation.

_____ 5. Inactivation of translation initiation factors blocks translation.

_____ 6. pre-mRNA splicing

# Fill in the Blanks

All cells of your body started out life with the same (7) _____ [p.210], because everyone arose by mitotic cell divisions from the same fertilized egg. And they all transcribe many of the same genes, because they are alike in most aspects of (8) _____ [p.210] and basic housekeeping activities.

In other ways, however, nearly all of your body cells become (9) _____ [p.210] in composition, structure and (10) _____ [p.210]. This process of cell (11) _____ [p.210] occurs during the development of multicelled organisms. Differences arise among cells that use a different subset of (12) _____ [p.210]. Specialized tissues and organs are the result.

# Choice

For each of the following, chose the stage of control over gene expression that is indicated by the statement. Answers may be used more than once. [p.211]

      a. controls at translation     b. controls after translation     c. controls before transcription
      d. control of transcript processing

_____ 13. The greatest range of controls occurs at this stage.

_____ 14. The modification of protein products occurs at this stage.

_____ 15. This stage dictates the timing of mRNA translation.

_____ 16. Histones are involved at this stage.

_____ 17. Base sequences at the untranslated ends of the mRNA act as zip codes.

## Matching

For each of the following, choose one of the terms from the list below. Some terms may be used more than once. [pp.212–215]

_____ 18. This occurs in female mammals and marsupials.

_____ 19. These code for regulatory proteins that are important in development.

_____ 20. The master genes of most eukaryotic organisms.

_____ 21. This forms a Barr body in the cell.

_____ 22. Ensures that equal amounts of gene products are present in each sex.

_____ 23. A form of selective gene expression.

a. X-chromosome inactivation
b. homeotic genes
c. dosage compensation
d. all of the above

---

# Self-Quiz

1. An arrangement where a promoter and a set of operators control more than one gene is called a _____. [p.217]
   a. translational unit
   b. gene
   c. operon
   d. enhancer
   e. none of the above

2. Which of the following acts as an activator protein for the lactose operon in _E. coli_? [p.216]
   a. lactose
   b. glucose
   c. ATP
   d. CAP
   e. all of the above

3. _____ bind to enhancer regulatory regions and speed up transcription. [p.210]
   a. activators
   b. repressors
   c. polymerases
   d. operators

4. The addition of a poly-A tail to an mRNA is an example of what level of control? [p.211]
   a. control after translation
   b. control at translation
   c. control at transcript processing
   d. control before transcription

5. Barr bodies are produced as a result of _____. [p.212]
   a. cell differentiation
   b. translational control
   c. X-chromosome inactivation
   d. selective gene expression

6. In multicelled organisms, cell differentiation occurs as a result of _____. [p.210]
   a. growth
   b. fertilization
   c. proto-oncogenes
   d. selective gene expression

7. Blocking an RNA polymerase's access to a gene is an example of which of the following controls? [p.210]
   a. control after translation
   b. control at translation
   c. control at transcript processing
   d. control before transcription

8. An experimental procedure in which genes are deleted from a wild-type organism to observe the effects on the organism is called a _____ experiment. [p.214]
   a. knockout
   b. knockdown
   c. homeotic
   d. X-inactivation
   e. none of the above

9. The emergence of embryonic tissues in predictable locations is called _____. [p.215]
   a. X-chromosome inactivation
   b. dosage compensation
   c. pattern formation
   d. selective gene expression

10. The master control genes of development in eukaryotic cells are called _____. [p.214]
   a. repressor genes
   b. homeotic genes
   c. proto-oncogenes
   d. activator genes

# Chapter Objectives/Review Questions

This section lists general and detailed chapter objectives that can be used as review questions. You can make maximum use of these items by writing answers on a separate sheet of paper. To check for accuracy, compare your answers with information given in the chapter or glossary.

1. Define the role of operators, enhancers, and promoters in gene control. [pp.210–211]
2. Define an operon. [p.217]
3. Identify the mechanisms of negative and positive control of the *lac* operon in *E. coli*. [pp.216–217]
4. Define cell differentiation. [p.210]
5. Recognize the various levels of control that are involved in gene expression. [pp.210–211]
6. Recognize the relationship between selective gene expression and cell differentiation. [pp.210–211]
7. Define a homeotic gene and the role of the homeodomain. [p.214]
8. Explain the process of X-chromosome inactivation and its relationship to dosage compensation. [p.212]
9. How are knockout experiments used to understand the process of gene control? [p.214]
10. What is pattern formation? [p.215]

# Chapter Summary

(1) _____ mechanisms govern when, how, and to what extent a cell's genes are (2) _____. They alter gene expression in response to changing (3) _____ both inside and (4) _____ the cell.

Diverse controls govern every step between gene (5) _____ and delivery of the final gene (6) _____ to a targeted location.

In multicelled species, (7) _____ genes guide the stage-by-stage development of new individuals. (8) _____ gene expression results in cell (9) _____. By this process, different cell lineages become specialized in composition, structure, and (10) _____.

The orderly, (11) _____ expression of certain genes is the basis of the body plan of complex (12) _____ organisms. In female mammals, most of the genes on one of the two X chromosomes are (13) _____ in every cell.

*Drosophila* research revealed how a (14) _____ body plan emerges. All cells in a developing embryo inherited the same genes, but they selectively (15) _____ or (16) _____ different fractions of those genes.

Prokaryotic gene controls govern responses to short-term changes in (17) _____ availability and other aspects of the (18) _____. The main gene controls bring about fast adjustments in rates of (19) _____.

# Integrating and Applying Key Concepts

1. In the past prescription and over-the counter drugs interacted primarily with receptors located on the surface of the cells. Many newer forms of drugs now actively target gene regulation at the transcriptional or translational level. Why are these drugs typically more effective than their earlier counterparts? What area of gene regulation do you think could be targeted by the next generation of drugs to make them safer and more effective?
2. The lac operon in *E. coli* is said to be an inducible operon, that is it is induced by its substrate lactose. Explain why this type of control mechanism is very efficient for the bacterial cell.

# 15

# STUDYING AND MANIPULATING GENOMES

## INTRODUCTION

The information that you have learned so far allows an almost complete understanding of the structure of DNA and the ways that DNA is used to produce proteins. You also now understand the processes involved with the cellular treatment of DNA. As scientists began to unravel these processes, new technology was developed. This technology allowed us to intentionally manipulate DNA and create organisms that contained desired DNA, often from an entirely different species. The development of this technology holds much promise for the medical field and the agricultural industry, in particular. However, the intentional manipulation of DNA is not without risk and controversy. This chapter first discusses some of the various technologies involved in the intentional manipulation of DNA. It then goes on to cover the potential benefits of genetic engineering, and the potential concerns.

## FOCAL POINTS

- Figure 15.2 [p.222] illustrates the use of enzymes in the creation of a recombinant DNA molecule.
- Figure 15.3 [p.222] shows a representative plasmid which is often used as a DNA vector.
- Figure 15.4 [p.223] animates the formation of a recombinant DNA molecule.
- Figure 15.7 [p.225] animates the polymerase chain reaction which is used to amplify DNA.

## Interactive Exercises

*Golden Rice or Frankenfood?* [pp.218–219]

## 15.1.  A MOLECULAR TOOLKIT [pp.222–223]

## 15.2.  FROM HAYSTACKS TO NEEDLES [pp.224–225]

*Selected Terms*

"Golden rice" [p.218], genetically modified organism (GMO) [p.221], *Haemophilus influenzae* [p.222], "sticky" end [p.222], genomic [p.224], Taq polymerase [p.225], *Thermus aquaticus* [p.225]

## Boldfaced Terms

[p.222] restriction enzyme _____

_____

[p.222] recombinant DNA _____

_____

[p.222] plasmid _____

_____

[p.222] DNA cloning _____

_____

[p.222] cloning vectors _____

_____

[p.223] clones _____

_____

[p.223] reverse transcriptase _____

_____

[p.223] cDNA _____

_____

[p.224] gene library _____

_____

[p.224] probe _____

_____

[p.224] nucleic acid hybridization _____

_____

[p.224] polymerase chain reaction (PCR) _____

_____

[p.225] primers _____

_____

*Plasmids*

If the plasmid shown below was cut with the restriction enzyme EcoR1,

1. How many fragments would be present?

2. Would the largest fragment contain an intact gene for ampicillin resistance?

3. How many HindIII sites would remain in the DNA?

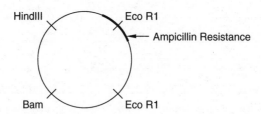

For each of the following, match the word or phrase with its correct definition.

_____ 4. reverse transcriptase [p.223]

_____ 5. restriction enzyme [p.222]

_____ 6. DNA ligase [p.222]

_____ 7. cDNA [p.223]

_____ 8. plasmid [p.222]

a. seals the nicks in a recombinant DNA molecule
b. a small circular piece of DNA located outside the chromosome in bacteria
c. a viral enzyme that uses mRNA to form cDNA
d. complementary DNA that is made from mRNA
e. a plasmid that accepts foreign DNA and allows it to be moved into a host cell
f. This cuts double-stranded DNA at a specific base sequence.

*Labeling*

Choose the correct description for each of the numbers in the diagram below. [p.223]

_____ 9.

_____ 10.

_____ 11.

_____ 12.

_____ 13.

_____ 14.

a. The DNA fragments and plasmid DNA are mixed together with DNA ligase.
b. A restriction enzyme cuts a specific base sequence in the DNA.
c. The DNA fragments have sticky ends.
d. Recombinant plasmids are formed.
e. The plasmid DNA has sticky ends.
f. A restriction enzyme cuts the same sequence in plasmid DNA.

## 15.3. DNA SEQUENCING [p.226]

## 15.4. ANALYZING DNA FINGERPRINTS [p.227]

### Boldfaced Terms

[p.226] DNA sequencing _____

_____

[p.226] gel electrophoresis _____

_____

[p.227] DNA fingerprint _____

_____

[p.227] short tandem repeats _____

_____

### Complete the Table

For each of the following, provide the role of the procedure in the study of molecular biology.

| Procedure | Role |
|---|---|
| gel electrophoresis [p.226] | 1. |
| DNA fingerprinting [p.227] | 2. |
| PCR [p.224] | 3. |
| nucleic acid hybridization [p.224] | 4. |
| DNA sequencing [p.227] | 5. |

### Matching

Match each of the following terms to its correct definition.

_____ 6. probe [p.224]

_____ 7. tandem repeats [p.227]

_____ 8. gene library [p.224]

_____ 9. *Thermus aquaticus* [p.225]

_____ 10. primers [p.225]

_____ 11. cDNA library [p.223]

a. synthetic nucleotide sequences that are used to identify the ends of a DNA fragment for PCR

b. a collection of bacterial cells that house cloned fragments of DNA

c. the heat-tolerant organism that provides the polymerase for PCR reactions

d. a radioisotope labeled fragment of DNA used in nucleic acid hybridization

e. cloned fragments that are obtained from mRNA

f. multiple copies of short DNA sequences, positioned one after another on the chromosome

12. Following is a DNA fingerprint from a crime scene. Please describe how DNA fingerprinting is used to convict suspects, and tell which of the following suspects is probably guilty of this crime.

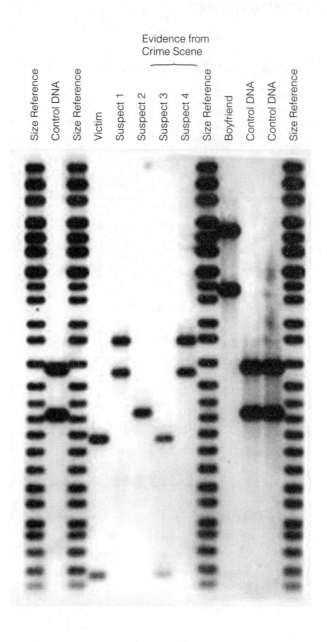

_____

_____

_____

_____

_____

## 15.5. THE RISE OF GENOMICS [pp.228–229]

## 15.6. GENETIC ENGINEERING [p.230]

### Selected Terms

Human Genome Project [p.228], structural genomics [p.229], comparative genomics [p.229], microarrays [p.229], *Eschericia coli* [p.230]

### The People and Institutions [p.228]

Describe the role of each of the following people or institutions in the sequencing of the human genome.

Walter Gilbert _____

NIH _____

James Watson _____

Craig Ventner _____

President Clinton and Tony Blair _____

### Boldfaced Terms

[p.229] genomics _____

_____

[p.229] DNA chips _____

_____

[p.230] genetic engineering _____

_____

[p.230] transgenic organism _____

_____

## 15.7. DESIGNER PLANTS [pp.230–231]

## 15.8. BIOTECH BARNYARDS [pp.232–233]

## 15.9. SAFETY ISSUES [p.233]

## 15.10. MODIFIED HUMANS? [p.234]

### Selected Words

*Agrobacterium tumefaciens* [p.230], *Bacillus thuringensis* [p.230], "Frankenfood"

### Boldfaced Terms

[p.231] xenotransplantation _____

_____

[p.234] gene therapy _____

_____

## True–False

If the statement is true, write "T" in the space provided. If false, correct the underlined word so that the statement becomes true.

_____ 1. Transferring a gene from one species to another is an example of <u>xenotransplantation</u>.

_____ 2. *Agrobacterium tumefaciens* is commonly used in the process of <u>human gene therapy</u>.

_____ 3. <u>DNA chips</u> are now used by researchers to rapidly screen genomes for genes of interest.

_____ 4. *Bacillus thuringensis* is commonly used as an <u>insecticide</u>.

## Choice

Choose which area of research is best described by each statement. Some answers may be used more than once.

       a. genetic engineering [p.230]    b. human gene therapy [p.234]    c. xenotransplantation [p.232]
       d. DNA chips [p.229]

_____ 5. the replacement of a defective gene with a healthy one to fight a specific disease

_____ 6. a source of drugs and protein for medical studies

_____ 7. used to study the expression of genes under specific conditions

_____ 8. the development of transgenic plants and animals

_____ 9. the movement of organs from one species to another

_____ 10. an array of the genes in a genome on a microscope slide

# Self-Quiz

1. The molecule cuts double-stranded DNA at specific locations. [p.222]
   a. plasmid
   b. DNA ligase
   c. restriction enzymes
   d. *Taq* polymerase
   e. none of the above

2. The movement of a gene from one species to another is an example of _____. [p.232]
   a. xenotransplantation
   b. genetic engineering
   c. human gene therapy
   d. eugenic engineering
   e. none of the above

3. A small, circular piece of bacterial DNA that contains only a few genes is called a _____. [p.222]
   a. ligase
   b. restriction enzyme
   c. tandem repeat
   d. plasmid
   e. cDNA

4. A DNA chip may contain thousands of genes in an area the size of a postage stamp. [p.229]
   a. true
   b. false

5. Which of the following is used to manufacture a cDNA library? [p.223]
   a. reverse transcriptase
   b. restriction enzymes
   c. DNA ligase
   d. *Taq* polymerase
   e. PCR

6. The _____ procedure can make billions of copies of a DNA sequence in a short period of time. [p.224]
   a. nucleic acid hybridization
   b. DNA fingerprinting
   c. automated DNA sequencing
   d. polymerase chain reaction
   e. genetic engineering

7. A radioisotope labeled probe is used in
   _____. [p.224]
   a. nucleic acid hybridization
   b. polymerase chain reaction
   c. automated DNA sequencing
   d. a cDNA library
   e. the use of restriction enzymes

8. _____ is involved with finding the evolutionary relationships among groups of organisms.
   [p.229]
   a. structural genomics
   b. functional genomics
   c. comparative genomics
   d. eugenic engineering
   e. all of the above

9. The separation of DNA fragments based on
   their size is called _____. [p.226]
   a. DNA fingerprinting
   b. polymerase chain reaction
   c. nucleic acid hybridization
   d. genetic engineering
   e. gel electrophoresis

10. Tandem repeats are used primarily for which
    of the following? [p.227]
    a. DNA fingerprinting
    b. nucleic acid hybridization
    c. polymerase chain reaction
    d. automated DNA sequencing
    e. all of the above

## Chapter Objectives/Review Questions

This section lists general and detailed chapter objectives that can be used as review questions. You can make maximum use of these items by writing answers on a separate sheet of paper. To check for accuracy, compare your answers with information given in the chapter or glossary.

1. Define recombinant DNA. [p.222]
2. Understand the difference in function between a DNA ligase and a restriction enzyme. [p.222]
3. Recognize how plasmids can be used as cloning vectors. [p.222]
4. Understand the relationship between cDNA and mRNA, and the differences between cDNA and genomic DNA. [p.223]
5. Understand the concept of a gene library and how it can be screened using nucleic acid hybridization. [p.224]
6. Understand the process of the polymerase chain reaction. [p.224]
7. How does gel electrophoresis separate DNA fragments? [p.226]
8. Why are tandem repeats important in the process of DNA fingerprinting? [p.226]
9. How can DNA fingerprinting be used to distinguish between individuals? [p.227]
10. Understand the goals of the Human Genome Project. [p.228]
11. What is the difference between structural and comparative genomics? [p.229]
12. What are the uses of a DNA chip? [p.229]
13. What is genetic engineering? [p.230]
14. What is xenotransplantation? [p.231]
16. How can gene therapy be used to fight human disease? [p.232]

## Chapter Summary

Researchers routinely make (1)_____ DNA by isolating, cutting, and joining DNA from different (2) _____. Plasmids and other (3) _____ can carry foreign DNA into host cells.

Researchers isolate and make copies of DNA in order to study it. (4) _____ copies particular (5) _____ of DNA in sufficient quantity for experiments and other practical purposes.

(6) _____ reveals the linear order of nucleotides in a fragment of DNA. A DNA (7) _____ is an individual's unique array of DNA sequences.

Genomics is the study of (8) _____. Comparisons of the genomes of different (9) _____ offer practical benefits.

Genetic engineering—the directed modification of an organism's (10) _____—is now used routinely in research and medical applications. It continues to raise many (11) _____ questions.

## Integrating and Applying Key Concepts

1. You have been assigned as part of a research team to produce a new transgenic plant. You have determined that a specific gene of interest needs to be copied from one plant, cloned, and then inserted into another species of plant. Using the procedures provided in this chapter, outline how that procedure could be performed.
2. A group of concerned citizens comes to you with questions about the safety of your genetically engineered plant. They say to you "genetic pollution is forever." What do you say to these citizens about the safety of your plant and why it should be produced? What are the possible benefits of genetically modified plants?

# 16

# EVIDENCE OF EVOLUTION

## INTRODUCTION

This chapter presents a brief history of evolutionary thought from the time of the ancient Greeks up to the discoveries of Darwin and Wallace. The chapter then presents supporting evidence of how natural selection has influenced life on earth. As you proceed through the chapter, note how each class of evidence supports the others and strengthens support for natural selection being the mechanism of evolutionary change.

## FOCAL POINTS

- The review of natural selection [p.245] relates Darwin's initial theory to what is currently known from genetics and molecular biology.
- Figure 16.12 [p.249] overlays major biological events with the geologic time scale and introduces the concepts of mass extinctions and adaptive radiation.
- Figure 16.12 [p.243] relates geologic time to a 12-hour clock.
- Figure 16.28 [p.261] provides a useful overview of the major groups of organisms.

---

## Interactive Exercises

*Measuring Time* [pp.238–239]

## 16.1. EARLY BELIEFS, CONFOUNDING DISCOVERIES [pp.240–241]

## 16.2. A FLURRY OF NEW THEORIES [pp.242–243]

## 16.3. DARWIN, WALLACE, AND NATURAL SELECTION [pp.244–245]

*Selected Words*

Chain of Being [p.240], "fluida" [p.242], *descent with modification* [p.244]

*Boldfaced Terms*

[p.240] species _____

_____

[p.240] biogeography _____

_____

[p.241] comparative morphology _____

_____

[p.241] fossils _____

_____

[p.242] evolution _____

_____

[p.242] catastrophism _____

_____

[p.243] theory of uniformity_____

_____

[p.245] natural selection _____

_____

## Matching

Choose the most appropriate answer for each term.

1. _____ comparative morphology [p.241]
2. _____ biogeography [p.240]
3. _____ fossils [p.241]
4. _____ Chain of Being [p.240]
5. _____ species [p.240]
6. _____ descent with modification

a. a theory that all species were formed at the same time and were linked together
b. each kind of being in the Chain of Beings
c. the study of patterns of distribution of species
d. traits of a common ancestor changed over time
e. the rock-like remains of past organisms
f. the study of body plans and structure among groups of organisms

## Choice

For each of the following, choose one of the individuals from the list below.

a. George Cuvier [p.242]    b. Jean Lamarck [pp.242–243]    c. Charles Lyell [p.243]
d. Aristotle [p.240]

7. _____ Theory of uniformity

8. _____ inheritance of acquired characteristics

9. _____ catastrophism

10. _____ Chain of Beings

11. _____ natural processes have more of an effect on Earth than did catastrophes

12. _____ force for change was an intense drive for perfection

13. _____ first to suggest that the environment is a factor in lines of descent

14. _____ one of the first naturalists

*Choice*

For each of the following, choose a term from the list below that best completes the sentence. A term may be used more than once. [pp.244–245]

        a. individual(s)    b. population(s)    c. alleles    d. natural selection

15. _____ Environmental agents of selection act on the range of variations, and the _____ may evolve as a result.

16. _____ Natural _____ have inherent reproductive capacity to increase numbers through successive generations.

17. _____ _____ share a pool of heritable information about traits, encoded in genes.

18. _____ _____ is the outcome of differences in reproduction among individuals of a population that vary in shared characteristics.

19. _____ No _____ can indefinitely grow in size.

20. _____ Variations in traits start with _____, slightly different molecular forms of genes that arise through mutations.

21. _____ Sooner or later, _____ end up competing for dwindling resources.


## 16.4. FOSSILS—EVIDENCE OF ANCIENT LIFE [pp.246–247]

## 16.5. DATING PIECES OF THE PUZZLE [pp.248–249]

*Selected Words*

*Fossil* [p.246], *a fossil record* [p.246], *trace* fossils [p.246]

*Boldfaced Terms*

[p.246] fossilization _____

_____

[p.247] stratification _____

_____

[p.248] half-life _____

_____

[p.248] radiometric dating _____

_____

[p.248] geologic time scale _____

_____

[p.249] macroevolution _____

_____

## Matching

Match each term with the correct statement.

1. _____ fossilization [p.246]
2. _____ fossil record [p.246]
3. _____ stratification [p.247]
4. _____ trace fossils [p.246]
5. _____ half-life [p.248]
6. _____ radiometric dating [p.248]
7. _____ geologic time scale [p.248]

a. the formation of sedimentary rocks
b. indirect evidence of past life, such as tracks and burrows
c. a vertical series of fossils that serves as a record of past life
d. the chronology of Earth's history
e. the use of parent to daughter elements to determine sample age
f. the amount of time required to reduce the amount of radioisotope in half
g. the process by which metal ions replace minerals in bones and hardened tissues

## Short Answer

8. Carbon-14 has a half-life of 5,370 years. Assume that a sample of organic material contains 4.0 grams of carbon-14. How many grams would be left after three half-lives? How many years would have elapsed?

_____

_____

_____

## Labeling

In the geologic time clock below, provide an example of an organism that appears at each indicated time. [p.249]

9. _____
10. _____
11. _____
12. _____
13. _____
14. _____

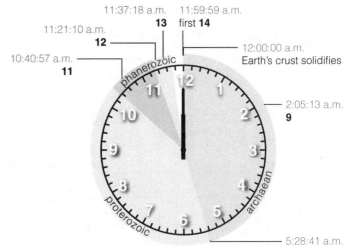

*Choice*

Refer to Figure 16.12 in the text [p.249]. For each of the following events, choose the most appropriate time period from the list provided. Some answers may be used more than once.

a. Cenozoic    b. Proteozoic    c. Archean    d. Paleozoic    e. Mesozoic

15. _____ origin of the vascular plants

16. _____ gymnosperms dominate the land plants

17. _____ origin of the angiosperms

18. _____ origin of the reptiles

19. _____ origin of the amphibians

20. _____ origin of eukaryotic cells

21. _____ origin of modern humans

22. _____ origin of life

23. _____ origin of the mammals

*TNOW:* For interactive exercises of this material, refer to Chapter 16: Art Labeling: Major Extinction Events and Chapter 16: Art Labeling: The Geologic Time Scale.

## 16.6. DRIFTING CONTINENTS, CHANGING SEAS [pp.250–251]

## 16.7. DIVERGENCES FROM A SHARED ANCESTOR [pp.252–253]

## 16.8. CHANGES IN PATTERNS OF DEVELOPMENT [pp.254–255]

## 16.9. CLUES IN DNA, RNA, AND PROTEINS [pp.256–257]

*Selected Words*

"Hot spots" [p.250], *morpho-* [p.252], *homo-* [p.252], "stem reptiles" [p.252], *Dlx* [p.254], *Hox* [p.254], *conserved* [p.256]

*Boldfaced Terms*

[p.250] Pangea _____

_____

[p.250] plate tectonics theory _____

_____

[p.250] Gondwana _____

_____

[p.252] homologous structures _____

_____

[p.252] morphological divergence _____

_____

[p.253] analogous structures _____

_____

[p.253] morphological convergence _____

_____

[p.256] molecular clock_____

_____

[p.257] mitochondrial DNA_____

_____

## Choice

For each of the following, identify the category of evidence to which the statement best applies.

> a. plate tectonics [pp.250–251]  b. morphological divergence and convergence [pp.252–253]
> c. patterns of development [pp.254–255]  d. DNA, RNA, and proteins [pp.256–257]

1. _____ molecular clocks that record the accumulation of neutral mutations

2. _____ homologous structures

3. _____ therapsid fossils in Africa and South America

4. _____ embryonic studies of early embryo stages in vertebrates

5. _____ studies of mitochondrial patterns of inheritance

6. _____ comparison of cytochrome-c protein sequences

7. _____ analogous structures

8. _____ the movement of continents over long periods of time

9. _____ examinations of conserved genes

10. _____ studies of *Hox* and *Dlx*

11. _____ studies of Gondwana and Pangea

12. _____ comparison of iron compasses in North America and Europe

## 16.10. ORGANIZING INFORMATION ABOUT SPECIES [pp.258–259]

## 16.11. HOW TO CONSTRUCT A CLADOGRAM [pp.260–261]

## 16.12. PREVIEW OF LIFE'S EVOLUTIONARY HISTORY [p.261]

### Selected Words

Taxonomy [p.258], *klados* [p.258], *sets within sets* [p.259], *ingroup* [p.260], *outgroup* [p.260], *relative* relatedness [p.260]

## Boldfaced Terms

[p.258] taxon _____

_____

[p.258] monophyletic group _____

_____

[p.258] six-kingdom classification system _____

_____

[p.258] three-domain system _____

_____

[p.258] cladistics _____

_____

[p.258] derived traits _____

_____

[p.259] evolutionary trees _____

_____

## Matching

Choose the most appropriate statement for each term. [pp.258–259]

1. _____ three-domain system
2. _____ taxon
3. _____ cladistics
4. _____ six-kingdom system
5. _____ derived traits
6. _____ evolutionary trees
7. _____ taxonomy
8. _____ monophyletic group

a. Bacteria, Archaea, Eukarya
b. Eubacteria, Archaea, Protists, Plants, Animals, Fungi
c. the science of naming and classifying species
d. these did not appear in the most recent ancestor
e. method of determining evolutionary relationships
f. a group of organisms
g. an ancestor and all of its descendants
h. diagrams presenting the best supported evolutionary relationships

## Cladogram Interpretation

For questions 9-13, refer to the following cladogram of five vertebrate taxa. [p.260]

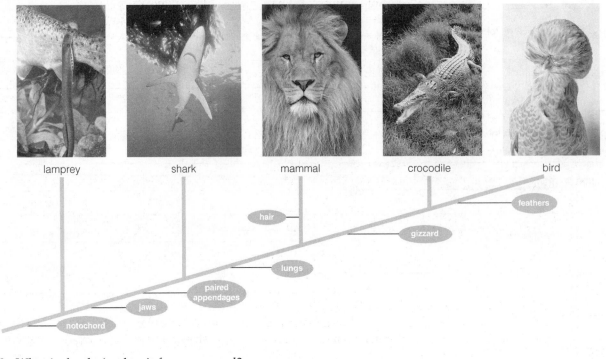

9. What is the derived trait for a mammal?_____

10. What group has jaws and paired appendages, but no lungs? _____

_____

11. What is the unique derived trait shared by birds and crocodiles? _____

_____

12. What taxon on the cladogram could be used as an outgroup? _____

13. Sharks and lampreys share what characteristics? _____

_____

## Self-Quiz

___ 1. Which of the following individuals
proposed the inheritance of acquired
characteristics? [p.242]
a. Cuvier
b. Aristotle
c. Darwin
d. Lamarck
e. Lyell

___ 2. Stratification is an important component in
the study of _____. [p.247]
a. comparative morphology
b. evolutionary trees
c. biogeography
d. fossil record

___ 3. Descent with modification is another name
for _____. [p.244]
a. natural selection
b. catastrophism
c. stratification
d. cladistics

___ 4. A half-life is _____. [p.248]
a. the time between two mass extinction
events
b. the time required for half of a radioactive
isotope to decay
c. half the lifespan of a species
d. a unit of the geological time scale

___ 5. The study of Pangea and Gondwana are
associated with _____. [pp.250–251]
a. biogeography
b. DNA, RNA and proteins
c. development
d. plate tectonics
e. fossil record

___ 6. Evolution of similar body parts in different
lineages is an example of _____.
a. catastrophism
b. morphological divergence
c. morphological convergence
d. speciation

___ 7. A trait that did not appear in the most recent
ancestor is called _____. [p.258]
a. homologous trait
b. analogous trait
c. derived trait
d. isolation trait
e. none of the above

___ 8. Which of the following represents the most
inclusive level of classification? [p.258]
a. species
b. genus
c. kingdom
d. order
e. phylum

___ 9. The scientific method for determining
evolutionary relationships is called
_____. [p.258]
a. cladistics
b. fossilization
c. uniformity
d. radiometric dating

___ 10. Which of the following was Charles Lyell's
contribution to the development of evolutionary thought? [p.243]
a. catastrophism
b. Chain of Beings
c. stratification
d. theory of uniformity
e. inheritance of acquired characteristics

# Chapter Objectives/Review Questions

1. Explain how biogeography, comparative morphology, and fossils contributed to the study of evolution.
[pp.240–241]
2. Give the basic premises of Cuvier's and Lamarck's theories. [p.242]
3. Explain the theory of uniformity and why it played an important role in the study of evolution. [p.243]
4. Describe the evidence that Darwin and Wallace used to develop their theories of natural selection.
[pp.244–245]
5. Explain the process of fossilization and the importance of stratification. [pp.247–248]
6. Explain radiometric dating. [p.248]
7. Understand the geologic time scale. [p.249]
8. Understand the importance of plate tectonics in studying evolution. [pp.250–251]
9. Distinguish between morphological convergence and divergence and give an example of each. [pp.252–253]
10. Recognize the importance of developmental studies in evolutionary research. [pp.254–255]

11. Define the term *molecular clock*. [p.256]
12. Understand the importance of mitochondrial DNA studies. [p.257]
13. Understand the differences between the six-kingdom and three-domain classification systems. [p.258]
14. Understand the concept of cladistics in evolutionary studies. [p.258]
15. Be able to interpret a cladogram. [p.260]

# Chapter Summary

Long ago, Western scientists started to catalog previously unknown species and think about their (1) _____ distribution. They discovered similarities and differences among major groups, including those represented as (2) _____ in layers of sedimentary rock.

Evidence of (3) _____, or changes in lines of descent, gradually accumulated. (4) _____ and Alfred Wallace independently developed a theory of (5) _____ to explain how (6) _____ trans that define each species evolve.

The fossil record offers (7) _____ evidence of past changes in lines of descent.

Correlating evolutionary theories with (8) _____ history helps explain the distribution of (9) _____, past and present.

Species of different (10) _____ often have similar body parts that may be evidence of descent from a shared (11) _____.

(12) _____ comparisons help us discover and confirm relationships among species and lineages.

Species are identified, named, and classified. Evolutionary (13) _____ diagrams are based on the premise that all species interconnect through shared ancestors—some remote, others recent. A current tree groups all organisms into three (14) _____: Bacteria, Archaea, and (15) _____.

# Integrating and Applying Key Concepts

1. In previous chapters, you defined a species as a group of interbreeding organisms that are reproductively isolated from other groups. What problem does this definition hold for scientists who study macroevolutionary processes? How would you propose that the definition be changed?
2. As a researcher, you have discovered two groups of plants on two isolated islands in the Pacific Ocean. These plants have some similar characteristics, but also some different ones. You are not sure whether they represent convergent or divergent evolution. How would you test this?

# 17

# PROCESSES OF EVOLUTION

## INTRODUCTION

This chapter examines what biologists know about how the process of evolution occurs. It presents a series of models that biologists use to examine evolutionary change over time. The chapter also introduces the concepts of speciation, adaptation, and extinction—all of which are potential outcomes of evolutionary change.

## FOCAL POINTS

- Figure 17.4 [p.269] presents the three models of natural selection.
- Figure 17.17 [p.278] illustrates the differences between prezygotic and postzygotic isolation mechanisms.
- Figure 17.23 [p.282] illustrates polyploidy in the history of modern wheat.

## Interactive Exercises

*Rise of the Super Rats* [pp.264–265]

## 17.1. INDIVIDUALS DON'T EVOLVE, POPULATIONS DO [pp.266–267]

*Selected Words*

Morphological traits [p.266], *morpho-* [p.266], physiological traits [p.266] *qualitative* differences [p.266], *quantitative* differences [p.266], *genotype* [p.266], *phenotype* [p.266], polymorphism [p.266], *polymorphos* [p.266]

*Boldfaced Terms*

[p.266] population _____

_____

[p.266] gene pool _____

_____

[p.266] alleles _____

_____

[p.266] lethal mutation _____

_____

[p.266] neutral mutation _____

_____

[p.267] allele frequencies _____

_____

[p.267] genetic equilibrium _____

_____

[p.267] microevolution _____

_____

## Matching

Match each of the following terms with the correct statement. [pp.266–267]

_____ 1. qualitative trait

_____ 2. quantitative trait

_____ 3. genotype

_____ 4. population

_____ 5. polymorphism

_____ 6. neutral mutation

_____ 7. morphological traits

_____ 8. lethal mutation

_____ 9. genetic equilibrium

_____ 10. gene pool

_____ 11. phenotype

_____ 12. microevolution

_____ 13. physiological traits

_____ 14. allele frequencies

a. a trait that has two or more different forms
b. a mutation that results in death for the organism
c. the observable characteristics of an individual
d. a mutation that has no influence on survival or reproduction
e. when a population is not evolving with respect to that gene
f. traits that are based on metabolic activities
g. traits that are based on the form of the species
h. a range of incrementally small variations in a trait
i. one group of individuals of the same species in a specified area
j. the sum of the genes in a population
k. small changes in the allele frequencies of a population over time
l. a gene that has three or more alleles at a frequency of more than 1%
m. an individual's complement of alleles
n. the relative abundance of alleles of a given gene among individuals of a population

## 17.2. WHEN IS A POPULATION NOT EVOLVING? [pp.268–269]

## 17.3. NATURAL SELECTION REVISITED [p.269]

### Selected Words

*Hardy-Weinberg equilibrium equation* [p.268], *directional* selection [p.269], *stabilizing* selection [p.269], *disruptive* selection [p.269]

### Boldfaced Terms

[p.269] natural selection _____

_____

## Population Genetics Problem

In a population of the fruit fly *Drosophila melanogaster*, the frequency of individuals with a certain autosomal recessive wing disorder is 16 percent. Using this information, calculate the following: [pp. 268–269]

1. The value of $q^2$ _____

2. The value of q _____

3. The value of p _____

4. The frequency of homozygous dominant individuals _____

_____

5. The frequency of heterozygous individuals _____

_____

6. Assuming that the population is in Hardy-Weinberg equilibrium, what will the frequency of the recessive mutation be in the gene pool in the next generation?
(HINT: This is represented by the value q.) _____

_____

*TNOW:* For an interactive exercise on this material, refer to Chapter 17: Art Labeling: Relationship between Genotype and Hardy-Weinberg Variables.

## Short Answer

7. List the five conditions of a stable population. [p.268] _____

_____

_____

_____

_____

_____

## Identification

Identify the type of selection in each of the graphs below. [p.269]

8. _____  9. _____  10. _____

8. _____

9. _____

10. _____

## 17.4. DIRECTIONAL SELECTION [pp.270–271]

## 17.5. SELECTION AGAINST OR IN FAVOR OF EXTREME PHENOTYPES [pp.272–273]

### Selected Words

Peppered moth [p.270], pocket mice [p.270], sociable weavers [p.272], black-bellied seedcracker [p.273]

### Boldfaced Terms

[p.270] directional selection _____

_____

[p.270] antibiotics _____

_____

[p.272] stabilizing selection _____

_____

[p.273] disruptive selection _____

_____

### Choice

For each of the following statements, choose the appropriate form of selection from the list below.

> a. stabilizing selection [p.272]      b. disruptive selection [p.273]
> c. directional selection [pp.270–271]

_____ 1. birth weight in human babies

_____ 2. the effects of predation on peppered moths

_____ 3. antibiotic resistance in bacteria

_____ 4. bill size in the black-bellied seedcracker

_____ 5. body weight in sociable weavers

_____ 6. favors the most common phenotype

_____ 7. allele frequencies shift in one direction

_____ 8. phenotypes at both ends of the range are favored

_____ 9. coat color in rock pocket mice

## 17.6. MAINTAINING VARIATION [pp.274–275]

## 17.7. GENETIC DRIFT—THE CHANCE CHANGES [pp.276–277]

## 17.8. GENE FLOW [p.277]

### Selected Words

*Dimorphos* [p.274], *balancing* selection [p.275], malaria [p.275], *probability* [p.276], *Ellis-van Creveld syndrome* [p.277], *emigration* [p.277], *immigration* [p.277]

## Boldfaced Terms

[p.274] sexual dimorphism _____

_____

[p.274] sexual selection _____

_____

[p.275] balanced polymorphism _____

_____

[p.276] genetic drift _____

_____

[p.276] bottleneck _____

_____

[p.276] founder effect _____

_____

[p.277] inbreeding _____

_____

[p.277] gene flow _____

_____

## Matching

Match each of the following definitions to the correct term. [pp.274–275]

_____ 1. a species that has distinct male and female phenotypes

_____ 2. environmental conditions favor heterozygous individuals

_____ 3. a drastic reduction in population size

_____ 4. the movement of alleles between populations

_____ 5. those better at securing mates are more successful genetically

_____ 6. when a very small population forms a new population

_____ 7. random change in allele frequencies over time

_____ 8. when two or more alleles persist at high frequencies in a population

a. sexual dimorphism
b. genetic drift
c. sexual selection
d. founder effect
e. balancing selection
f. balanced polymorphism
g. bottleneck
h. gene flow

## True–False

If the statement is true, place a "T" in the blank. If the statement is false, correct the underlined term so that the statement is correct.

_____ 9. The random changes of allele frequencies associated with genetic drift frequently lead to a <u>gain</u> in genetic diversity. [p.276]

_____ 10. <u>Sexual selection</u> helps maintain the $Hb^A/Hb^S$ heterozygous condition in humans. [p.275]

_____ 11. Bottleneck and founder effects are examples of <u>genetic drift</u>. [p.276]

_____ 12. <u>Gene flow</u> counteracts the effects of mutation, natural selection, and genetic drift. [p.277]

_____ 13. The high rate of Ellis-van Creveld syndrome in Amish populations is due to gene flow. [p.277]

_____ 14. Inbreeding <u>increases</u> the genetic diversity of a population. [p.277]

# 17.9. REPRODUCTIVE ISOLATION [pp.278–279]

# 17.10. ALLOPATRIC SPECIATION [pp.280–281]

# 17.11. OTHER SPECIATION MODELS [pp.282–283]

### Selected Words

Species [p.278], *reproductive isolation* [p.278], *prezygotic* [p.278], *postzygotic* [p.278], *allo-* [p.280], *patria* [p.280], *sym-* [p.282], *auto*polyploid [p.282], *allo*polyploids [p.282]

### Boldfaced Terms

[p.278] speciation_____

_____

[p.278] reproductive isolating mechanisms_____

_____

[p.280] allopatric speciation _____

_____

[p.280] archipelago _____

_____

[p.282] sympatric speciation _____

_____

[p.282] polyploid _____

_____

[p.283] parapatric speciation _____

_____

## Labeling

Provide the indicated label for each of the numbered items in the diagram below. [p.278]

1. _____

2. _____

3. _____

4. _____

5. _____

6. _____

7. _____

8. _____

9. _____

**Different species!**

**1 isolating mechanisms**

a.

**2**: Individuals of different species reproduce at different times.

**3**: Individuals cannot mate or pollinate because of physical incompatibilities.

**4**: Individuals of different species ignore or do not get the required cues for sex.

**5**: Individuals of different species live in different places and never do meet up.

**They interbreed anyway.**

b.

**6**: Reproductive cells meet up, but no fertilization occurs.

**Zygotes form, but . . .**

**7 isolating mechanisms**

c.

**8**: Hybrid embryos die early, or new individuals die before they can reproduce.

**9**: Hybrid individuals or their offspring do not make functional gametes.

No offspring, sterile offspring, or weak offspring that die before reproducing

## Choice

For each of the following, choose the appropriate model of speciation from the list below.

a. allopatric speciation [pp.280–281]    b. sympatric speciation [pp.282–283]
c. parapatric speciation [p.283]

_____ 10. the formation of hybrids is an example

_____ 11. new species form within the home range of an existing species

_____ 12. physical separation between populations

_____ 13. polyploidy is a leading cause

_____ 14. an archipelago is an example

_____ 15. occurs when a population extends across a broad region with diverse habitats

_____ 16. an example is the formation of modern wheat

## 17.12. MACROEVOLUTION [pp.284–285]

## 17.13. FOR THE BIRDS [p.286]

## 17.14. ADAPTATION TO WHAT? [pp.286–287]

### Selected Words

*Niche* [p.284], "adaptation" [p.286], *short-term* adaptation [p.286]

### Boldfaced Terms

[p.284] gradualism_____

_____

[p.284] punctuated equilibrium_____

_____

[p.284] adaptive radiation _____

_____

[p.284] key innovation _____

_____

[p.284] coevolution _____

_____

[p.285] extinction _____

_____

[p.285] mass extinctions _____

_____

[p.286] evolutionary adaptation _____

_____

### Matching

Match each term to the correct definition.

_____ 1. punctuated equilibrium [p.284]

_____ 2. mass extinctions [p.285]

_____ 3. evolutionary adaptation [p.286]

_____ 4. niche [p.284]

_____ 5. gradualism [p.284]

_____ 6. extinction [p.285]

_____ 7. coevolution [p.284]

_____ 8. key innovation [p.284]

_____ 9. adaptive radiation [p.294]

a. evolution occurs over brief time spans, followed by period of little or no change
b. a burst of divergences from a single lineage
c. a structural or functional modification that allows an individual or species to exploit a habitat more efficiently
d. a trait that improves an individual's odds of surviving in a given habitat
e. evolution occurs by slight changes over long time spans
f. two species that evolve together over time
g. the loss of a species
h. a certain way of life
i. a loss of a significant number of lineages

# Self-Quiz

___ 1. The sum of all of the genes of a population is called _____. [p.266]
a. the gene pool
b. microevolution
c. a bottleneck
d. macroevolution
e. none of the above

___ 2. If the alleles frequencies for a specific gene are not changing, the population is said to be _____ for that gene. [p.267]
a. evolving
b. going extinct
c. in genetic drift
d. in genetic equilibrium

___ 3. Which of the following is not correct regarding a stable population? [p.268]
a. there is no mutation
b. the population size is small
c. there is equal reproductive ability
d. mating is random
e. there is no gene flow

___ 4. Antibiotic and insecticide resistance are examples of _____ selection. [p.269]
a. sexual
b. directional
c. stabilizing
d. disruptive

___ 5. The form of selection that tends to preserve intermediate phenotypes is known as _____ selection. [p.272]
a. balancing
b. directional
c. stabilizing
d. disruptive

___ 6. The preservation of a heterozygous condition in a population is called _____. [p.275]
a. genetic drift
b. gene flow
c. stabilizing selection
d. a balanced polymorphism
e. sexual dimorphism

___ 7. The random changes in allele frequencies over time is called _____. [p.276]
a. gene flow
b. equilibrium
c. genetic drift
d. balancing selection

___ 8. Which of the following is not a prezygotic isolation mechanism? [p.278]
a. hybrid sterility
b. temporal isolation
c. behavioral isolation
d. ecological isolation
e. all of the above are prezygotic

___ 9. Which of the following forms of speciation is the result of geographic separation of two species? [p.280]
a. parapatric
b. allopatric
c. sympatric
d. dispatric

___ 10. The rate of evolutionary change that is characterized by rapid periods of change followed by long periods of little or no change is called _____. [p.284]
a. genetic equilibrium
b. punctuated equilibrium
c. genetic drift
d. balanced polymorphism
e. macroevolution

---

# Chapter Objectives/Review Questions

1. Describe the difference between a qualitative and quantitative trait. [p.266]
2. Be able to define gene pool, genotype, and phenotype. [p.266]
3. Distinguish between a lethal and neutral mutation. [p.266]
4. Understand the concept of genetic equilibrium. [p.267]
5. Understand the characteristics of a stable population. [p.268]
6. Be able to use the Hardy-Weinberg equation to determine allele and genotype frequencies in a population. [pp.268–269]

7. Understand the three models of selection and be able to give an example of each. [pp.269–273]
8. Understand the significance of sexual selection. [p.274]
9. Define a balanced polymorphism and give an example. [p.275]
10. Understand how genetic drift, gene flow, and inbreeding influence allele frequencies in a population. [pp.276–277]
11. Know the major forms of prezygotic and postzygotic isolation. [pp.278–279]
12. Distinguish between allopatric, sympatric, and parapatric speciation and give an example of each. [pp.280–283]
13. Recognize the importance of polyploidy as a sympatric speciation mechanism. [p.282]
14. Distinguish between punctuated equilibrium and gradualism. [p.284]
15. Understand the importance of adaptive radiation and coevolution of species. [p.284]
16. Be able to define evolutionary adaptation. [p.286]

# Chapter Summary

(1) _____ evolve. Individuals of a population differ in which (2) _____ they inherit, and thus in (3) _____.

Over generations, any allele may increase or decrease in (4) _____ in a population. Such shifts occur by the

microevolutionary processes of (5) _____, natural selection, (6) _____ drift, and gene (7)_____.

Sexually reproducing species consist of one or more populations of individuals that (8) _____ successfully

under (9) _____ conditions, produce (10) _____ offspring, and are reproductively (11) _____ from other

species. The origin of new species varies in details and duration. Typically, it starts after gene (12) _____ ends

between parts of a population. Microevolutionary events occur (13) _____ and lead to genetic (14) _____ of

the subpopulations. Such divergences are reinforced as (15) _____ isolation mechanisms evolve.

Genetic change above the population level is called (16) _____. Recurring patterns of macroevolution include

preadaptation, (17) _____ radiations, coevolution, and (18) _____.

An evolutionary (19) _____ is a heritable aspect of form, function, (20) _____, or development that increases

an individual's capacity to (21) _____ and reproduce in a particular (22) _____.

# Integrating and Applying Key Concepts

1. Do you think that it would be possible for a viral infection to completely eliminate our species from the planet? What type of evolutionary force would this represent? How would the species genetically respond to this?
2. Why is it important for scientists to assign mathematical models, such as the Hardy-Weinberg equation, to natural phenomena? What are the benefits and limitations to doing this?

# 18

# LIFE'S ORIGIN AND EARLY EVOLUTION

## INTRODUCTION

This chapter provides a quick tour through the history of life, starting with the conditions on a very young Earth more than 4 billion years ago. As you progress through this chapter, take note of the major evolutionary events that occurred as life changed to adapt to changes in the environment of the planet.

## FOCAL POINTS

- Figure 18.7c [p.295] provides a schematic view of the major chemical changes that may have occurred to produce living cells.
- Figure 18.12 [pp.300–301] provides an overview of the major milestones in the history of life.

## Interactive Exercises

*Looking for Life in All the Wrong Places* [pp.290–291]

## 18.1. IN THE BEGINNING [pp.292–293]

## 18.2. HOW DID CELLS EMERGE? [pp.294–295]

*Selected Words*

*Thermus aquaticus* [p.290], polymerase chain reaction (PCR) [p.290], bioprospecting [p.290], *Cyanidium caldarium* [p.290], nanobes [p.291]

*Boldfaced Terms*

[p.292] big bang _____

_____

[p.295] protocell _____

_____

[p.295] RNA world _____

_____

[p.295] ribozymes _____

_____

## Matching

Match the statement to the appropriate term. [pp.290–291]

1. _____ organisms that are ⅒ the size of normal bacteria but still contain DNA

2. _____ a "heat-loving" bacteria that makes PCR possible

3. _____ the process that caused a revolution in biotechnology

4. _____ a red alga that lives in hot acidic springs

5. _____ exploration for organisms in extreme environments

a. *Thermus aquaticus*
b. *Cyanidium caldarium*
c. bioprospecting
d. polymerase chain reaction
e. nanobes

## Choice

For each of the following, choose the most appropriate stage of physical/chemical evolution of life.

a. conditions on the early Earth [pp.292–293]  b. origin of the molecules of life [p.293]
c. origin of agents of metabolism [p.294]  d. origin of the plasma membrane [pp.294–295]
e. origin of self-replicating genetic systems. [p.295]

6. _____ Protocells that were little more than membrane sacs protecting information-storing templates and various metabolic agents of the environment

7. _____ Simple systems of RNA, enzymes, and coenzymes have been created in the lab.

8. _____ Four billion years ago the Earth was a thin-crusted inferno.

9. _____ One hypothesis has simple organic molecules forming in outer space.

10. _____ Life may have started as an RNA world.

11. _____ Atmosphere was a mix of hydrogen, nitrogen, carbon monoxide, and carbon dioxide.

12. _____ Simple metabolic pathways evolved at hydrothermal vents.

13. _____ The Stanley Miller experiments demonstrate that some amino acids could form when atmospheric gases were stimulated by lightening.

14. _____ Metabolism first formed on clay-rich tidal flats.

15. _____ Ribozymes in an RNA world

## Labeling

Provide the missing term for each of the numbered items in the diagram below. [p.295]

16. _____

17. _____

18. _____

19. _____

20. _____

21. _____

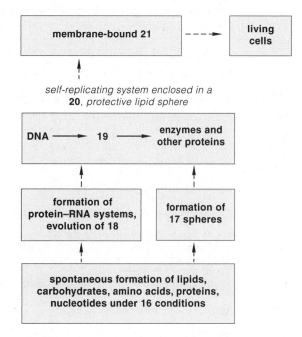

## 18.3. THE FIRST CELLS [pp.296–297]

## 18.4. WHERE DID ORGANELLES COME FROM? [pp.298–299]

## 18.5. TIME LINE FOR LIFE'S ORIGIN AND EVOLUTION [pp.300–301]

### Selected Words

*Bangiomorpha pubescens* [p.296], *endo-* [p.298], *symbiosis* [p.298], *Amoeba proteus* [p.299]

### Boldfaced Terms

[p.296] stromatolites _____

_____

[p.298] endosymbiosis _____

_____

### Matching

Match each of the following with the correct statement.

1. _____ endosymbiosis [p.298]

2. _____ *Bangiomorpha pubescens* [pp.296–297]

3. _____ eukaryotic cells [pp.296–297]

4. _____ stromatolites [pp.296–297]

5. _____ prokaryotic cells [pp.296–297]

a. an interaction of two organisms to the benefit of both
b. these are fossilized remains of early autotrophs
c. the earliest form of cells
d. one of the earliest sexually reproducing species
e. these originated in the Proteozoic in response to increased oxygen levels

## Short Answer

6. List three outcomes of the formation of an oxygen-rich atmosphere. [p.296]

_____

_____

_____

## Choice

For each of the following, choose the most appropriate category from the list below. [pp.298–299]

        a. origin of the nucleus and ER     b. origin of the chloroplast     c. origin of the mitochondria

7. _____ allowed for the protection of the genetic material

8. _____ bacterial origins of these organelles are supported by studies of *Amoeba proteus*

9. _____ cyanobacteria origins of these are supported by studies of the glaucophytes

10. _____ these made the production of ATP by aerobic pathways possible

11. _____ evolved in response to an oxygen-rich atmosphere

12. _____ formed areas in which specific substances could be concentrated

## Labeling

Match each number in the diagram below with the correct event in the history of life. [pp.300–301]

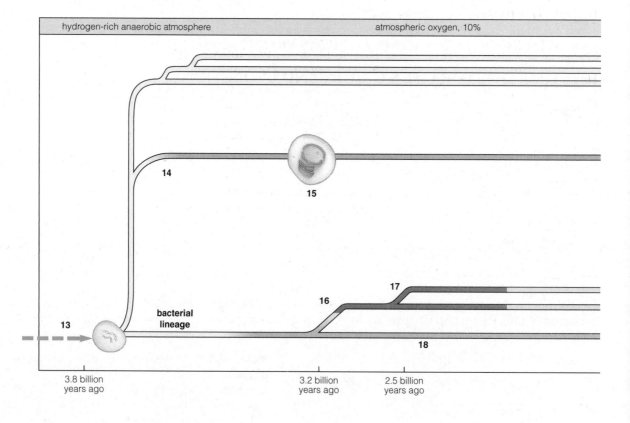

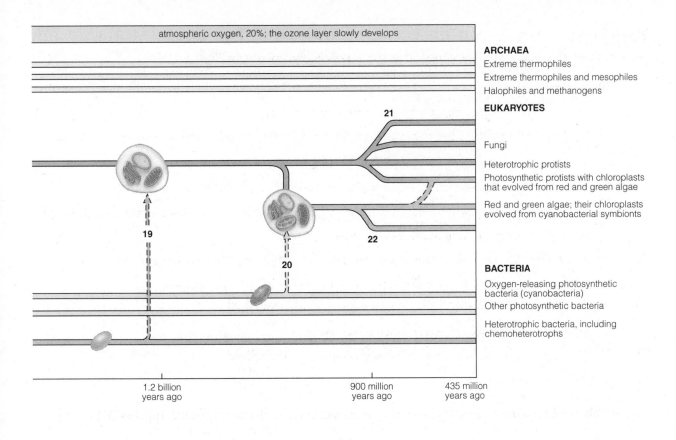

atmospheric oxygen, 20%; the ozone layer slowly develops

**ARCHAEA**
Extreme thermophiles
Extreme thermophiles and mesophiles
Halophiles and methanogens

**EUKARYOTES**

Fungi

Heterotrophic protists
Photosynthetic protists with chloroplasts that evolved from red and green algae

Red and green algae; their chloroplasts evolved from cyanobacterial symbionts

**BACTERIA**
Oxygen-releasing photosynthetic bacteria (cyanobacteria)
Other photosynthetic bacteria

Heterotrophic bacteria, including chemoheterotrophs

1.2 billion years ago

900 million years ago

435 million years ago

13. _____

14. _____

15. _____

16. _____

17. _____

18. _____

19. _____

20. _____

21. _____

22. _____

a. start of aerobic respiration
b. cyclic pathway of photosynthesis
c. noncyclic pathway of photosynthesis
d. origin of the prokaryotes
e. lineage leading to animals
f. endosymbiotic origin of the mitochondria
g. endomembrane system and nucleus
h. lineage leading to plants
i. ancestors of eukaryotes
j. endosymbiotic origin of the chloroplasts

*TNOW:* For an interactive exercise of this material, refer to Chapter 18: Art Labeling: Major Events in the Evolution of Life, Part A and Chapter 18: Art Labeling: Major Events in the Evolution of Life, Part B.

## Self-Quiz

___ 1. The endosymbiotic theory explains which of the following? [p.298]
   a. the evolution of mitochondria and chloroplasts in eukaryotes
   b. the beginning of self-replicating systems
   c. the formation of the plasma membrane
   d. the formation of the nucleus
   e. none of the above

___ 2. Which of the following was absent from the atmosphere of the early Earth? [p.292]
   a. carbon dioxide
   b. carbon monoxide
   c. nitrogen
   d. oxygen
   e. all of the above were present

___ 3. Which of the following systems evolved first? [pp.300–301]
   a. cyclic photosynthesis
   b. noncyclic photosynthesis
   c. aerobic respiration
   d. origins of the chloroplast

___ 4. Which of the following is a "heat-loving" bacteria that made PCR possible? [p.290]
   a. *Thermus aquaticus*
   b. *Cyanidium caldarium*
   c. *Ameoba proteus*
   d. *Bangiomorpha pubescens*

___ 5. What is a stromatolite? [p.296]
   a. the first organism with a nucleus
   b. a nanobe
   c. the first sexually reproducing organism
   d. fossilized remains of ancient autotrophs

___ 6. The development of oxygen in the atmosphere was initially a benefit to living organisms. [p.292]
   a. true
   b. false

___ 7. The RNA world is associated with which of the following? [p.295]
   a. origin of the mitochondria
   b. origin of the eukaryotes
   c. origin of agents of metabolism
   d. origin of self-replicating systems
   e. none of the above

___ 8. Which of the following is probably not located where early organic materials may have been formed? [p.293]
   a. in space on meteorites
   b. near hydrothermal vents
   c. in the atmosphere
   d. on land

___ 9. Which of the following processes probably occurred first on sun-baked tidal flats? [p.294]
   a. origin of self-replicating systems
   b. origin of agents of metabolism
   c. origin of DNA
   d. origin of the nucleus and ER

___ 10. An oxygen-rich atmosphere _____. [p.296]
   a. allowed the formation of the ozone layer
   b. prevented spontaneous formation of life
   c. Allowed the evolution of aerobic respiration
   d. All of the above

## Chapter Objectives/Review Questions

1. What is the significance of organisms such as *Thermus aquaticus* and *Cyanidium caldarium?* [p.290]
2. What is the significance of the big bang? [p.292]
3. List the major components of the Earth's early atmosphere. [p.292]
4. What is the significance of Stanley Miller's experiments? [p.293]
5. What is the role of tidal flats and hydrothermal vents in the evolution of life? [p.293]
6. Explain how metabolic pathways may have initially evolved. [p.294]
7. What were the first plasma membranes made of? [pp.294–295]
8. Explain what is meant by an RNA world? [p.295]
9. What is a stromatolite? [p.296]
10. What effects did the accumulation of oxygen in the atmosphere have on early life? [p.296]
11. What is the significance of *B. pubescens* and *A. proteus*? [pp.296, 299]

12. Why did the nucleus and ER evolve in eukaryotic organisms? [p.298]
13. Explain the importance of the endosymbiotic theory. [pp.298–299]
14. Place important events in the evolution of life in chronological order. [pp.300–301]

---

# Chapter Summary

When Earth first formed about (1) _____ years ago, conditions were too harsh to support life. Over time, its crust cooled, (2) _____ formed, and (3) _____ compounds of the sort now found in living cells may have formed spontaneously or arrived in (4) _____.

Laboratory (5) _____ and advanced computer simulations support the hypothesis that forerunners of living cells arose through known physical and (6) _____ processes, such as the tendency of lipids to assemble into (7)_____-like structures when mixed with water.

The first cells probably were (8) _____ prokaryotes. Some gave rise to (9) _____, others to archaeans and to ancestors of (10) _____ cells. (11) _____ bacteria started releasing oxygen into the (12) _____. Oxygen accumulated over time and became a global (13) _____ pressure.

The nucleus, ER, and other (14) _____ organelles are defining features of eukaryotic cells. Some may have evolved from infoldings of the (15) _____. Mitochondria and (16) _____ appear to be descended from bacterial cells that became modified after taking up permanent residence in host cells.

A (17) _____ for milestones in the history of life offers insight into shared connections among all (18) _____.

---

# Integrating and Applying Key Concepts

1. Some scientists have suggested that it may be possible to transform Mars into an Earth-like planet using microorganisms from Earth. Mars has a thin carbon dioxide atmosphere, but no oxygen. Assuming that water exists in sufficient quantities, what organisms would you first introduce on the planet? Given the history of the Earth, how long would it take to establish an oxygen-rich atmosphere?
2. Jupiter's moon Europa, and Saturn's moon Titan are believed to hold important clues about conditions of the early Earth. Why would it be important for us to go to these locations in order to understand the evolution of life?

# 19

# PROKARYOTES AND VIRUSES

## INTRODUCTION

"Out of sight, out of mind" is often people's mindset when it comes to the myriad microorganisms with which we share this planet. We tend to overlook the tiny organisms that outnumber us by the billions. These tiny organisms that aid us in digestion, decomposition, and vitamin K and pharmaceutical production; that help us remain healthy; and that are vital in environmental cycling of elements are often thought of only when they make us sick. In reality, it is the small minority of these organism that are disease causing, or pathogenic. The vast majority of microbes are not only beneficial, but necessary, to us. This chapter describes some of the microorganisms in our environment and our interactions with them. Two domains of life, the Bacteria and the Archaea, are discussed in this chapter. The metabolic diversity as well as the pathogenic potential of these prokaryotes is explained. Finally, the chapter examines the viruses, the viroids, and the prions; mere chemicals that have come to have important implications in the realm of human health.

## FOCAL POINTS

- Table 19.1 [p.306] reviews the characteristics of the prokaryotic cell.
- Figure 19.4 [p.307] animates prokaryotic fission, the process by which prokaryotic cells divide.
- Table 19.2 [p.312] summarizes the steps in the viral replication cycle.
- Figure 19.14 [p.313] shows the replication of the Herpes simplex virus, a common animal virus that infects humans.
- Table 19.3 [p.315] lists the eight deadliest infectious diseases and their impact on human health.

---

## Interactive Exercises

*West Nile Virus Takes Off* [pp.304–305]

### 19.1. CHARACTERISTICS OF PROKARYOTIC CELLS [pp.306–306]

*Selected Terms*

*Pro-* [p.306], *karyon* [p.306], *photo-* [p.306], *chemo-* [p.306]

*Boldfaced Terms*

[p.306] autotrophs _____

_____

[p.306] heterotrophs _____

_____

[p.307] prokaryotic chromosome _____

_____

[p.307] prokaryotic fission _____

_____

[p.307] horizontal gene transfer _____

_____

[p.307] conjugation _____

_____

[p.307] plasmid _____

_____

[p.307] numerical taxonomy _____

_____

## Labeling

On the following diagram of a prokaryotic cell, label each of the indicated cellular components. Then, match each structure to its correct function. [p.306]

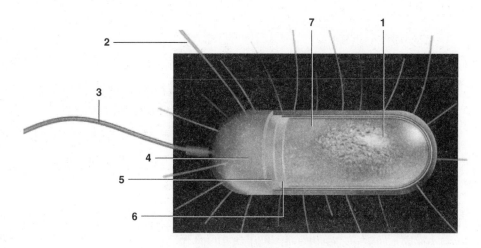

1. _____ ( )
2. _____ ( )
3. _____ ( )
4. _____ ( )
5. _____ ( )
6. _____ ( )
7. _____ ( )

a. this protects the cell from phagocytes
b. the genetic material of a prokaryote
c. this is involved in bacterial conjugation
d. this helps the cell resist rupturing as internal pressure increases
e. this contains the ribosomes, the site of protein synthesis
f. this rotates like a propeller for locomotion
g. the membrane of the cell

## Matching

For each of the following, choose the type of metabolic diversity that is best indicated by the statement. Some answers may be used more than once. [p.306]

_____ 8. use another organism as its energy and carbon source

_____ 9. use the sun's energy, but break down organic compounds for carbon

_____ 10. use photosynthesis and carbon dioxide

_____ 11. oxidize inorganic compounds such as iron and sulfur

_____ 12. these can either be saprophytic or parasitic

a. photoheterotroph
b. chemoheterotroph
c. photoautotrophs
d. chemoautotrophs

## Fill in the Blanks

Most often, a prokaryotic cell reproduces by a division mechanism called (13) _____ [p.307]. Briefly, a parent cell replicates its (14) _____ [p.307] molecule, and the replica docks at the plasma membrane next to the parent molecule. New (15) _____ and _____ [p.307] get added between them and so moves the two apart. The membrane and (16) _____ [p.307] extend into the cell's midsection and divide it into two (17) _____ [p.307] daughter cells.

## Draw

Draw the three typical shapes of bacterial cells [p.306]:

18.


19.


20.



## 19.2. THE BACTERIA [pp.308–309]

## 19.3. THE ARCHAEANS [pp.310–311]

### Selected Terms

*Aquifex* [p.308], *Thermus aquaticus* [p.308], *Spirulina* [p.308], *Anabaena* [p.308], *Escherichia coli* [p.308], *Neisseria gonorrheae* [p.308], *Vibrio cholerae* [p.308], *Heliobacter pylori* [p.304], *Thiomargarita namibiensis* [p.308], *Rhizobium* [p.308], *Chlamydia trachomatis* [p.308], *Lactobacillus* [p.309], *L. acidophilus* [p.309], *Clostridium* [p.309], *Bacillus* [p.309], *Bacillus anthracis* [p.309], *Clostridium tetani* [p.309], *C. botulinium* [p.309], *Borrelia burgdorferi* [p.309], *Nanoarchaeum equitans* [p.311], *Sulfolobus* [p.311]

### Boldfaced Terms

[p.308] cyanobacteria _____

_____

[p.308] nitrogen fixation _____

_____

[p.308] proteobacteria _____

_____

[p.308] pathogen _____

_____

[p.308] disease _____

_____

[p.308] chlamydias _____

_____

[p.309] gram-positive bacteria _____

_____

[p.309] endospores _____

_____

[p.309] spirochetes _____

_____

[p.310] methanogens _____

_____

[p.310] extreme halophiles _____

_____

[p.310] extreme thermophiles _____

_____

## Matching

Choose the correct term for each of the following. Some answers may be used more than once. [pp.308–309]

_____ 1. Also known as the gram-negative bacteria

_____ 2. Some of these produce endospores to survive hostile conditions.

_____ 3. These are usually parasites due to their inability to manufacture ATP.

_____ 4. These have a spring-like appearance and are all motile.

_____ 5. Photoautotrophs that cycle key nutrients in ecosystems

_____ 6. The most diverse group of bacteria

_____ 7. The walls of this group can be stained purple using a specific dye.

a. proteobacteria
b. chlamydias
c. spirochetes
d. cyanobacteria
e. gram-positive bacteria

## Complete the Table

For each of the following species, provide the general classification of bacteria to which the species belongs. Choose from cyanobacteria, gram-positive bacteria, proteobacteria, chlamydia, spirochetes, and Archaea. [pp.308–311]

| Species | Classification |
|---|---|
| Anabaena | 8. |
| E. coli | 9. |
| Spirulina | 10. |
| Lactobacillus acidophilus | 11. |
| Sulfolobus | 12. |
| N. gonorrheae | 13. |
| Rhizobium | 14. |
| Chlamydia trachomatis | 15. |
| Vibrio cholerea | 16. |
| Heliobacter pylori | 17. |
| Borrelia bergdorferi | 18. |
| Bacillus anthracis | 19. |
| Nanoarchaeum equitans | 20. |
| Clostridium botulinum | 21. |

## Choice

For each of the following, choose which group of Archaea bacteria is best described by the statement. Some answers may be used more than once. [pp.310–311]

        a. extreme halophiles      b. methanogens      c. extreme thermophiles      d. all of the above

_____ 22. These live in or near deep-sea hydrothermal vents.

_____ 23. Are strict anaerobes.

_____ 24. These are the source of DNA polymerases in PCR reactions.

_____ 25. Uses a special protein, bacteriorhodopsin, to generate ATP.

_____ 26. Some of these form symbiotic relationships in the guts of termites.

## 19.4. THE VIRUSES [pp.312–313]

## 19.5. VIROIDS AND PRIONS [pp.314–315]

### Selected Terms

*Herpes simplex* [p.312]

### Boldfaced Terms

p.312] virus _____

_____

[p.312] lytic pathway _____

_____

[p.312] lysogenic pathway _____

_____

[p.314] viroid _____

_____

[p.314] prions _____

_____

### Fill in the Blanks

Today, we define a virus as a noncellular (1) _____ [p.312] particle that consists of DNA or (2) _____ [p.312] enclosed within a (3) _____ [p.312] coat. The coat protects the genetic material as the virus journeys to a new (4) _____ [p.312] cell. (5) _____ [p.312] viruses have sheaths, (6) _____ _____ [p.312], and other accessory structures attached to the (7) _____ [p.312]. Some of the coat proteins on each type of virus chemically recognize and bind to specific (8) _____ [p.312] on host cells.

## Labeling

In the diagram below, choose the most appropriate label for each number. [p.313]

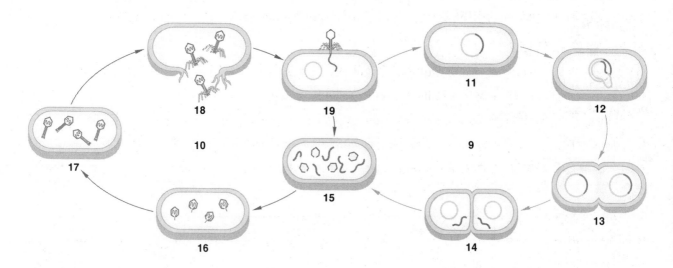

9. _____

10. _____

11. _____

12. _____

13. _____

14. _____

15. _____

16. _____

17. _____

18. _____

19. _____

a. Viral DNA integrates into the host chromosome.
b. After cell division, each daughter cell has recombinant DNA.
c. The viral DNA directs the host cell to make viral proteins and replicate viral DNA.
d. Lysis of the host cell
e. Virus injects genetic material into a suitable host cell.
f. The lysogenic pathway
g. Tail fibers and other parts are added to the coats.
h. Viral proteins are assembled into coats.
i. The viral DNA is excised from the bacterial chromosome.
j. The lytic pathway
l. The bacterial chromosome with the integrated viral DNA is replicated.

## Choice

For each of the following statements, choose one of the following answers.

　　　a. viroids [p.314]　　b. prions [p.314]　　c. viruses [pp.309–310]　　d. none of the above

_____ 20. circles of RNA with no protein coat or protein-encoding genes

_____ 21. is classified as a living organism

_____ 22. infectious proteins

_____ 23. infectious DNA or RNA contained within a protein coat

_____ 24. the etiologic agent of "mad-cow" disease

_____ 25. resemble the introns of eukaryotic DNA

## 19.6. EVOLUTION AND INFECTIOUS DISEASES [p.315]

### Selected Terms

*Epidemic* [p.315], *pandemic* [p.315], *Ebola* [p.315], *Streotococcus pneumoniae* [p.315], *S. mitis* [p.315]

### Fill in the Blanks [p.315]

(1) _____ is the invasion of a cell or multicelled body by a pathogen. (2) _____ follows when the pathogen multiplies and metabolic activities of its descendants interfere with the body activities.

It takes direct or indirect contact with an infected person or their secretions for a (3) _____disease to spread. (4) _____ diseases, such as whooping cough, occur irregularly among very (5) _____ people. (6) _____ diseases, such as tuberculosis, are more or less always present in a population but at low constant levels.

During an (7) _____, a disease quickly spreads though part of a population for just a limited time, then it subsides. (8) _____ refers to epidemics of the same disease breaking out in several countries in the same interval of time. (9) _____ and (10) _____ are two pandemics that have occurred during the last 25 years.

Ebola is a very fatal disease among humans, but the virus has no impact on monkeys because humans are a _____ (11) host for this virus. (12)_____ was the first antibiotic to be manufactured synthetically. This drug prompted optimism about the future of infectious diseases. Unfortunately, antibiotic (13) _____ is becoming a public health emergency, evidenced by the fact that the common childhood bacterium, *Streptococcus pneumoniae*, is now becoming resistant to (14)_____. This resistance was obtained through (15) _____ gene transfer from *S. mitis*.

---

# Self-Quiz

1. Which viral pathway has the genetic material from the virus integrating into the bacterial chromosome? [p.313]
   a. lysogenic pathway
   b. lytic pathway
   c. biogenic pathway
   d. auto-assembling pathway

2. Which of the following would not be present in a prokaryotic cell? [p.306]
   a. ribosomes
   b. polysaccharide coating
   c. membrane-bound organelles
   d. cell wall

3. The most diverse group of bacteria on the planet are the _____. [pp.308–309]
   a. cyanobacteria
   b. chlamydias
   c. proteobacteria
   d. gram-positive bacteria
   e. Archaea

4. Which of the following is classified as a living organism? [pp.312–314]
   a. prions
   b. viroids
   c. viruses
   d. None of the above are considered living organisms.

5. All of the following may be found in a virus, except _____. [pp.312–313]
   a. a protein coat
   b. genetic material
   c. ribosomes
   d. an envelope
   e. tail fibers

6. A disease that occurs in several countries of the world at the same time is called a _____. [p.315]
   a. sporadic outbreak
   b. epidemic
   c. endemic outbreak
   d. pandemic

7. Methanogens, extreme halophiles, and extreme thermophiles are all examples of which of the following bacterial groups? [pp.310–311]

    a. proteobacteria
    b. cyanobacteria
    c. spirochetes
    d. gram-positive
    e. Archaea

8. Which of the following gets their energy from photosynthesis and their carbon from carbon dioxide? [p.306]

    a. chemoheterotrophs
    b. chemoautotrophs
    c. photoheterotrophs
    d. photoautotrophs

9. A infectious protein is called a _____. [p.314]

    a. viroid
    b. virus
    c. prion
    d. archaean

10. The use of a sex pilus to transfer genetic information between bacteria is called _____. [p.307]

    a. lateral gene transfer
    b. asexual reproduction
    c. bacterial fission
    d. bacterial conjugation

## Chapter Objectives/Review Questions

This section lists general and detailed chapter objectives that can be used as review questions. You can make maximum use of these items by writing answers on a separate sheet of paper. To check for accuracy, compare your answers with information given in the chapter or glossary.

1. Understand the physical structure of a typical prokaryote. [p.306]
2. Know the major forms of metabolic diversity for the prokaryotes, and the energy and carbon source for each. [p.306]
3. Understand the processes of bacterial conjugation, prokaryotic fission, and lateral gene transfer as they relate to bacterial genetics. [p.307]
4. Recognize the major groups of bacteria. [pp.308–309]
5. Understand the relationship of the Archaea to the bacteria and eukaryotes. [pp.310–311]
6. Understand the environments in which the Archaea are found. [pp.310–311]
7. Know the major components of a typical virus. [p.312]
8. Be able to label the key steps of the lysogenic and lytic pathways and know the steps in animal virus replication. [p.313]
9. Define a viroid and recognize its possible evolutionary source. [p.314]
10. Understand the difference between an infection and a disease, and between epidemic and pandemic. [p.315]
11. Define antibiotic resistance. [p.315]
12. Recognize the significance of prions in disease. [p.314]

## Chapter Summary

We divide all prokaryotic cells into the domains (1) _____ and (2) _____. These single-celled organisms are structurally simple; they do not have a (3) _____ or the diverse cytoplasmic (4) _____ found in most eukaryotic cells. Collectively, they show great (5) _____ diversity.

Bacteria and archaeans alone reproduce by prokaryotic (6) _____. Exchanges of chromosomal DNA and (7) _____ DNA often occur within and among species.

Bacteria are the most studied (8) _____ species. They are the most abundant and widely distributed organisms. Archaeans, discovered more recently, are less well known. Many are adapted for (9) _____ environments.

Viruses are noncellular particles that cannot (10) _____ themselves without taking over the metabolic machinery of a host cell. A (11) _____ coat encloses their DNA or RNA and a few enzymes. Some viruses become enveloped in membrane when they bud from host cells.

Viroids and (12) _____ are even simpler than viruses. Viroids are short sequences of infectious (13) _____. Prions are infectious misfolded versions of normal (14) _____.

An immense variety of (15) _____, or disease-causing agents, infect human hosts. Pathogens and their hosts have been coevolving by way of (16) _____.

## Integrating and Applying Key Concepts

1. Imagine that you are working on a NASA project in exobiology, which is the study of potential life outside of the Earth. You send a probe to Europa, one of Jupiter's moons, and discover a new species of bacteria. Europa is too far away from the Sun to get any solar input, so what type of metabolism could it have? What group of Earth bacteria do you think it would be most like? Defend your answer.

2. Bacterial conjugation involves the transfer of genetic material through a sex pilus, but it is not considered to be a forma of bacterial reproduction. Why is this not reproduction? What purpose does this serve the bacteria that carry on conjugation?

3. Antibiotic resistance is becoming a public health emergency in the United States and abroad. Do your own research to learn why the following activities are beneficial (and of absolute necessity) in the fight against antibiotic resistance.

   a. Do not take antibiotics for the cold virus.

   b. Take the entire antibiotic dose in your prescription.

   c. Do not share antibiotics.

   d. Avoid eating meat that has been fed antibiotics.

4. How does conjugation (discussed in the previous section) play a role in antibiotic resistance?

# 20

# PROTISTS—THE SIMPLEST EUKARYOTES

## INTRODUCTION

This chapter begins our exploration of the eukaryotic organisms. The protists are a group of organisms that, although often unicellular, has a tremendous impact on human life. Many of these simple organisms are photosynthetic, producing oxygen and removing carbon dioxide from the atmosphere. Members of this group of organisms also are often the base of aquatic food chains. Many industrial, food, and cosmetic products are made from these organisms. Not all protists are so beneficial to us, however. The more notable organisms in this group are responsible for some of the world's most deadly and prevalent diseases. Malaria, for example, infects more than 3 million people worldwide and is caused by protists of the genus *Plasmodium*. This chapter surveys this large and biologically important group of organisms and prepares the student for the study of the more complex eukaryotic species in the following chapters.

## FOCAL POINTS

- Figure 20.2 [p.320] illustrates the typical protist life cycles.
- Figure 20.5 [p.321] shows the organelles of a typical protist, *Euglena*.
- Figure 20.7 [p.323] animates the protist, *Paramecium*, and its organelles.
- Figure 20.9 [p.324] depicts the life cycle of the protist that causes malaria, *Plasmodium*.
- Figure 21.13 [p.327] illustrates the life cycle of *Chlamydomonas*.
- Figure 20.15 [p.328] animates the life cycle of the red algae, *Porphyra*. This is a precursor to the alternation of generations in the plant life cycles that you will read about in the next chapter.
- Table 20.1 [p.330] provides a comparison of prokaryotes with eukaryotes.

---

## Interactive Exercises

*Tiny Critters, Big Impacts* [pp.318–319]

### 20.1.  AN EVOLUTIONARY ROAD MAP [p.320]

### 20.2.  EVOLUTIONARILY ANCIENT FLAGELLATES [p.321]

*Selected Terms*

*Primary endosymbiosis* [p.320], *secondary endosymbiosis* [p.320], *Giardia  lamblia* [p.321], *Trichomonas vaginalis* [p.321], *Trypanosoma brucei* [p.321], *T. cruzi* [p.321], *Leishmania donovani* [p.321]

## Boldfaced Terms

[p.320] protists _____

_____

[p.321] flagellated protozoans _____

_____

[p.321] pellicle _____

_____

[p.321] vector _____

_____

[p.321] euglenoids _____

_____

## Fill in the Blanks [p.320]

Of all existing species, protists are most likely the first (1) _____ cells. Unlike prokaryotes, protist cells have a
(2) _____. Most also have mitochondria, (3) _____ and many have (4) _____. Until recently, protists were
defined mostly by what they were not—not bacteria, not plants, not fungi, not animals. They were lumped
into a (5) _____ meant to signify an evolutionary crossroad between (6) _____ and "higher" forms of life.

"Protists" are being sorted out into a number of (7) _____ groups.

## Diagram

Match the letters on the diagram of the *Euglena* with the following structures. [p.321]

8. long flagellum _____
9. chloroplast _____
10. Golgi body _____
11. eyespot _____
12. pellicle _____
13. endoplasmic reticulum _____
14. mitochondrion _____
15. nucleus _____
16. Because the *Euglena* has a chloroplast, would it be a heterotroph or an autotroph? _____

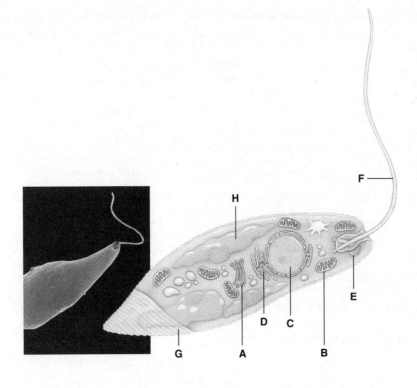

## 20.3. SHELLED AMOEBAS [p.322]

## 20.4. ALVEOLATES [p.322]

## 20.5. MALARIA AND THE NIGHT-FEEDING MOSQUITOES [p.324]

### Selected Terms

*Alveolus* [p.322], *Paramecium* [p.322], *Karenia brevis* [p.323], *Plasmodium* [p.322], *Anopheles* [p.324], *Artemisia* [p.324]

### Boldfaced Terms

[p.322] foraminiferans _____

_____

[p.322] radiolarians _____

_____

[p.322] plankton _____

_____

[p.322] ciliates _____

_____

[p.322] binary fission _____

_____

[p.322] dinoflagellates _____

_____

[p.323] algal blooms _____

_____

## Labeling

For each of the following statements, choose the correct letter on the diagram of *Plasmodium's* lifecycle. [p.324]

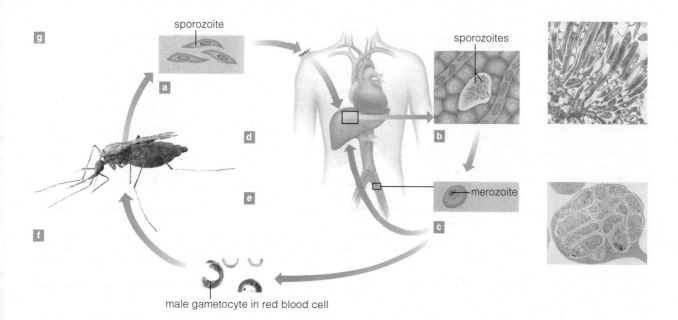

male gametocyte in red blood cell

_____ 1. *Plasmodium* zygotes develop in the gut of female mosquitoes.

_____ 2. Sporozoites asexually reproduce in the liver cells.

_____ 3. Some of the merozoites enter the liver and cause more malaria episodes.

_____ 4. Mosquito bites a human and the bloodstream carries the sporozoites to the liver.

_____ 5. Some merozoites develop into male and female gametophytes and enter the bloodstream.

_____ 6. Female mosquito sucks blood from an infected human.

_____ 7. The merozoites invade red blood cells and reproduce asexually.

## Choice

For each of the following, choose the appropriate group of protists that is described by the statement.

        a. ciliated protozoans [p.322]    b. flagellated protozoans [p.321]    c. apicomplexans [pp.323–24]
        d. euglenoids [p.321]           e. dinoflagellates [pp.322–323]    f. shelled amoebas [p.322]

_____ 8. Body plan includes cellulose plates beneath the plasma membrane and two flagella.

_____ 9. Species that cause malaria belong to this group.

_____ 10. A species of this group may cause cramping and severe diarrhea in humans.

_____ 11. Some members of this group cause leishmaniasis.

_____ 12. Often contain two nuclei, and can reproduce both sexually and asexually.

_____ 13. May cause red tide in coastal marine environments.

_____ 14. These may cause bioluminescence in marine environments.

_____ 15. Possess a unique microtubular device to penetrate a host cell.

_____ 16. Are a component of plankton in the ocean.

_____ 17. Contain an eyespot that directs the organism toward light.

_____ 18. Gave a shell of calcium carbonate and live on the ocean floor.

_____ 19. Part of marine plankton; secrete a glassy, silica shell.

_____ 20. Causes Trichomoniasis, one of the most common, curable sexually transmitted diseases.


## 20.6. SINGLE-CELLED STRAMENOPHILES [p.325]

### Selected Terms

*Phytophthora ramorum* [p.325], *P. infestans* [p.325], *phytoplankton* [p.325]

### Boldfaced Terms

[p.325] oomycotes _____

_____

[p.325] diatoms _____

_____

[p.325] coccolithophores _____

_____

[p.325] golden algae _____

_____


## 20.7. BROWN ALGAE [p.326]

## 20.8. GREEN ALGAE [pp.326–327]

## 20.9. RED ALGAE [p.328]

### Selected Terms

*Macrocystis* [p.326], *Chlamydomonas* [p.326], *Volvox* [p.326], *Ulva* [p.326], *Cladophora* [p.326], *Chlorella* [p.326], *Chara* [p.327], *Porphyra* [p.328]

### Boldfaced Terms

[p.326] brown algae _____

_____

[p.326] chlorophytes _____

_____

[p.326] charophytes _____

_____

[p.326] red algae _____

_____

## Choice

For each of the following, choose the correct group of protists that is being described by the statement. Some statements may have more than one answer.

> a. brown algae [p.326]  b. oomycotes [p.325]  c. green algae [p.326]  d. red algae [p.328]

_____ 1. Extracts of these are used commercially in foods and cosmetics.

_____ 2. The group that is the most similar to plants.

_____ 3. Heterotrophs that absorb nutrients that form during their life cycle.

_____ 4. The diatoms are a member of this group.

_____ 5. Chloroplasts in these organisms may have evolved from cyanobacteria.

_____ 6. The deepest living of the marine algae.

_____ 7. Chalk and limestone deposits are the remains of these organisms.

_____ 8. Many members of this group act as decomposers in aquatic habitats.

_____ 9. Contain both the chlorophytes and charophytes.

_____10. The group that is the most similar to the fungi.

## Complete the Table

For each of the species listed below, indicate the group of protists to which it belongs. Choose from green algae, red algae, brown algae, oomycotes.

| Species | Group |
|---|---|
| Volvox [p.326] | 11. |
| Porphyra [p.328] | 12. |
| Chlamydomonas [p.326] | 13. |
| Chara [p.326] | 14. |
| Phytopthora [p.321] | 15. |

## 20.10.  AMOEBOID CELLS AT THE CROSSROADS [p.329]

### Selected Terms

*Entamoeba histolytica* [p.329], *plasmodial slime molds* [p.329], *cellular slime molds* [p.329], *Dictyostelium discoideum* [p.329], *Amoeba proteus* [p.329], Physarum [p.329]

## Boldfaced Terms

[p.329] amoebas _____

_____

[p.329] slime mold _____

_____

## Complete the Table

For each of the species listed below, indicate the group of protists to which it belongs. Choose from amoeboid protozoans, dinoflagellates, euglenoids, apicomplexans, flagellated protozoans, ciliated protozoans, brown algae, red algae, green algae. List a disease that it causes, if any. If the organism is not pathogenic, write "NP" in the last column.

| Species | Group | Disease |
|---|---|---|
| Paramecium [p.322] | 1. | 11. |
| Plasmodium [p.323] | 2. | 12. |
| Chlamydomonas [p.326] | 3. | 13. |
| Giardia lamblia [p.321] | 4. | 14. |
| Entamoeba histolytica [p.329] | 5. | 15. |
| Trypanosoma brucei [p.321] | 6. | 16. |
| Volvox [p.326] | 7. | 17. |
| Trichomonas vaginalis [p.321] | 8. | 18. |
| Physarum [p.329] | 9. | 19. |
| Euglena [p.321] | 10. | 20. |

# Self-Quiz

1. Red tide and neurotoxic shellfish poisoning are caused by _____. [p.323]
   a. apicomplexans
   b. amoeboid protozoans
   c. dinoflagellates
   d. chlorophytes
   e. ciliated protozoans

2. The chloroplasts of which of the following is most closely related to the cyanobacteria? [p.328]
   a. green algae
   b. euglenoids
   c. brown algae
   d. red algae
   e. diatoms

3. The malaria-causing parasite *Plasmodium* belongs to which of the following? [p.324]
   a. flagellated protozoans
   b. ciliated protozoans
   c. dinoflagellates
   d. oomycota
   e. apicomplexans

4. Calcium carbonate and silica deposits at the bottom of ancient oceans are the remnants of _____. [p.325]
   a. eugleoids
   b. brown algae
   c. oomycotes
   d. chrysophytes
   e. chlorophytes

5. Charophytes and chlorophytes are most closely related to _____. [p.326]
   a. plants
   b. animals
   c. fungi
   d. endophytic fungi
   e. none of the above

6. Which of the following is present in a typical protist?
   a. a nucleus
   b. more than one chromosome
   c. cytoskeleton
   d. ribosomes
   e. all of the above

7. A multicelled, eukaryotic, heterotrophic saprobe would belong to which of the following classifications? [p.326]
   a. animal
   b. plant
   c. protist
   d. fungi
   e. none of the above

8. Charophytes and chlorophytes are most closely related to _____. [p.323]
   a. plants
   b. animals
   c. fungi
   d. endophytic fungi
   e. none of the above

9. Which of the following may form a symbiotic relationship with a fungus to form a lichen? [p.330]
   a. red algae
   b. brown algae
   c. green algae
   d. diatoms
   e. euglenoids

10. Which of the following is present in a typical protist?
    a. a nucleus
    b. more than one chromosome
    c. cytoskeleton
    d. ribosomes
    e. all of the above

## Chapter Objectives/Review Questions

This section lists general and detailed chapter objectives that can be used as review questions. You can make maximum use of these items by writing answers on a separate sheet of paper. To check for accuracy, compare your answers with information given in the chapter or glossary.

1. List the general characteristics of a protist. [p.320]
2. Distinguish between flagellated protozoans, euglenoids, amoeboid protozoans, and ciliated protozoans. [pp.321–323, 329]
3. _____ is the protist responsible for giardiasis. [p.321]
4. Understand the life cycle of *Plasmodium*. [p.324]
5. Understand the relationship between the dinoflagellates and red tide and neurotoxic shellfish poisoning. [p.323]
6. Know the protist source of calcium carbonate and silica deposits. [p.325]
7. Understand the economic significance of the brown and red algae. [pp.326–327]
8. Understand the relationship of the red algae to the cyanobacteria. [p.322]
9. Recognize the similarities of the green algae to the plants. [p.326]
10. Understand the life cycle of a green algae. [p.327]

## Chapter Summary

Protists include many lineages of single-celled (1) _____organisms and their closest (2) _____ relatives. Gene sequencing and other methods are clarifying how protest lineages are related to one another and to plants, (3) _____, and (4) _____.

Two of the earliest lineages of eukaryotic cells are known informally as flagellated (5) _____. The foraminiferans and (6) _____ are another ancient lineage.

Ciliated protozoans, (7) _____, and apicomplexans are single-celled (8) _____, predators, and parasites with a unique layer of tiny (9) _____ under their plasma membrane.

Chrysophytes, diatoms, and brown algae are (10) _____ stramenophiles, most of which are (11) _____. The colorless, parasitic (12) _____ are an offshoot of the stramenophile lineage.

Red and green algae are (13) _____ single cells and multicelled forms. One lineage of multicelled green algae is the closest living relatives of land (14) _____.

A great variety of amoeboid species formally classified as separate lineages are now united as the (15) _____. They are the closest living protistan relatives of (16) _____ and animals.

## Integrating and Applying Key Concepts

1. Some scientists do not believe that the protistans should possess their own kingdom. Assume that you have been asked to reassign the major group of protistans outlined in this chapter into the plant, animal, and fungi kingdoms. Where would you place each group and why? What groups represent a special problem in reclassification? Why?
2. You read in the text that malaria kills a child every 30 seconds. Because this disease is so prevalent and so devastating, many authorities are beginning to suggest the reintroduction of DDT use into malaria endemic regions. DDT was banned in the United States because of its environmental toxicity. Do you feel that DDT should be used to control the mosquitoes that carry malaria? Why or why not?
3. Prepare an evolutionary tree, similar to the one shown in Figure 20.3 of your text, that shows the relationship and organization of the protists in this chapter. Use only the following words in your tree: flagellates, brown algae, diatoms, *Giardia lamblia*, *Trypanosoma brucei*, *Leishmania donovani*, alveolates, shelled amoebas, foraminiferans, radiolarans, paramecium, ciliates, dinoflagellates, *Karenia brevis*, apicomplexans, *Plasmodium*, oomyccotes, coccolithotrophs, *Macrocystis*, green algae, *chlamydomonas*, Red algae, amoebozoans, *Amoeba proteus*, slime molds, *Physarum*, *Trichomonas vaginalis*.

# 21

# PLANT EVOLUTION

## INTRODUCTION

Plants are the reason we, as a species, survive. All animal life on earth, and certainly *Homo sapiens*, is dependent upon plants for oxygen and food. Plants have the unique and vital capacity to produce sugar from carbon found in the atmosphere and water in the environment. Thankfully, in the process they generate oxygen for us to breathe. Plants are a tremendously successful and diverse group of organisms that have been on Earth 475 million years. During this time, the atmosphere and climate of the earth have undergone drastic changes, and the plants have responded to these changes. This chapter discusses the four major groups of plants and their evolution. The life cycle of these four groups are described in detail, as are the structures that are vital to the life cycles of these multicellular, eukaryotic organisms.

## FOCAL POINTS

- Figure 21.2 [p.334] introduces the generalized life cycle for land plants and identifies the difference in sporophytes and gametophytes.
- Figure 21.5 [p.336] animates the life cycle of a moss.
- Figure 21.9 [p.339] illustrates the life cycle of a fern.
- Figure 21.15 [p.343] shows the life cycle of a conifer.
- Figure 21.19 [p.346] diagrams the life cycle of an angiosperm.
- Table 21.1 [p.348] compares the major plant groups.

## Interactive Exercises

*Beginnings and Endings* [pp.232–233]

## 21.1. EVOLUTIONARY TRENDS AMONG PLANTS [pp.334–335]

*Selected Words*

*Cooksonia* [p.334], *Psilophyton* [p.334]

*Boldfaced Terms*

[p.334] gametophyte _____

_____

[p.334] sporophyte _____

_____

[p.334] cuticle _____

_____

[p.334] stomata _____

_____

[p.334] vascular tissues _____

_____

[p.334] xylem _____

_____

[p.334] phloem _____

_____

[p.334] lignin _____

_____

[p.335] pollen grain _____

_____

[p.335] seed _____

_____

## Matching

Match each term with its description.

1. ___ stomata [p.334]
2. ___ lignin [p.334]
3. ___ phloem [p.334]
4. ___ xylem [p.334]
5. ___ cuticle [p.334]
6. ___ gametophyte [p.335]
7. ___ sporophytes [p.335]
8. ___ spores [p.335]
9. ___ pollen grains [p.335]
10. ___ seed [p.335]

a. a waxy coat that conserves water inside shoots on hot, dry days
b. a haploid body that produces haploid gametes
c. vascular tissue that distributes water
d. develops from a diploid zygote into multicelled vegetative bodies that produce spores by way of meiosis
e. resting structures, typically walled, that help a new generation wait out harsh environmental conditions; germinate and develop into gametophytes
f. embryo sporophyte packaged in nutritive tissues and a tough, waterproof coat
g. an opening in the cuticle used to maintain the balance of $CO_2$ and water intake
h. vascular tissue that distributes sugars
i. an organic compound in the cell walls of many plants that lends support
j. formed from small microspores in gymnosperms and angiosperms; a protective wall encloses a few cells that will develop into a mature, sperm-bearing male gametophyte

## 21.2. THE BRYOPHYTES—NO VASCULAR TISSUES [pp.336–337]

### Selected Words

*Polytrichumm* [p.337], *Sphagum* [p.337]

### Boldfaced Terms

[p.336] bryophytes _____

_____

[p.337] rhizoids _____

_____

[p.337] embryophytes _____

_____

[p.337] peat bogs _____

_____

## True–False

If the statement is true, write "T" in the blank. If the statement is false, correct it by writing the correct word(s) for the underlined word(s) in the answer blank.

_____ 1. Mosses, liverworts, and hornworts are all <u>avascular</u> bryophytes.

_____ 2. Bryophytes <u>have</u> true leaves, stems, and roots. [p.336]

_____ 3. Most bryophytes have <u>rhizomes</u> that absorb and attach gametophytes to the soil. [p.336]

_____ 4. In mosses, gametes form by mitosis at <u>gametophyte</u> shoot tips. [p.336]

_____ 5. The moss <u>gametophyte</u> is a stalk and a jacketed structure in which haploid spores develop. [p.336]

_____ 6. When organic remains of peat mosses accumulate over time, they become compressed into exceedingly moist mats called <u>peat bogs</u>. [p.337]

_____ 7. The metabolic products of bryophytes have a(n) <u>basic</u> pH and restrict the growth of bacterial and fungal decomposers. [p.337]

_____ 8. Because of antiseptic properties and absorbency of <u>peat</u>, World War I doctors used it for bandages. [p.337]

_____ 9. <u>Algae</u> are the simplest land plants to display a water conserving waxy cuticle on above-ground parts, a cellular jacket around the gamete-producing parts, and gametophytes are large and don't draw nutrients from sporophytes as they do in land plants. [p.336]

## Dichotomous Choice

For each statement below, refer to the corresponding numbers on these sketches of the moss life cycle (also see Figure 21.5, p.336 in the text). Circle the words in parentheses that make each statement correct. [p.338]

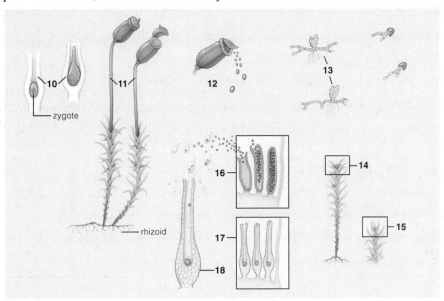

10. The zygote shown is (haploid/diploid) and represents the future (gametophyte/sporophyte) plant.

11. The moss plants shown above the line represent (gametophyte/sporophyte) plants, and their tissues are (haploid/diploid); the moss plants shown below the dotted line represent (gametophyte/sporophyte) plants, and their tissues are (haploid/diploid).

12. (Mitosis/Meiosis) is occurring in the capsule-shaped structure, and the cells being released from it are (haploid gametes/haploid spores).

13. The two structures shown represent young (gametophyte/sporophyte) plants, and their tissues are (haploid/diploid).

14. The moss plant shown here represents a (gametophyte/sporophyte) plant.

15. The moss plant shown here represents a (gametophyte/sporophyte) plant.

16. The cells being released from this structure are (haploid sperms/diploid spores).

17. The cells at the base of the flask-shaped structures are (haploid spores/haploid eggs).

## 21.3. SEEDLESS VASCULAR PLANTS [pp.340–341]

### Selected Words

*Lycopodium* [p.338], *Selaginella* [p.338], *Psilotum* [p.338], *Equisetum* [p.338]

### Boldfaced Terms

[p.338] rhizome _____

_____

[p.340] sori _____

_____

[p.340] epiphytes _____

_____

### Choice

Choose from the following four options to fill in the blanks below:

      a. lycophytes    b. horsetails    c. ferns    d. applies to a, b, and c

_____ 1. *Equisetum* [p.338]

_____ 2. Club mosses, the best-known types, range from the tundra to the tropics. [p.340]

_____ 3. Seedless vascular plants [pp.340–341]

_____ 4. Rust-colored patches, the sori, are on the lower surface of most fronds. [p.339]

_____ 5. Includes tropical tree types and species growing as epiphytes [p.339]

_____ 6. American colonists and pioneers traveling west gathered these stems for use as disposable pot scrubbers. [p.338]

_____ 7. Leaves (fronds) are divided into leaflets. [p.341]

_____ 8. When chamber walls pop open, spores are catapulted through the air. [p.339]

_____ 9. Grow near many streams, railroad tracks, roads, and vacant lots [p.338]

_____ 10. Sporophytes usually have rhizomes, hollow stems, and microphylls that look like scales. [p.338]

_____ 11. Their sporophytes are adapted to conditions on land. [pp.338–339]

_____ 12. *Lycopodium* [p.338]

_____ 13. Young leaves develop in a coiled pattern that looks a bit like a fiddlehead and are uncoiled by maturity. [p.339]

_____ 14. When spores germinate, they give rise to free-living gametophytes no bigger than one millimeter to one centimeter across. [p.338]

_____ 15. Requires environmental water to complete the sexual phase of the life cycle [p.338]

_____ 16. After germination, each spore gives rise to a heart-shaped gametophyte, a few centimeters across. [p.339]

## Dichotomous Choice

For each statement below, refer to the corresponding numbers on these sketches of the fern life cycle (also see Figure 21.9, p.339 in the text). Circle the words in parentheses that make each statement correct. [p.339]

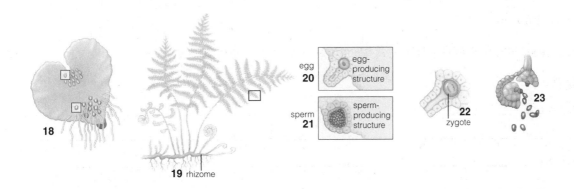

17. This structure represents the (gametophyte/sporophyte) plant; its tissues are (haploid/diploid).

18. The entire structure represents the (gametophyte/sporophyte) plant; its tissues are (haploid/diploid).

19. The egg shown is (haploid/diploid); (mitosis/meiosis) gave rise to the egg.

20. The sperm-producing structure is located on the (gametophyte/sporophyte) plant; its tissues are (haploid/diploid).

21. The zygote is (haploid/diploid) and is located within the egg-producing structure found on the (gametophyte/sporophyte) plant.

22. This structure functions in (sperm production/spore production) and the cells being released from it are (haploid/diploid); the structure itself is located on the (gametophyte/sporophyte) plant.

## 21.4. ANCIENT CARBON TREASURES [p.340]

## 21.5. THE RISE OF SEED-BEARING PLANTS [p.341]

### Selected Words

*Lepidodendron* [p.340], *Calamites* [p.340], *Homo erectus* [p.341], *Aloe vera* [p.341]

### Boldfaced Terms

[p.340] coal _____

_____

[p.341] pollination _____

_____

[p.343] megaspores _____

_____

[p.343] ovule _____

_____

## Matching

Match each term with its description.

1. ___ carboniferous forests [p.340]
2. ___ *Lepidodendron*, a coal fomer [p.340]
3. ___ *Calamites*, a coal former [p.340]
4. ___ coal [p.340]
5. ___ nonrenewable source of energy [p.340]
6. ___ two spore types in seed-bearing plants [p.341]
7. ___ pollination [p.341]
8. ___ ovules [p.341]
9. ___ the rise of seed plants [p.341]
10. ___ seed coat [p.341]
11. ___ edible treasures from flowering plants [p.341]

a. the arrival of pollen on female reproductive parts
b. energy source that cannot be replaced, such as coal
c. coordinates with structural modifications, such as a water-conserving cuticle and two spore types
d. derived from the ovule tissues to protect the embryo
e. giant club mosses growing in Carboniferous forests that rose to nearly forty meters
f. formed by compression into great seams through heat and pressure on peat
g. female reproductive structures that mature into seeds
h. vast, represented by giant club mosses and giant horsetails
i. leaves, fruits, seeds, stems, roots
j. a giant horsetail growing in Carboniferous forests
k. microspores and megaspores

## 21.6. GYMNOSPERMS—PLANTS WITH "NAKED" SEEDS [pp.342–343]

### Selected Words

*Gymnos* [p.342], *sperma* [p.342], *evergreen* [p.342], *deciduous* [p.342], *Ginkgo biloba* [p.342], *Ephedra* [p.342], *Welwitschia* [p.342]

### Boldfaced Terms

[p.342] gymnosperms _____

_____

[p.342] conifers _____

_____

[p.342] cones _____

_____

[p.342] cycads _____

_____

[p.342] ginkgo _____

_____

[p.342] gnetophytes _____

_____

## Choice

Choose from the following five options to fill in the blanks below. [pp.342–343]

a. cycads   b. ginkgos   c. gnetophytes   d. conifers   e. all gymnosperms (includes a, b, c, and d)

_____ 1. The thick, fleshy seeds of female trees smell terrible.

_____ 2. Includes pines, redwoods, and the bristlecone pine

_____ 3. Only a single species survives, the maidenhair tree.

_____ 4. Includes *Welwitschia* and *Ephedra*

_____ 5. Form pollen-bearing cones and seed-bearing cones on separate plants; many wild species face extinction.

_____ 6. Includes firs, spruces, junipers, cypresses, larches, and podocarps

_____ 7. Ovules

_____ 8. Male trees are now widely planted; they have attractive, fan-shaped leaves and resist insects, disease, and air pollutants.

_____ 9. Ovules and seeds are exposed on spore-producing structures.

_____ 10. The sporophyte has a deep taproot and a woody stem with strobili; two strap-shaped leaves may grow to five meters long and split lengthwise repeatedly during growth.

_____ 11. Most species are "evergreen" trees and shrubs with needlelike or scale-like leaves; a few are deciduous.

_____ 12. Includes conifers, cycads, ginkgos, and gnetophytes

## Dichotomous Choice

For each statement below, refer to the corresponding number on these sketches of the pine life cycle (also see Figure 21.15, p.343 in the text). Circle the words in parentheses that make each statement correct. [p.345]

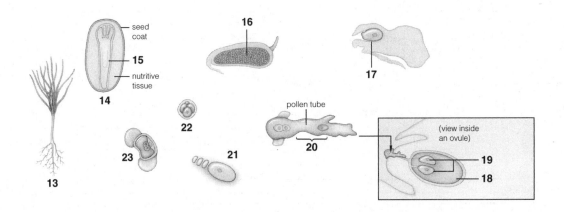

13. This plant is a (gametophyte/sporophyte) and its tissues are (haploid/diploid).
14. The entire structure shown is a (seed/fruit).
15. This structure represents a young (gametophyte/sporophyte) plant; its tissues are (haploid/diploid).
16. The entire structure shown is a(n) (egg sac/pollen sac) and may be found as part of the pine (flower/cone).
17. This structure is the (egg/ovule) and it is (haploid/diploid).
18. The inside cellular mass of this structure represents the (male gametophyte/female gametophyte) plant; its cells are (haploid/diploid).
19. These structures are (eggs/sperms) and they are (haploid/diploid); they were produced by (mitosis/meiosis).
20. The cellular portion of this structure is known as the (male gametophyte/female gametophyte) plant; it is (haploid/diploid).
21. The largest cell shown is a (microspore/megaspore); it was produced by (mitosis/meiosis) and is (haploid/diploid).
22. The quartet of cells shown are (microspores/megaspores); each is (haploid/diploid) and will develop into a(n) (ovule/pollen grain).
23. This structure is a (megaspore/pollen grain) and through pollination, it will be moved to a (male cone/female cone).

## 21.7.  ANGIOSPERMS—THE FLOWERING PLANTS [pp.344–345]

## 21.8.  FOCUS ON A FLOWERING PLANT LIFE CYCLE [p.346]

### Selected Words

*Angio* [p.344], *sperma* [p.344], *Eucalyptus* [p.345]

### Boldfaced Terms

[p.344] angiosperm _____

_____

[p.344] flower _____

_____

[p.344] coevolution _____

_____

[p.344] pollinators _____

_____

[p.345] magnoliids _____

_____

[p.345] eudicots _____

_____

[p.345] monocots _____

_____

## Matching

Match each term with its description.

1. _____ flower [p.344]
2. _____ coevolution [p.344]
3. _____ pollinators [p.344]
4. _____ ovary [p.344]
5. _____ examples of magnoliids [p.345]
6. _____ examples of eudicots [p.345]
7. _____ examples of monocots [p.345]
8. _____ stamen [p.344]
9. _____ carpel [p.344]
10. _____ cotyledons [p.347]

a. cabbages, daisies, most other herbaceous plants, most flowering shrubs and trees, and cacti
b. microspores form here
c. magnolias, avocados, nutmeg, and pepper plants
d. seed leaves
e. agents that deliver pollen of one species to female parts of the same species
f. a specialized reproductive shoot found only in angiosperms
g. megaspores form here
h. orchids, palms, lilies, and grasses such as rye, sugarcane, corn, rice, wheat, and many other highly valued plants
i. where ovules and seeds develop; also usually an immature fruit
j. refers to two or more species evolving jointly because of their close ecological interactions

## Dichotomous Choice

For each statement below, refer to the corresponding numbers on these sketches of the flowering plant life cycle (monocot) (also see Figure 21.19, p.346 in the text). Circle the words in parentheses that make each statement correct. [p.346]

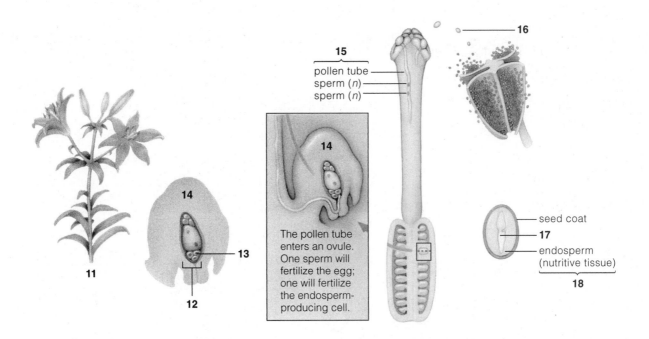

11. The plant is a (gametophyte/sporophyte); its tissues are (haploid/diploid).
12. The reduced plant is a (female gametophyte/female sporophyte); its cells are (haploid/diploid).
13. The single cell is a(n) (egg/spore); it is (haploid/diploid), and it was produced by (mitosis/meiosis).
14. This structure is called an (ovule/ovary); its cells are (haploid/diploid) and it later can develop into a(n) (fruit/seed).

15. The reduced plant is a (male gametophyte/male sporophyte); its cells are (haploid/diploid), and they were produced by (mitosis/meiosis).

16. The small cell is a (released spore/released pollen grain) and it is (haploid/diploid).

17. This structure is a(n) (embryo gametophyte plant/embryo sporophyte plant); its cells are (haploid/diploid), and it was produced by (mitosis/meiosis) of a (haploid zygote/diploid zygote).

18. The entire structure is called a (seed/fruit); it is a matured (ovary/ovule).

---

# Self-Quiz

___ 1. The _____ is *not* a trend in the evolution of plants. [p.334]
   a. evolution of roots, stems, and leaves
   b. shift from homosporous plants to heterosporous plants
   c. shift from diploid to haploid dominance
   d. development of xylem and phloem
   e. development of cuticles and stomata

___ 2. Plants possessing xylem and phloem are called _____ plants. [p.334]
   a. gametophytes
   b. nonvascular
   c. vascular
   d. sporophytes
   e. seedless

___ 3. Existing nonvascular plants do *not* include _____. [p.338]
   a. horsetails
   b. mosses
   c. liverworts
   d. hornworts

___ 4. _____ are *not* seedless vascular plants. [pp.338–339, 342]
   a. lycophytes
   b. gymnosperms
   c. horsetails
   d. *Equisetum* plants
   e. ferns

___ 5. In horsetails, lycopods, and ferns, _____. [pp.338–339]
   a. Spores give rise to gametophytes.
   b. The dominant plant is a gametophyte.
   c. The sporophyte bears sperm- and egg-producing structures.
   d. The gametophyte bears sperm- and egg-producing structures.
   e. Both a and d are correct.

___ 6. _____ are seed plants. [pp.341–346]
   a. Cycads
   b. Ginkgos
   c. Conifers
   d. Angiosperms
   e. All of the above

___ 7. The diploid stage progresses through this sequence: _____. [pp.336, 339, 343, 346]
   a. gametophyte ——> male and female gametes
   b. spores ——> sporophyte
   c. zygote ——> sporophyte
   d. zygote ——> gametophyte
   e. spores ——> gametes

___ 8. In which of the following is the gametophyte the dominant plant? [pp.336, 339, 343, 346]
   a. seedless vascular plants
   b. gymnosperms
   c. bryophytes
   d. angiosperms
   e. ferns

___ 9. Microspores give rise to _____. [p.341]
   a. megaspores
   b. female gametophytes
   c. male cones
   d. pollen grains
   e. embryos

___ 10. Magnoliids, eudicots, and monocots are groups of _____. [p.345]
   a. gymnosperms
   b. bryophytes
   c. club mosses
   d. lycophytes
   e. angiosperms

# Chapter Objectives/Review Questions

1. Define *deforestation* and cite the damage it has done to the world's forests. [pp.332–333]
2. Be able to list the trends that have occurred in the evolution of plants. [pp.334–335]
3. List the groups included in the modern bryophytes and describe their major features. [pp.336–337]
4. Be familiar with the major features of the moss life cycle. The _____ plant is the dominant form in this life cycle. [p.336]
5. The main lineages of seedless vascular plants are lycophytes, horsetails, and _____; describe the habitats and major features of each. [pp.338–339]
6. Be familiar with the major features of the fern life cycle. The _____ plant is the dominant form in this life cycle. [p.339]
7. The major carbon treasure left to us by vast Carboniferous forests was _____. [p.340]
8. Define the following as features that developed to allow the rise of seed-bearing plants: *microspores, pollination, megaspores,* and *ovules.* [p.341]
9. Be familiar with representatives and the features of conifers, cycads, ginkgos, and gnetophytes. [pp.342–343]
10. Be familiar with the major features of the pine life cycle. The _____ plant is the dominant form in this life cycle. [p.343]
11. The outstanding feature of the angiosperms is the _____, a specialized reproductive shoot. [p.344]
12. Define *coevolution, pollinator, magnoliids, eudicots,* and *monocots.* [pp.344–345]
13. Be familiar with the major features of the angiosperm life cycle. The _____ plant is the dominant form in this life cycle. [p.344–347]

# Chapter Summary

The earliest known plants date from (1) _____ million years ago. Ever since then, environmental changes have triggered (2) _____, adaptive (3) _____, and extinctions. Structural and (4) _____ adaptations of lineages are responses to some of the changes.

Bryophytes are (5) _____, with no internal pipelines to conduct water and solutes through the plant body. A (6) _____-producing stage dominates their life cycle, and sperm reach the eggs by swimming through droplets or films of (7) _____.

Lycophytes, whisk ferns, horsetails, and ferns have (8) _____ tissues but do not produce (9) _____. A large spore-producing body that has (10) _____ vascular tissues dominates the life cycle. As with bryophytes, (11) _____ swim through water to reach eggs.

Gymnosperms and, later, angiosperms, radiated into higher and (12) _____ environments. The packaging of male gametes in (13) _____ grains and embryo sporophytes in (14) _____ contributed to the expansion of these groups into new habitats.

Angiosperms alone make (15) _____, which wind, water, and animals help pollinate. In distribution and (16) _____, angiosperms are the most successful group of plants. Nearly all plant species that we rely upon for food are (17) _____.

## Integrating and Applying Key Concepts

1. Prepare a table using the four column titles: "Plant group? Dominant plant? Vascular tissue present? and Seeds Present?" List the following plant groups in the leftmost column: bryophytes, lycophytes, horsetails, ferns, gymnosperms, and angiosperms. Then, fill in the information indicated by the column titles for each plant group. Check your work against the text. List evolutionary trends you can discern after looking over the information.
2. Discuss why an angiosperm, such as an oak tree, can reach great heights, but a bryophyte never grows taller than 8 inches tall.

# 22

# FUNGI

## INTRODUCTION

Somewhere between the plants and the animals are the fungi. While this group of organisms shares characteristics with both, molecular studies have shown us that this diverse group of organisms is more closely related to animals than to plants. The fungi exist worldwide and are important economically, ecologically, and medically. Fungi are important in the food industry in the production of bread, beer, and cheese, for example. Fungi are the most numerous and adept decomposers. And fungi do cause many plant and animal diseases. This chapter describes these organisms and their life cycles, with proper emphasis on their importance and diversity.

## FOCAL POINTS

- Figure 22.4 [p.353] animates the life cycle of a black bread mold, a representative fungus.
- Figure 22.7 [p.355] shows the life cycle of a typical club fungus.
- Figure 22.8 [p.356] illustrates a lichen growing on a birch tree.

## Interactive Exercises

*Food, Forest, Fungi* [pp.350–351]

## 22.1. CHARACTERISTICS OF FUNGI [p.352]

*Boldfaced Terms*

[p.352] fungi _____

_____

[p.352] chytrids _____

_____

[p.352] microsporidians _____

_____

[p.352] mycelium _____

_____

[p.352] hypha _____

_____

## 22.2. ZYGOMYCETES—THE ZYGOTE FUNGI [p.353]

## 22.3. ASCOMYCETES—THE SAC FUNGI [p.354]

## 22.4. BASIDIOMYCETES—THE CLUB FUNGI [p.355]

### Selected Terms

*Rhizopus* [p.353], *R. nigrans* [p.353], *R. oryzae* [p.353], *R. stolonifer* [p.353], *Candida* [p.354], *Saccharomyces cervisiae* [p.354], *Arthobotrys* [p.354], *Penicillium chrysogenum* [p.354], *Aspergillus* [p.354], *Neurospora crassa* [p.354], *Agaricus bisporus* [p.355], *Armillaria ostoyae* [p.355]

### Boldfaced Terms

[p.353] zygomycetes _____

_____

[p.354] sac fungi_____

_____

[p.355] club fungi_____

_____

### Completion

For each of the fungi listed, fill in the table and list what type of fungi they are (zygomycete, sac, club) and list any health, economical, or ecological importance for that organism.

| Fungi | Zygomycete, sac or club | Importance |
|---|---|---|
| *Rhizopus stolonifer* | 1. | 7. |
| *Agaricus bisporus* | 2. | 8. |
| *Penicillium chrysogenum* | 3. | 9. |
| *Arthrobotrys* | 4. | 10. |
| *Aspergillus* | 5. | 11. |
| *Candida* | 6. | 12. |

## 22.5. THE FUNGAL SYMBIONTS [pp.356–357]

## 22.6. AN UNLOVED FEW [p.357]

### Selected Terms

*Phythophthora* [p.356], *Cryphonectria parasitica* [p.356], *Histoplasma capsulatum* [p.357], *Penicillium* [p.357], *Aspergillusa* [p.357], *Claviceps purpurea* [p.357]

## Boldfaced Terms

[p.356] lichens _____

_____

[p.356] mutualism _____

_____

[p.356] mycorrhiza _____

_____

## Labeling

For each number choose the corresponding label from the diagram below. [p.355]

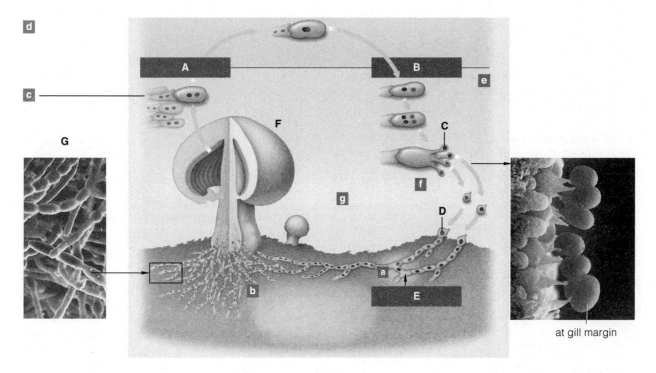

at gill margin

_____ 1. meiosis

_____ 2. a hypha

_____ 3. cytoplasmic fusion

_____ 4. a mushroom or fruiting body

_____ 5. cells with two nuclei form on the gills

_____ 6. a spore

_____ 7. nuclear fusion

*Choice*

For each of the following statements, choose the group that is best described by the statement.

a. lichens [p.330]     b. zygomycetes [p.328]     c. sac fungi [pp.328–329]     d. mycorrhiza [p.330]

_____ 8. Most food-spoiling fungi belong to this group.

_____ 9. This group has the ability to colonize hostile environments.

_____ 10. Baking yeast belongs to this group.

_____ 11. Bread mold belongs to this group.

_____ 12. A fungus and a photoautotroph in a symbiotic relationship

_____ 13. Truffles are this type of fungi.

_____ 14. A fungus in a mutualistic relationship with the roots of a plant.

## Self-Quiz

1. A fungus that forms a mutualistic relationship with the root of a plant is called a _____. [p.356]
   a. lichen
   b. oomycota
   c. endophytic symbiont
   d. mycorrhiza
   e. none of the above

2. A single cellular filament of a fungus is called a _____. [p.352]
   a. hypha
   b. mycelium
   c. spore
   d. fruiting body
   e. mycorrhiza

3. A multicelled, eukaryotic, heterotrophic saprobe would belong to which of the following classifications? [p.356]
   a. animal
   b. plant
   c. protist
   d. fungi
   e. none of the above

4. Which of the following may form a symbiotic relationship with a fungus to form a lichen? [p.356]
   a. red algae
   b. brown algae
   c. green algae
   d. diatoms
   e. euglenoids

5. A mat of absorptive filaments of a fungus is called a _____. [p.352]
   a. hypha
   b. mycelium
   c. spore
   d. fruiting body
   e. mycorrhiza

## Chapter Objectives/Review Questions

This section lists general and detailed chapter objectives that can be used as review questions. You can make maximum use of these items by writing answers on a separate sheet of paper. To check for accuracy, compare your answers with information given in the chapter or glossary.

1. Identify the main characteristics of a fungus. [p.352]
2. Understand the basic anatomy of a fungus. [p.352]
3. Understand the life cycle of a fungus. [p.353]
4. Distinguish between the various groups of fungi. [pp.353–355]
5. Understand the symbiotic relationships of the endophytic fungi, lichens, and mycorrhizae. [p.356]

# Chapter Summary

Fungi are single-celled and multicelled (1) _____ more closely related to (2) _____ than to plants. They feed by secreting digestive enzymes into (3) _____ matter in their surroundings, then absorbing the released nutrients. Multicelled species form a mesh of absorptive (4) _____, some of which intertwine as spore-producing structures.

(5) _____ fungi, (6) _____ fungi, and (7) _____ fungi are three major groups. Others include chytrids, and ancient lineages with flagellated spores, and the (8) _____ parasites called microsporidians.

(9) _____ formation dominates fungal life cycles. Zygote fungi make thick-walled zygotes that give rise to (10) _____ spores. Many sac fungi and club fungi make complex (11) _____-bearing structures. In sac fungi, (12) _____ spores form in a sac. In club fungi, they form at the tip of a club-shaped (13) _____.

Many fungi live on, in, or with other (14) _____. Some live inside plant leaves, stems, or roots. Others form (15) _____ by living with algae or cyanobacteria.

A minority of fungi are (16) _____. Certain species cause diseases in crop plants and (17) _____.

# 23

# ANIMAL EVOLUTION—THE INVERTEBRATES

## INTRODUCTION

The organisms that make up the world of animals are defined by several shared characteristics, including multicellularity, heterotrophic feeding in all, and sexual reproduction and tissue formation in most. While all animals share these common characteristics, they have evolved many different and sometimes unique adaptations that allow them to survive in various environments and conditions. This chapter begins our exploration of the animal world with a look at the invertebrates, the animals with no backbone. From the sponges to the insects and on to the spiny skinned echinoderms, you will learn about the characteristics of these diverse animals that make them able to survive, make them unique, make them important to us, and sometimes give them the ability to cause disease.

## FOCAL POINTS

- Table 23.1 [p.362] shows the major animal groups covered in Chapters 23 and 24.
- Figure 23.3 [p.363] animates the differences between body cavity arrangements in animals.
- Figure 23.6 [p.364] illustrates a proposed family tree for the major animal groups.
- Figure 23.12 [p.367] animates the life cycle of a colonial hydroid.
- Figure 23.13 [p.368] shows the organ systems of a planaria.
- Figure 23.15 [p.369] shows the life cycle of a beef tapeworm.
- Figure 23.16 [p.370] depicts the body segments of the annelids.
- Figure 23.18 [p.371] illustrates the earthworm body plan.
- Figure 23.20 [p.372] illustrates the aquatic snail body plan.
- Figure 23.21 [p.373] illustrates the body plan of the cephalopods.
- Figure 23.22 [p.374] illustrates the roundworm body plan.
- Figure 23.26 [p.376] shows a spider's internal organization.
- Figure 23.28 [p.377] illustrates the clawed lobster body plan.
- Figure 23.30 [p.378] gives examples of insect appendages.
- Figure 23.31 [p.378] shows different patterns of insect development.

# Interactive Exercises

*Old Genes, New Drugs* [pp.360–361]

## 23.1. ANIMAL TRAITS AND TRENDS [pp.362–363]

## 23.2. ANIMAL ORIGINS AND EARLY RADIATIONS [p.364]

### Selected Terms

*Conotoxins* [p.362], *incomplete* digestive system [p.362], *complete* digestive system, *peritoneum* [p.362], *Trichoplax adhaerens* [p.364]

### Boldfaced Terms

[p.362] animals _____

_____

[p.362] epithelium _____

_____

[p.362] connective tissues _____

_____

[p.362] ectoderm _____

_____

[p.362] endoderm _____

_____

[p.362] mesoderm _____

_____

[p.362] radial symmetry _____

_____

[p.362] bilateral symmetry _____

_____

[p.362] cephalization _____

_____

[p.362] gut _____

_____

[p.363] coelom _____

_____

[p.363] segmentation_____

_____

[p.364] placozoan_____

_____

## True–False

If the statement is true, write "T" in the space provided. If false, correct the underlined word so that the statement is correct. [p.362]

_____ 1. All animals are multicelled and <u>do not</u> possess cell walls.

_____ 2. Animals are <u>autotrophs</u>.

_____ 3. Animals typically reproduce sexually, and <u>many</u> reproduce asexually as well.

_____ 4. Most animals are <u>motile</u> during all or part of the life cycle.

## Matching

Choose the most appropriate description for each of the following terms. [pp.362–363]

_____ 5. vertebrates

_____ 6. bilateral symmetry

_____ 7. cephalization

_____ 8. endoderm

_____ 9. invertebrates

_____ 10. ectoderm

_____ 11. mesoderm

_____ 12. placozoan

_____ 13. gut

_____ 14. radial symmetry

_____ 15. coelom

_____ 16. connective tissues

_____ 17. segmentation

_____ 18. pseudocoel

a. the tissue layer that is the source of muscles and other organs; it is between the ecto- and endoderm
b. the site of food digestion
c. cells that bind tissues together and offer structural support
d. the division of the body into repeated interconnecting units
e. animals that possess a backbone
f. a body arrangement in which many appendages and organs are paired
g. a body cavity between the gut and body wall
h. the tissue layer that develops into the gut lining
i. the most common form of animal, does not have a backbone
j. a body form in which the parts are arranged around a central axis
l. a reduced, unlined coelom, present in some early animals
m. the development of the anterior end into a distinct head
n. animals with no true tissues; only one known species—*Trichoplax adhaerans*
o. the tissue layer that lines the body surfaces

## Fill in the Blanks [p.364]

How did animals arise? Perhaps the ancestor of the earliest animals was a (19) _____ _____. Each of these organisms has a ring of absorptive structures (microvilli) around the base of a (20) _____. Over time, some of the colonial cells have shown (21) _____ of labor. This is evidenced by the fact that cells on the outside of the colony are involved with (22) _____ while cells on the insides are specialized for (23) _____. Comparisons of (24) _____ sequences also point to choanoflagellates as the group most likely to have been involved in the early stages of (25) _____ evolution. Animals first lived in the (26) _____, and in the (27) _____ there was an explosion in animal diversity such that all major animal lineages evolved within 40 million years.

## Labeling

Choose the most appropriate label for each of the numbers in the diagram below. [p.356]

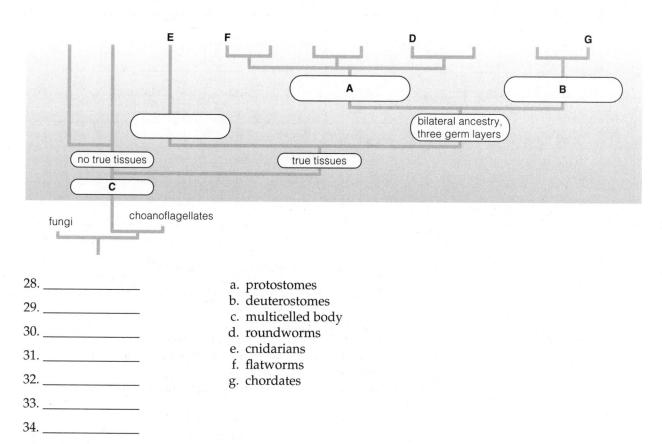

28. _____
29. _____
30. _____
31. _____
32. _____
33. _____
34. _____

a. protostomes
b. deuterostomes
c. multicelled body
d. roundworms
e. cnidarians
f. flatworms
g. chordates

## 23.3. SPONGES—SUCCESS IN SIMPLICITY [p.365]

## 23.4. CNIDARIANS—SIMPLE TISSUES, NO ORGANS [pp.366–367]

## 23.5. FLATWORMS—SIMPLE ORGAN SYSTEMS [pp.368–369]

### Selected Terms

*Anthozoans* [p.366], *medusozoans* [p.366], *Obelia* [p.367], *Physalia* [p.367], *Schistosoma* [p.368], *Taenia saginata* [p.369]

### Boldfaced Terms

[p.365] sponges _____

_____

[p.365] larva _____

_____

[p.366] cnidarians _____

_____

[p.367] nerve net _____

_____

[p.367] hydrostatic skeleton _____

_____

[p.367] nematocysts _____

_____

[p.368] organs _____

_____

[p.368] organ system _____

_____

[p.368] flatworms _____

_____

[p.368] pharynx _____

_____

[p.368] nerve cords _____

_____

[p.368] ganglia _____

_____

[p.369] proglottids _____

_____

## Matching

For each statement, choose the correct group of animals to which the statement best applies. Answers may be used more than once.

      a. sponges [p.365]     b. cnidarians [pp.366–367]     c. flatworms [pp.368–369]
      d. all of the above

_____ 1. Possess nematocysts.

_____ 2. Possess a primitive nervous system called a nerve net.

_____ 3. The first animal with organs.

_____ 4. Have no symmetry, tissues, or organs.

_____ 5. All members of this group have an ectoderm, an endoderm, and a mesoderm.

_____ 6. May reproduce asexually by fragmentation.

_____ 7. The majority are bilateral.

_____ 8. Body forms are called polyps and medusas.

_____ 9. Many of this group have radial symmetry.

_____ 10. Members of this group may utilize definitive and intermediate hosts.

_____ 11. Lack a body coelom.

## Labeling

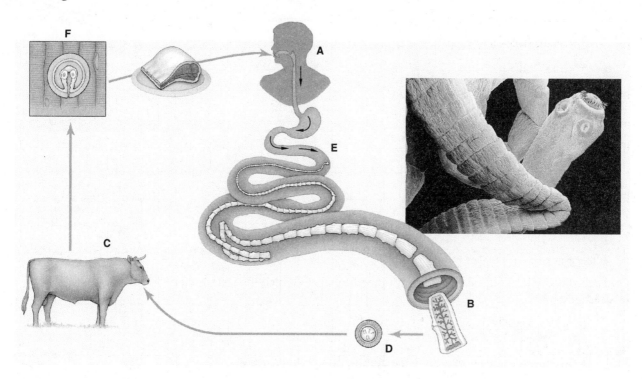

This is the life cycle of the beef tapeworm. Match the following statement with the proper label from the picture above.

12. the definitive host _____

13. the larvae, with the inverted scolex of future tapeworm _____

14. the intermediate host _____

15. scolex _____

16. fertilized egg _____

17. sexually mature proglottid _____

## 23.6. ANNELIDS—SEGMENTS GALORE [pp.370–371]

## 23.7. THE PLIABLE MOLLUSKS [pp.372–373]

## 23.8. ROUNDWORMS [p.374]

### Selected Terms

Oligo- [p.370], poly- [p.370], *Hirudo medicinalis* [p.370], gastropods [p.372], chitons [p.372], bivalves [p.372], cephalopods [p.372], *Caenorhabditis elegans* [p.374], *Trichinella spiralis* [p.374], *Ascaris lumbricoides* [p.374], *Wucheria bancrofti* [p.374], *Enterobius vermicularis* [p.374]

### Boldfaced Terms

[p.370] annelids _____

_____

[p.370] closed circulatory system _____

_____

[p.370] nephridium _____

_____

[p.372] mollusks_____

_____

[p.374] roundworms _____

_____

[p.374] molting _____

_____

## Matching

For each statement, choose the correct group of animals to which the statement best applies. Answers may be used more than once.

> a. mollusks [pp.372–373]    b. annelids [pp.370–371]    c. roundworms [p.374]
> d. none of the above

_____ 1. Members of this group have the ability of jet propulsion.

_____ 2. Has a false coelum packed with reproductive organs.

_____ 3. Many of this group have a radula.

_____ 4. Cephalopods belong to this group.

_____ 5. Are responsible for diseases such as trichinosis and elephantitis.

_____ 6. Possess multiple hearts and a closed circulatory system.

_____ 7. Possess an extension of their body called a mantle.

_____ 8. Leeches belong to this group.

_____ 9. Also known as the nematodes.

_____ 10. The most segmented of all the animals.

## Matching

Match each of the following body parts to its correct function.

_____ 11. nerve cords [p.370]

_____ 12. chaetae [p.370]

_____ 13. mantle [p.372]

_____ 14. radula [p.372]

_____ 15. brain [p.370]

_____ 16. nephridia [p.370]

_____ 17. gill [p.372]

a. clusters of nerve cell bodies that control activity

b. regulates the composition and volume of body fluids

c. a cloak-like extension of the body mass

d. lines of communication that help the brain coordinate activities

e. chitin-reinforced bristles that assist in movement, present in most annelids

f. a tongue-like organ that assists in feeding

g. respiratory organ found in most mollusks

## Labeling

Provide the name of each indicated body part in the diagram below. [p.364]

18. _____

19. _____

20. _____

21. _____

22. _____

23. _____

24. _____

25. _____

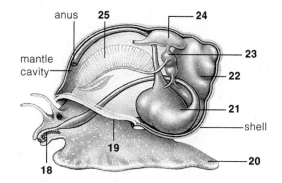

## Complete the Table [p.374]

| Species | Disease |
|---------|---------|
| Trichinella spiralis | 26. |
| Enterobius vermucularis | 27. |
| Ascaris lumbricoides | 28. |
| Wuchereria bancrofti | 29. |

## 23.9.  WHY SUCH SPECTACULAR ARTHROPOD DIVERSITY? [p.375]

## 23.10.  SPIDERS AND THEIR RELATIVES [p.376]

## 23.11.  A LOOK AT THE CRUSTACEANS [p.377]

## 23.12.  A LOOK AT INSECT DIVERSITY [pp.378–379]

## 23.13.  UNWELCOME ARTHROPODS [p.380]

## 23.14.  THE SPINY SKINNED ECHINODERMS [p.381]

### Selected Terms

Trilobites [p.375], chelicerates [p.375], crustaceans [p.375], uniraminarians [p.375], arthropod [p.375], *Cyclommatus* [p.379], *Diabrotica vigifera* [p.379], *Loxosceles* [p.380], *Latrodectus* [p.380], *Ixodes scapularis* [p.380], *Borrelia burgdorferi* [p.380], *Rickettsia rickettsii* [p.380]

## Boldfaced Terms

[p.375] arthropods _____

_____

[p.375] exoskeleton _____

_____

[p.375] eyes _____

_____

[p.375] antennae _____

_____

[p.375] metamorphosis _____

_____

[p.375] arthropods _____

_____

[p.376] malpighian tubules _____

_____

[p.378] metamorphosis _____

_____

[p.381] echinoderm _____

_____

[p.381] water-vascular system _____

_____

## True–False

Write "T" in the space provided if the statement is true. If the statement is false, correct the underlined word so that the statement is correct. [p.375]

_____ 1. The arthropods are <u>bilateral</u> animals with a <u>closed</u> circulatory system.

_____ 2. In many arthropods, there is a <u>metamorphosis</u> of body form between the embryo and adult stages.

_____ 3. Arthropods utilize a hardened <u>endoskeleton</u> for support.

_____ 4. The <u>jointed</u> appendages of the arthropods allow for the diverse evolution of this group.

_____ 5. Arthropods have a complete digestive system and a <u>pseudocoel</u>.

## Matching

For each of the following statements, identify the group(s) of arthropods that is best described. Some statements may have more than one answer.

a. crustaceans [p.377]     b. spiders and their relatives [p.376]     c. insects [p.378]
d. all arthropods

_____ 6. These are marine organisms that are frequently consumed by humans.

_____ 7. Possess special appendages called chelicerae

_____ 8. Body is divided into three distinct parts—head, thorax, and abdomen.

_____ 9. Undergo periodic molts to shed their exoskeleton

_____ 10. In terms of number of species, the largest group on the planet

_____ 11. Gills and trachea are the respiratory structures.

_____ 12. Specialize in predatory or parasitic life styles

_____ 13. Includes the copepods and pillbugs

_____ 14. Malpighian tubules are used for waste excretion.

_____ 15. Includes the horseshoe crabs and mites

_____ 16. Winged invertebrates

## Fill in the Blanks [p.378]

In most adult insects, the (17) _____ body is divided into three distinct parts: a head, (18) _____, and (19) _____ [p.370]. The head has paired sensory (20) _____ and paired mouthparts that are specialized for biting, chewing and other tasks. The (21) _____ has three pairs of legs and, usually, (22) _____ pairs of wings. Insects are the only winged (23) _____

Insects have a (24) _____ digestive system divided into a foregut, a midgut where food is (25) _____ and a hindgut where (26) _____ is reabsorbed. A system of (27) _____ disposes of wastes and maintains (28) _____ concentrations in body fluid.

## True–False

Write "T" in the space provided if the statement is true. If the statement is false, correct the underlined word so that the statement is correct. [p.381]

_____ 29. The water-vascular system of Echinoderms is used for <u>locomotion</u>.

_____ 30. Most echinoderms are <u>bilateral</u> as adults.

_____ 31. Echinoderms have interlocking <u>calcium carbonate</u> plates on their surface.

_____ 32. Echinoderms have a <u>centralized</u> nervous system.

## Complete the Table

Complete the table below by filling in the group of animal and the effect on human health for each of the species listed. [p.380]

| Animal | Group | Impact on Health |
|--------|-------|------------------|
| Loxosceles | 33. | 38. |
| Ixodes scapularis | 34. | 39. |
| Latrodectus | 35. | 40. |
| American dog tick carrying Rickettsia rickettsii | 36. | 41. |
| Mosquitoes | 37. | 42. |

## List

List the six key adaptations that contributed to the evolutionary success of the arthropods:

43. _____

44. _____

45. _____

46. _____

47. _____

48. _____

# Self-Quiz

1. Nematocysts are present in the _____. [p.367]
   a. cnidarians
   b. flatworms
   c. roundworms
   d. echinoderms

2. Which of the following is a characteristic of an animal? [p.362]
   a. multicelled
   b. aerobic heterotrophs
   c. most are motile
   d. most reproduce sexually
   e. all of the above are correct.

3. A water-vascular system is a characteristic of the _____. [p.381]
   a. echinoderms
   b. crustaceans
   c. poriferans
   d. cnidarians

4. Which of the following is the first group to develop organ systems? [p.368]
   a. roundworms
   b. flatworms
   c. cnidarians
   d. arthropods

5. Nephridia and malpighian tubules are involved in _____. [pp.370, 376]
   a. nervous system function
   b. locomotion
   c. waste removal
   d. reproduction
   e. none of the above

6. _____ symmetry allows for the process of cephalization. [p.362]
   a. radial
   b. axial
   c. perpendicular
   d. bilateral
   e. none of the above

7. Which of the following represents the earliest known form of animals? [p.365]
   a. poriferans
   b. cnidarians
   c. ediacarans
   d. arthropods

8. A complete lack of symmetry is a characteristic of the _____. [p.365]
   a. cnidarians
   b. echinoderms
   c. poriferans
   d. arthropods
   e. flatworms

9. You have identified an organism with true tissues and an organ with no segments or coelom. To which of the following groups could this organism belong? [p.356]
   a. cnidarians
   b. flatworms
   c. arthropods
   d. chordates
   e. annelids

10. These are the most segmented organisms in the animal kingdom. [p.370]
    a. annelids
    b. roundworms
    c. flatworms
    d. echinoderms

11. Which of the following tissue layers is correctly matched to its correct function? [p.362]
    a. ectoderm—source of muscles and other organs.
    b. mesoderm—lines the body surfaces
    c. endoderm—forms the lining of the gut
    d. none of the above are matched correctly
    e. all of the above are matched correctly

12. The only winged invertebrates are the _____. [p.378]
    a. spiders
    b. crustaceans
    c. cephalopods
    d. insects
    e. Echinoderms

13. Cephalopods belong to which of the following groups? [p.372]
    a. annelids
    b. mollusks
    c. roundworms
    d. arthropods

14. The body cavity of an animal is called the _____. [p.363]
    a. peritoneum
    b. gut
    c. protostome
    d. coelom
    e. mesoderm

15. Jointed appendages are a characteristic of the _____. [p.375]
    a. echinoderms
    b. arthropods
    c. poriferans
    d. mollusks

## Chapter Objectives/Review Questions

This section lists general and detailed chapter objectives that can be used as review questions. You can make maximum use of these items by writing answers on a separate sheet of paper. Fill in answers where blanks are provided. To check for accuracy, compare your answers with information given in the chapter or glossary.

1. List the general characteristics of all animals. [p.362]
2. The majority of animals are _____ (vertebrates, invertebrates). [p.362]
3. Explain how bilateral symmetry leads to cephalization. [p.362]
4. Distinguish between a gut and coelum. [pp.362–363]
5. List the primary difference between a protostome and a deuterostome. [p.364]
6. Define segmentation. [p.363]
7. Recognize the general organization of the animal kingdom, and the key events in the evolution of the animal body form. [p.364]

8. Identify the key characteristics of a poriferan. [p.365]
9. Identify the key characteristics of a cnidarian. [pp.366–367]
10. Explain how organs can be combined into an organ system. [p.368]
11. Distinguish between an intermediate and definitive host. [p.368]
12. List the advantages of segmentation. [p.370]
13. Explain the purpose of the nephridia. [p.370]
14. Identify the key characteristics of a mollusk. [pp.372–373]
15. Understand the relationship of roundworms to human diseases. [p.374]
16. Identify the key characteristics of all arthropods. [p.375]
17. What is metamorphosis and how does it relate to insect development? [p.375]
18. Distinguish among crustaceans, cheicerates, and insects. [pp.376–379]
19. Identify the key characteristics of an echinoderm. [p.381]
20. Give the purpose of the water vascular system. [p.381]

## Complete the Table

| Group | Some Representatives | Existing Species |
| --- | --- | --- |
| [p.364] 21. | *Trichoplax*; simplest animal | 1 |
| Poriferans | [p.365] 22. | 8,000 |
| [p.366] 23. | Hydrozoans, jellyfishes, corals, sea anemones | 11,000 |
| Flatworms | [p.368] 24. | 15,000 |
| [p.370] 25. | Leeches, earthworms | 15,000 |
| Mollusks | [p.372] 26. | 110,000 |
| Arthropods | [p.375] 27. | 1,000,000 |
| [p.374] Roundworms | 28. | 20,000 |
| [p.381] 29. | Sea stars, sea urchins | 6,000 |

# Chapter Summary

Animals are multicelled (1) _____ that ingest other organisms, grow and (2) _____ through a series of stages, and actively move about during all or part of the (3) ____ cycle. Cells of most animals form (4) _____ and extracellular matrixes.

The earliest animals were small and structurally (5) _____. Their descendants evolved larger bodies with a more (6) _____ structure and greater (7) _____ among specialized parts.

Animals' body plans vary. Bodies may or may not show (8) _____. There may or may not be an internal (9) _____ cavity, a head, or division into (10) _____. An early divergence gave rise to two major branches: protostomes and (11) _____.

Sponges and placozoans have no body (12) _____ or true (13) _____. Cnidarians are (14) _____ symmetrical, with two tissue layers and a gelatinous matrix between the two.

Most animals show (15) _____ symmetry. Bilateral animals have tissues, organs, and organ (16) _____. All adult tissues arise from two or three simple layers that form in early (17) _____.

In diversity, numbers, and distribution, (18) _____ are the most successful animals. In the seas, (19) _____ are the dominant arthropod lineage; on land, (20) _____ rule.

Echinoderms are on the same branch of the animal family tree as the (21) _____. They are (22) _____ with bilateral ancestors, but adults now have a decidedly (23) _____ body plan.

## Integrating and Applying Key Concepts

Juvenile hormone disruptors are becoming an important weapon in the war against pest insects. These chemicals stop the insect from molting. How could this be used to control pest populations? If a farmer finds a pest insect on his crop and applies these chemicals, will he immediately decrease the damage to his crops? Why or why not?

# 24

# ANIMAL EVOLUTION—THE VERTEBRATES

## INTRODUCTION

The diversity of life on Earth is further reflected in the animal world, and in this chapter that diversity is illustrated through a study of the vertebrate animals. This chapter discusses, in a sequential, organized way, the evolution of vertebrate animals from the seas in which they began to the present day humans. The relationships between such varied animals as the fish and the human being are discussed and the innovations that took place at each evolutionary branch are pointed out. The varied animals all share some physical features that have been maintained throughout the ages. Again, the ability of species to adapt to various environments is illustrated, but what happens when a group of animals does not respond quickly enough to environmental changes is also discussed. Such was the case, we believe, with the dinosaurs.

## FOCAL POINTS

- Figure 24.7 [p.388] compares the gill supporting structures in fishes.
- Figure 24.8 [p.389] shows where living vertebrates fit into the chordate family tree.
- Figure 24.9 [p.390] diagrams the body plan of a perch, a bony fish.
- Figure 24.14 [p.394] shows the family tree of the amniotes.
- Figure 24.16 [p.396] animates the body plan of a crocodile, a representative reptile.
- Figure 24.19 [p.398] diagrams the components of a bird egg.
- Figure 24.21 [p.399] shows a bird skeleton and the main flight muscles.
- Table 24.1 [p.402] organizes primate classification.
- Figure 24.29 [p.404] presents samples of fossilized hominoid skulls.

## Interactive Exercises

*Interpreting and Misinterpreting the Facts* [pp.384–385]

## 24.1.  THE CHORDATE HERITAGE [pp.386–387]

## 24.2.  EVOLUTIONARY TRENDS AMONG THE VERTEBRATES [pp.388–389]

*Selected Terms*

*Dromaeosaurus* [p.384], *Archaeopteryx* [p.384], *Rhopalaea crassa* [p.386]

## Boldfaced Terms

[p.386] chordates _____

_____

[p.386] notochord _____

_____

[p.386] vertebrates _____

_____

[p.386] tunicates _____

_____

[p.386] filter feeders _____

_____

[p.386] lancelets _____

_____

[p.387] craniates _____

_____

[p.388] jaws _____

_____

[p.388] vertebrae _____

_____

[p.388] gills _____

_____

[p.389] lungs _____

_____

## Labeling

Using Figure 24.1, fill in the matching letter of the following structures:

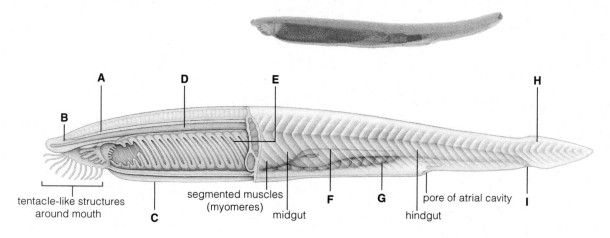

tentacle-like structures around mouth

segmented muscles (myomeres)

midgut

hindgut

pore of atrial cavity

1. eyespot _____
2. epidermis _____
3. aorta _____
4. anus _____
5. pharynx with gill slits _____
6. notochord _____
7. tail extending past the anus _____
8. gonad _____
9. dorsal, tubular nerve cord _____
10. Which four letters represent the four defining traits of a chordate?

## Choice

Choose one of the following adaptations for each of the statements. Some answers may be used more than once. [pp.388–389]

  a. jaws   b. vertebrae   c. fins   d. gills   e. lungs

_____ 11. This was the earliest of the trends, and it sparked the development of additional trends.
_____ 12. These assist in stabilizing and guiding the body through the water.
_____ 13. These have a large surface area to exchange gases with the environment.
_____ 14. The evolution of this overcame the use of armor by the ostracoderms.
_____ 15. This led to the development of more efficient circulatory systems in land animals.
_____ 16. The placoderms were the first group to possess this structure.
_____ 17. These are internally moistened sacs for gas exchange.

## Labeling

Label each of the indicated groups on the diagram below. [p.381]

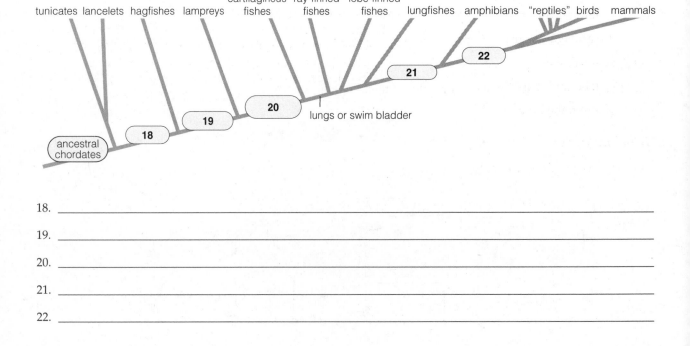

18. _____

19. _____

20. _____

21. _____

22. _____

## 24.3. JAWED FISHES AND THE RISE OF TETRAPODS [pp.390–391]

## 24.4. AMPHIBIANS—THE FIRST TETRAPODS ON LAND [pp.392–393]

## 24.5. VANISHING ACTS [p.393]

### Selected Terms

*Latimaria* [p.391], *Latimeria* [p.391], *Acanthostega* [p.391], *Ichthyostega* [p.391], *Eusthenoptheron* [p.391], *Notophthalmus* [p.392], *Rana aurora* [p.393], *Ribeiroia* [p.393]

### Boldfaced Terms

[p.390] swim bladder _____

_____

[p.390] cartilaginous fishes _____

_____

[p.391] ray-finned fishes _____

_____

[p.391] lobed-finned fishes _____

_____

[p.391] lung fishes _____

_____

[p.392] amphibians _____

_____

### Matching

Match each of the following groups to its correct description. [pp.390–391]

_____ 1. tetrapods

_____ 2. cartilaginous fish

_____ 3. lungfishes

_____ 4. lobe-finned fish

_____ 5. ray-finned fish

a. the most diverse fish in terms of body color, size and shape
b. Sharks and rays are the major representatives of this group.
c. The Coelacanth is the last remaining type of this group.
d. possess both gills and small pouches that serve as lungs
e. four-legged walkers

### Labeling [p.390]

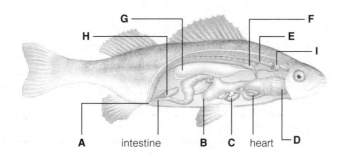

Fill in the matching letter of the following structures:

6. swim bladder _____

7. ovary _____

8. gills _____

9. kidney _____

10. stomach _____

11. nerve cord _____

12. cloaca _____

13. liver _____

14. brain _____

## 24.6. THE RISE OF AMNIOTES [pp.394–395]

## 24.7. SO LONG, DINOSAURS [p.395]

## 24.8. PORTFOLIO OF MODERN "REPTILES" [pp.396–397]

## 24.9. BIRDS—THE FEATHERED ONES [pp.398–399]

### Selected Terms

*Lystrosaurus* [p.394], *Maiasaura* [p.394], *Temnodotosaurus* [p.394], repto- [p.396], *Sphenodon* [p.397], *Archaeopteryx* [p.398], *Confuciusornis sanctus* [p.398]

### Boldfaced Terms

[p.394] amniotes _____

_____

[p.394] synapsids _____

_____

[p.395] dinosaurs _____

_____

[p.395] K-T asteroid impact hypothesis _____

_____

[p.396] cloaca _____

_____

[p.398] birds _____

_____

[p.398] migration _____

_____

## Matching

Match each of the following descriptions to the correct term. [pp.394–395]

_____ 1. Produce eggs that have four membranes, have scaly skins and kidneys.

_____ 2. The intense heat produced by a large asteroid impact.

_____ 3. Amniote group that includes the reptiles and birds.

_____ 4. This group has basic amniote features, but is not a bird or mammal.

_____ 5. The amniotes that include mammals and their extinct relatives.

_____ 6. The event at the end of the Cretaceous that ended the reign of the dinosaurs.

a. synapsids
b. K-T asteroid impact theory
c. global broiling hypothesis
d. sauropsids
e. reptiles
f. amniotes

## Complete the Table

Provide the missing information for each of the following groups of reptiles. [pp.396–397]

| Group | Key Characteristics | # Species Remaining |
|---|---|---|
| 7. | Most advanced reptiles; closest relatives of the birds | 23 |
| 8. | The most diverse group of reptiles | 4,710 |
| 9. | Flexible skull bones and jaws; carnivores | 2,955 |
| tuatara | Resemble modern amphibians in brain and locomotion. | 10. |
| turtles | 11. | 305 |

## Labeling [p.399]

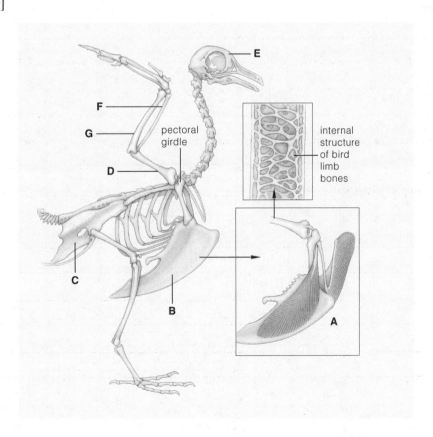

pectoral girdle

internal structure of bird limb bones

Fill in the matching letter of the following structures:

12. radius _____

13. ulna _____

14. sternum _____

15. pelvic girdle _____

16. humerus _____

17. skull _____

18. flight muscles _____


## 24.10. THE RISE OF MAMMALS [pp.400–401]

## 24.11. FROM EARLY PRIMATES TO HOMINIDS [pp.402–403]

## 24.12. EMERGENCE OF EARLY HUMANS [pp.404–405]

## 24.13. EMERGENCE OF MODERN HUMANS [pp.406–407]

### Selected Terms

Mamma [p.400]; prosimians [p.402]; anthropoids [p.402]; hominoids [p.402]; enhanced daytime vision [p.402]; upright walking [p.402]; better grips [p.402]; *prehensile* movements [p.403]; *opposable* movements [p.403]; modified jaws and teeth [p.403]; brain, behavior, and culture [p.403]; *Sahelanthropus tchadensis* [p.404]; *Orrorion tugenensis* [p.404]; *Ardipithecus ramidus* [p.404]; *Australpithecus afarensis* [p.404]; Paranthropus [p.404]; homo [p.405]; *Homo habilus* [p.405]; *Homo erectus* [p.405]; *Homo sapiens* [p.406]

### Boldfaced Terms

[p.400] mammals _____

_____

[p.400] monotremes _____

_____

[p.400] marsupials _____

_____

[p.400] eutherians _____

_____

[p.401] placenta _____

_____

[p.402] primates _____

_____

[p.402] hominids _____

_____

[p.402] foramen magnum _____

_____

[p.403] culture _____

_____

[p.404] australopiths _____

_____

[p.405] humans _____

_____

[p.406] multiregional model _____

_____

[p.406] replacement model _____

_____

## True–False

If the statement is true, write "T" in the space provided. If false, correct the underlined word so that the statement is correct. [pp.400–401]

_____ 1. Two of the unique characteristics of <u>mammals</u> are their ability to make hair and excrete milk.

_____ 2. Humans are descended from a group of synapsids called the <u>therapsids</u>.

_____ 3. <u>Therapsids</u>, marsupials, and eutherians represent the three lineages of mammals.

_____ 4. Following the Cenozoic, both monotremes and <u>placental</u> mammals were isolated on Australia.

_____ 5. <u>Placental</u> mammals had a competitive edge over marsupials due to a higher metabolic rate, precise control over body temperature, and a more efficient method of nourishing embryos.

_____ 6. The aardvark, spiny anteater, and giant anteater are examples of <u>divergent</u> evolution.

## Fill in the Blanks [pp.402–407]

The five trends that define the lineage leading to humans were set in motion among the first (7) _____. First, a reliance on the sense of smell became less important than (8) _____ [p.394] vision. Second, skeletal changes promoted (9) _____ —upright walking—which freed (10) _____ for novel tasks. Third, bone and muscle changes led to diverse (11) _____ motions. Fourth, (12) _____ became less specialized to accommodate an omnivorous diet. Fifth, the evolution of the (13) _____, behavior, and (14) _____ became interlocked. Culture is the sum of a social group's (15) _____ patterns, passed on through the generations by (16) _____ and symbolic behavior. In brief, uniquely human traits emerged by modification of (17) _____ that had evolved earlier, in ancestral forms.

## Matching

Match each of the following groups to its correct description. [pp.404–407]

_____ 18. *Homo habilis*

_____ 19. *Australopithecus aferensis*

_____ 20. *Paranthropus*

_____ 21. *Homo erectus*

_____ 22. *Homo sapiens*

a. the "handy-man"
b. indisputably, the first bipedal hominid
c. The multiregional and replacement models attempt to explain the origin of this species.
d. "upright man"
e. one of the australopiths; wide face and large molars; died out 1.2 million years ago

# Self-Quiz

1. Which of the following groups is an immediate ancestor of the tetrapods? [pp.390–391]
   a. cartilaginous fish
   b. ray-finned fish
   c. lungfish
   d. Coelacanths
   e. none of the above

2. Bipedalism and culture are found in all _____. [pp.402–403]
   a. synapsids
   b. vertebrates
   c. dinosaurs
   d. hominids
   e. primates

3. Which of the following is a not a characteristic of all chordates? [p.386]
   a. gills slits in a pharynx
   b. vertebrae made of calcium
   c. a tail that extends past the anus
   d. a notochord
   e. a nerve cord parallel to the gut and notochord

4. Reptiles, mammals, and birds are all _____. [p.394]
   a. amniotes
   b. examples of organisms with four-chambered hearts in the entire group
   c. warm-blooded
   d. examples of sauropsids

5. The evolution of the _____ in fish set the stage for other key innovations leading to tetrapods. [p.390]
   a. brain
   b. lungs
   c. jaws
   d. fins
   e. legs

6. The term "handy-man" refers to _____. [p.405]
   a. all hominids
   b. *Homo habilus*
   c. *Australopithecus aferensis*
   d. *Homo sapiens*
   e. all primates

7. Tunicates and lancelets are _____. [p.386]
   a. fish
   b. vertebrates
   c. amphibians
   d. invertebrate chordates
   e. hominids

8. The first tetrapods on land were the _____. [p.392]
   a. therapsids
   b. amphibians
   c. synapsids
   d. tunicates
   e. Coelacanths

9. The K-T asteroid impact theory attempts to explain the extinction of the _____. [p.395]
   a. Australopithecus species
   b. dinosaurs
   c. tuataras
   d. monotremes

10. *Archaeopteryx* represents the first of the _____. [p.398]
    a. tetrapods
    b. lungfish
    c. vertebrate chordates
    d. feathered reptiles
    e. hominids

# Chapter Objectives/Review Questions

1. List the four key characteristics of all chordates. [p.386]
2. Understand the difference between the invertebrate and the vertebrate chordates. [pp.386–387]
3. List the key innovations in the development of the vertebrate body plan. Be able to place these innovations in correct chronological order on an evolutionary diagram. [pp.388–389]
4. Distinguish between the cartilaginous and the bony fishes. [pp.390–391]
5. List the key characteristics of an amphibian. [pp.392–393]
6. Explain why amphibian species are on the decline. [p.393]
7. Explain the significance of an amniote. [p.394]
8. List two hypotheses that have been advanced to account for the extinction of the dinosaurs. [p.395]
9. Give the key characteristics of the major reptile groups. [pp.396–397]
10. Explain how birds are adapted for flight. [pp.398–399]
11. List the key characteristics of mammals. [pp.400–401]
12. List the three major groups of mammals. [pp.400–401]
13. Distinguish between a primate and a hominoid. [pp.402–403]
14. List the five trends in primate development that define the lineage to humans. [pp.402–403]
15. Give the significance of the australopiths. [p.404]
16. Recognize the key species in the evolution of the hominoids. [p.404]
17. Understand the major hypotheses on the origins of *Homo sapiens*. [p.406]

---

# Chapter Summary

A unique set of four traits characterizes the (1) _____: a supporting rod (notochord), a dorsal (2) _____ cord, a pharynx with (3) _____ in the wall, and a (4)_____ extending past an anus. Certain invertebrates and all (5) _____ belong to this group.

In some vertebrate lineages, a (6) _____ replaced the notochord as the partner of (7) _____ used in motion. Jaws evolved, sparking the evolution of novel (8) _____ organs and (9) _____ expansions. On land, lungs replaced (10) _____, and more efficient blood circulation enhanced (11) _____ exchange. Fleshy fins with skeletal supports evolved into (12) _____, which are now typical of vertebrates on land.

Vertebrates first evolved in the (13) _____, where lineages of cartilaginous and bony (14) _____ persist. Of all vertebrates, modern bony fishes show the most (15) _____. Mutations in master genes that control body plans were pivotal in the rise of aquatic (16) _____ and their move onto dry land.

As a group, the (17) _____—known informally as the reptiles, birds, and (18) _____—are vertebrate lineages that radiated into nearly all habitats on (19) _____.

Primates that were ancestral to the (20) _____ lineage became physically and behaviorally adapted to change in global (21) _____ and available (22) _____. Behavioral and cultural flexibility helped humans disperse from (23) _____ throughout the world.

# 25

# PLANTS AND ANIMALS—COMMON CHALLENGES

## INTRODUCTION

Plants and animals do not appear, at first glance, to be all that similar. On closer inspection, however it is clear that both plants and animals face many of the same challenges. For example, both plants and animals must regulate their internal environments, fight off disease, and undergo growth and development. This chapter discusses many of the processes common to plants and animals during their life cycles. Emphasis is placed on the concept of homeostasis, which is critical to the survival of all multicellular organisms. After completing this chapter, you should understand that many essential life functions are shared by both plants and animals and both types of organisms are able to accomplish these processes very effectively.

## FOCAL POINTS

- Figure 25.2 [p.412] illustrates the anatomy of a tomato plant.
- Figure 25.8 [p.416] shows the essential components of the regulation of body temperature, a negative feedback system.
- Figure 25.13 [p.420] diagrams the generalized signal transduction pathway and how this pathway would be used to induce apoptosis.

## Interactive Exercises

*A Cautionary Tale* [pp.410–411]

## 25.1. LEVELS OF STRUCTURAL ORGANIZATION [pp.412–413]

## 25.2. RECURRING CHALLENGES TO SURVIVAL [pp.414–415]

*Selected Terms*

*Anatomy* [p.411], *physiology* [p.411], *quantitative* terms [p.412], *qualitative* terms [p.412]

*Boldfaced Terms*

[p.412] tissue _____

_____

[p.412] organ _____

_____

[p.412] organ system _____

_____

[p.412] growth _____

_____

[p.412] development _____

_____

[p.413] extracellular fluid (ECF) _____

_____

[p.413] homeostasis _____

_____

[p.414] diffusion _____

_____

[p.414] active transport _____

_____

[p.415] habitat _____

_____

## Matching

Choose the correct statement for each of the following terms. [pp.412–415]

_____ 1. organ

_____ 2. tissue

_____ 3. organ system

_____ 4. development

_____ 5. internal environment

_____ 6. homeostasis

_____ 7. growth

a. the maintenance of the internal environment within tolerable limits

b. a structure that contains at least two tissues

c. the increase in cell size, number, or volume

d. a succession of stages in which tissues, organs, or organ systems form

e. the body fluids not located within a cell

f. a community of cells and intercellular substances that are interacting in one or more tasks

g. two or more organs interacting physically or chemically

## Fill in the Blanks [p.413]

To stay alive, plants and animal cells must be bathed in a (8) _____ that delivers nutrients and carries away metabolic (9) _____. In this they are no different from free-living single cells. But each plant or animal has thousands to many trillions of living cells that must draw (10) _____ from the fluid bathing them and dump wastes into it.

Body fluid not inside cells—(11) _____ fluid—is an internal environment. Changes in its composition and (12) _____ affect cell activities. The type and number of (13) _____ are vital; they must be kept at concentrations compatible with (14) _____. It makes no difference whether the plant or animal is simple or complex, it requires a (15) _____ fluid environment for all of its living cells. This concept is central to understanding how plants and animals work.

## True–False

If the statement is true, write "T" in the space provided. If false, correct the underlined word so that the statement is correct. [pp.414–415]

_____ 16. The process of active transport moves molecules <u>against</u> concentration gradients.

_____ 17. Active and passive transport helps to maintain the <u>external</u> environment and metabolism by adjusting the kinds, amounts, and movement of substances.

_____ 18. The overall size and shape of a cell is determined by the <u>surface-to-volume ratio</u>.

_____ 19. An <u>ecosystem</u> is defined as the place where a species normally lives.

# 25.3. HOMEOSTASIS IN ANIMALS [pp.416–417]

# 25.4. DOES HOMEOSTASIS OCCUR IN PLANTS? [pp.418–419]

# 25.5. HOW CELLS RECEIVE AND RESPOND TO SIGNALS [pp.420–421]

## Selected Terms

Intensify [p.417], _Lupinus arboreus_ [p.418]

## Boldfaced Terms

[p.416] interstitial fluid _____

_____

[p.416] sensory receptors _____

_____

[p.416] integrator _____

_____

[p.416] effectors _____

_____

[p.416] negative feedback mechanisms _____

_____

[p.417] positive feedback mechanisms _____

_____

[p.418] system acquired resistance _____

_____

[p.418] compartmentalization _____

_____

[p.419] circadian rhythm _____

_____

[p.420] apoptosis _____

_____

## Matching

Match each of the following statements with its correct term. [p.416]

_____ 1. the fluid portion of the blood

_____ 2. processes information concerning stimuli

_____ 3. carries out responses to stimuli

_____ 4. fluid that fills the spaces between cells and tissues

_____ 5. detect specific forms of energy called stimuli

a. interstitial fluid
b. effector
c. sensory receptor
d. plasma
e. integrator

## Choice

Indicate whether the statement is associated with negative (N) or positive (P) feedback. [pp.416–417]

_____ 6. After a setpoint is reached, a response is triggered to reverse an activity.

_____ 7. An example is the movement of a signal along a neuron.

_____ 8. A chain of events that intensify change from the original condition.

_____ 9. An example is the internal thermostat of mammals.

## Fill in the Blanks [p.418]

Most trees protect their internal environment by (10) _____ off the area around wounds. They unleash phenols and other (11) _____ compounds. Many secrete (12) _____. The heavy flow of gooey compounds saturates and protects bark and (13) _____ at an attack site. It can also seep into soil around roots.

Some (14) _____ are so potent that they also kill cells of the tree itself. As a result, (15) _____ form around injured, infected, or (16) _____ sites, and new tissues grow over them. The plant response to attack is called (17) _____

## Labeling

Label the diagram with the appropriate letter.

a. A signal binds to a receptor, usually at the cell surface.
b. Changes alter the cell's metabolism, gene expression, or rate of division.
c. Binding of the signal brings about changes in cell properties, activities, or both.

## Self-Quiz

1. The maintenance of the internal body within acceptable limits is called _____. [p.413]
   a. integration
   b. apoptosis
   c. homeostasis
   d. circadian rhythm

2. An _____ is responsible for the detection of a specific form of energy, or stimuli. [p.416]
   a. integrator
   b. effector
   c. sensory receptor
   d. all of the above

3. _____ refers to the increase in the number, size, or volume of cells. [p.412]
   a. Growth
   b. Development
   c. Apoptosis
   d. Integration

4. Programmed cell death is called _____. [p.420]
   a. compartmentalization
   b. apoptosis
   c. growth
   d. development

5. The place where a species normally lives is called its _____. [p.415]
   a. home
   b. ecosystem
   c. habitat
   d. community

6. A _____ has at least two tissues organized in certain proportions and patterns to perform one or more common tasks. [p.412]
   a. organism
   b. organ system
   c. organ
   d. tissue
   e. cell

7. An intensification of a change by a pathway would be an example of a positive feedback mechanism. [p.417]
   a. true
   b. false

8. The overall limits on the size of a cell are determined by its _____. [p.414]
   a. surface-to-volume ratio
   b. number of receptors
   c. degree of compartmentalization
   d. number of tissue layers

## Chapter Objectives/Review Questions

This section lists general and detailed chapter objectives that can be used as review questions. You can make maximum use of these items by writing answers on a separate sheet of paper. Fill in answers where blanks are provided. To check for accuracy, compare your answers with information given in the chapter or glossary.

1. Distinguish among a tissue, an organ, and an organ system. [p.412]
2. Compare the processes of growth and development. [p.412]
3. Define an internal environment. [p.413]
4. Explain what is meant by homeostasis and whether it interacts with the internal or external environment. [p.413]
5. How does the surface-to-volume ratio limit cell size? [p.414]
6. Define a habitat. [p.415]
7. Trace the movement of a signal from the sensory receptor, to the integrator, and then the effector. Define the role of each. [p.416]
8. Distinguish between positive and negative feedback and give an example of each. [pp.416–417]
9. Explain how plants conduct homeostasis. [p.418]
10. Define a circadian rhythm and how it relates to homeostasis. [p.419]
11. Define apoptosis. [p.420]

# Chapter Summary

(1) _____ is the study of body form at one or more levels of structural organization, from molecules to cells, (2) _____, and organ systems. (3) _____ is the study of patterns and processes by which the body functions in its (4) _____. The structure of most body parts correlates with current or past functions, and it emerges during (5) _____ and development.

Animals and plants must exchange (6) _____ with their environment, (7) _____ materials throughout a body, maintain the volume and composition of their (8) _____ environment, and coordinate (9) _____ activities. They must also respond to threats and to variations in available (10) _____.

Extracellular (11) _____ bathes all living cells in a multicelled body. Cells, tissues, (12) _____, and organ systems contribute to maintaining this internal environment within a range that individual (13) _____ can tolerate.

(14) _____ is the process of maintaining favorable operating conditions in the body's internal environment. Negative and positive (15) _____ systems contribute to homeostasis.

Cells of tissues and organs communicate by secreting (16) _____ molecules into the extracellular fluid, and by selectively responding to (17) _____ secreted by other cells.

# Integrating and Applying Key Concepts

1. How might an understanding of apoptosis benefit modern medicine? How could it be used to fight disease and increase longevity?

# 26

# PLANT TISSUES

## INTRODUCTION

This chapter investigates the different types of tissues in plants. These tissue types are very different from those found in the animals, thus special care should be taken to learn the function and location of the key tissue types.

## FOCAL POINTS

- Figure 26.2 [p.426] illustrates the different tissue types in a plant.
- Figure 26.3 [p.427] diagrams the differences between a monocot and eudicot.
- Figure 26.14 [p.433] illustrates the structure of a leaf.
- Figure 26.22 [p.437] diagrams the structure of a woody stem.

## Interactive Exercises

*Droughts Versus Civilization* [pp.424–425]

### 26.1. COMPONENTS OF THE PLANT BODY [pp.426–427]

*Selected Words*

"Seed leaves" [p.426], *apical* meristems [p.427], *lateral* meristems [p.427]

*Boldfaced Terms*

[p.426] shoots _____

_____

[p.426] roots _____

_____

[p.426] ground tissue system _____

_____

[p.426] vascular tissue system _____

_____

[p.426] dermal tissue system _____

_____

[p.426] cotyledons _____

_____

[p.426] meristems _____

_____

[p.427] primary growth _____

_____

[p.427] secondary growth _____

_____

## Short Answer

1. List the ways that drought can affect plants. [p.425]

_____

_____

_____

## Identification

Use the terms in exercises 2–6 to identify each part of this illustration. Then, complete each exercise by noting the letter of each term's description in the parentheses. [p.426]

2. _____ ground tissues (   )

3. _____ vascular tissues (   )

4. _____ dermal tissues (   )

5. _____ shoot system (   )

6. _____ root system (   )

a. aboveground, includes stems, leaves, and flowers (reproductive shoots)
b. typically grows below ground, anchors aboveground parts, absorbs water and dissolved minerals, cells store and release food, and anchors the aboveground parts
c. serves basic functions, such as food and water storage
d. protects all exposed plant surfaces
e. threads through ground tissue and delivers water and solutes through the plant

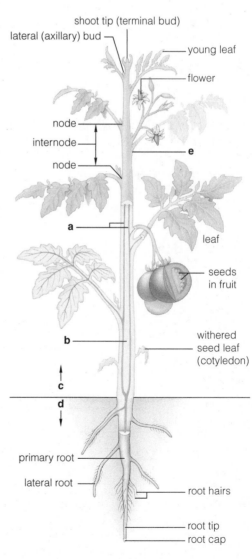

*TNOW:* For an animated exercise of this material, refer to Chapter 26: Art Labeling: Body Plan of a Tomato Plant.

*Short Answer*

7. Explain the difference between primary and secondary growth with regards to meristems. [pp.426–427]

_____

_____

_____

*Complete the Table*

8. There are two classes of flowering plants, monocots and eudicots. Complete the table by supplying the information requested. [p.427]

| Class | Number of Cotyledons | Number of Floral Parts | Leaf Venation | Pollen Grains | Vascular Bundles |
|---|---|---|---|---|---|
| a. Monocots | | | | | |
| b. Eudicots | | | | | |

## 26.2. COMPONENTS OF PLANT TISSUES [pp.428–429]

*Selected Words*

*Radial* sections [p.428], *tangential* sections [p.428], *transverse* sections [p.428], *sclerids* [p.428], *vessel members* [p.428], *tracheids* [p.428], *sieve-tube members* [p.429], *sieve plates* [p.429], *companion cells* [p.421]

*Boldfaced Terms*

[p.428] parenchyma _____

_____

[p.428] collenchyma _____

_____

[p.428] sclerenchyma _____

_____

[p.428] xylem _____

_____

[p.429] phloem _____

_____

[p.429] epidermis _____

_____

[p.429] cuticle _____

_____

## Labeling

Label the indicated components of the diagram below. [p.429]

1. _____

2. _____

3. _____

4. _____

5. _____

6. _____

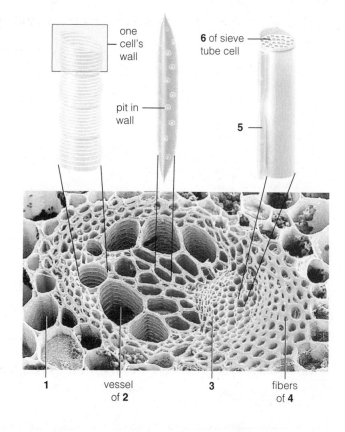

one cell's wall

pit in wall

**6** of sieve tube cell

**5**

1

vessel of **2**

3

fibers of **4**

## Choice

For each of the following statements, choose the appropriate form of plant tissue from the list below. [pp.428–429]

a. xylem    b. phloem    c. epidermis    d. cuticle

7. _____ conserves water and protects against pathogens

8. _____ the main cells are sieve-tube members

9. _____ conducts water and dissolved mineral ions

10. _____ secretes the cuticle

11. _____ consists of vessel members and tracheids

12. _____ rich in a waxy substance called cutin

13. _____ first dermal tissue to form on a plant

14. _____ conducts sugars and other organic solutes

15. _____ companion cells assist in its function

## 26.3. PRIMARY STRUCTURE OF SHOOTS [pp.430–431]

## 26.4. A CLOSER LOOK AT LEAVES [pp.432–433]

## Selected Words

*Axillary buds* [p.422], *cortex* [p.431], *pith* [p.431], *deciduous* species [p.432], *evergreen* plants [p.432], *palisade* mesophyll [p.433], *spongy* mesophyll [p.433]

## Boldfaced Terms

[p.430] terminal buds _____

_____

[p.430] lateral buds _____

_____

[p.431] vascular bundles _____

_____

[p.432] mesophyll _____

_____

[p.433] veins _____

_____

## Labeling

Label each numbered structure in these illustrations of a eudicot stem. [p.430]

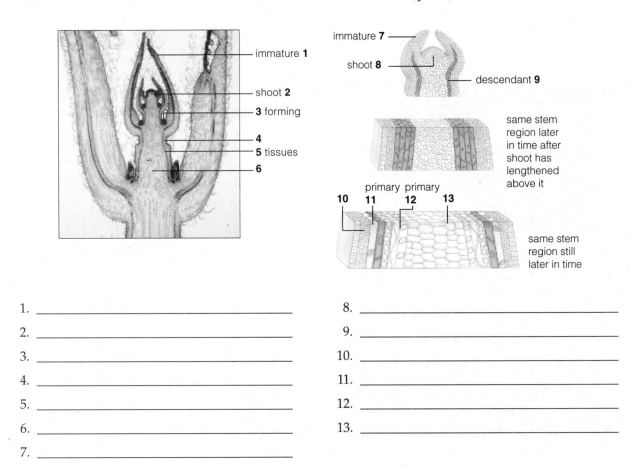

1. _____
2. _____
3. _____
4. _____
5. _____
6. _____
7. _____

8. _____
9. _____
10. _____
11. _____
12. _____
13. _____

## Labeling

Name each of the genetically dictated, internal, cross-sectional stem patterns as either eudicot or monocot. [p.431]

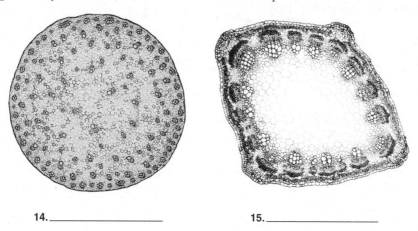

14. _____  15. _____

## Labeling

Name each numbered structure in these illustrations of leaf structure and organization. [p.433]

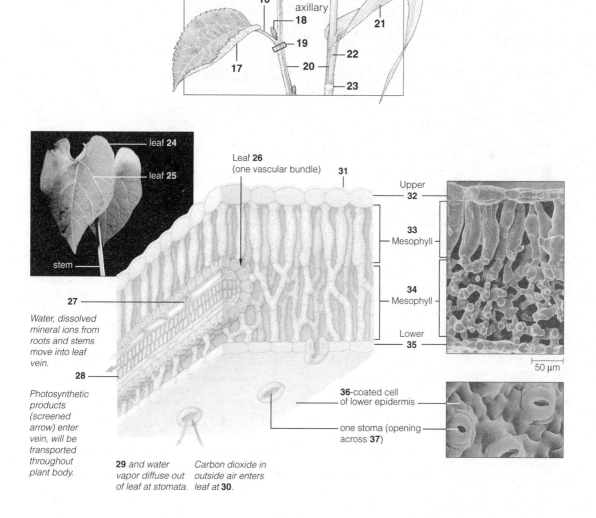

16. _____

17. _____

18. _____

19. _____

20. _____

21. _____

22. _____

23. _____

24. _____

25. _____

26. _____

27. _____

28. _____

29. _____

30. _____

31. _____

32. _____

33. _____

34. _____

35. _____

36. _____

37. _____

## 26.5.  PRIMARY STRUCTURE OF ROOTS [pp.434–435]

### Selected Words

Primary root [p.434], pericycle [p.435]

### Boldfaced Terms

[p.434] taproot system _____

_____

[p.434] fibrous root system _____

_____

[p.435] roots hairs _____

_____

[p.435] vascular cylinder _____

_____

[p.435] endodermis _____

_____

### Labeling

Label the numbered parts of these illustrations. (*Note:* Some numbers are used more than once; they refer to the same structure.) [pp.434–435]

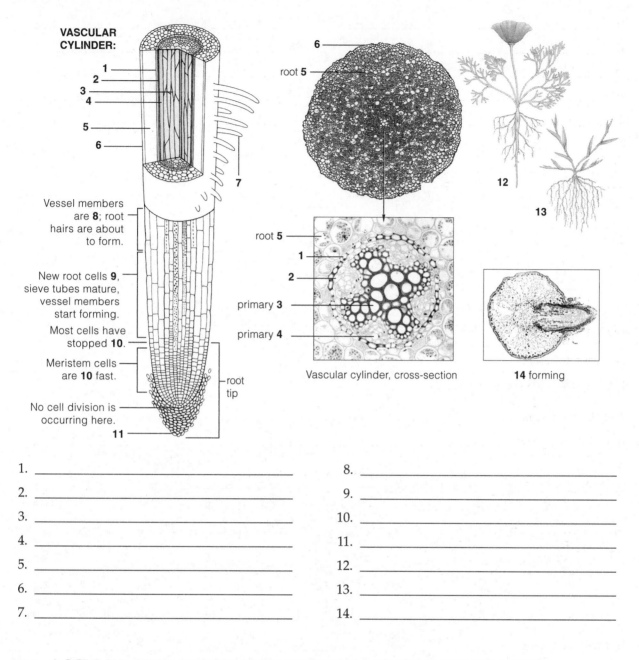

VASCULAR CYLINDER:

1
2
3
4
5
6
7

Vessel members are **8**; root hairs are about to form.

New root cells **9**, sieve tubes mature, vessel members start forming.

Most cells have stopped **10**.

Meristem cells are **10** fast.

No cell division is occurring here.

11

root tip

6
root **5**

root **5**
1
2
primary **3**
primary **4**

Vascular cylinder, cross-section

12
13

**14** forming

1. _____
2. _____
3. _____
4. _____
5. _____
6. _____
7. _____

8. _____
9. _____
10. _____
11. _____
12. _____
13. _____
14. _____

## 26.6. ACCUMULATED SECONDARY GROWTH—THE WOODY PLANTS [pp.436–437]

### Selected Words

*Annuals* [p.436], *biennials* [p.436], *perennials* [p.436], *primary* growth [p.436], *secondary* growth [p.436], *early* wood [p.429], *late* wood [p.437], "tree ring" [p.437]

### Boldfaced Terms

[p.436] vascular cambium _____

_____

[p.436] cork cambium _____

_____

[p.436] cork cambium _____

_____

[p.436] bark _____

_____

[p.437] cork _____

_____

[p.437] heartwood _____

_____

[p.437] sapwood _____

_____

[p.437] hardwood _____

_____

[p .437] softwood _____

_____

## Labeling

Label each part of these illustrations, which show woody stems and primary and secondary growth. [p.437]

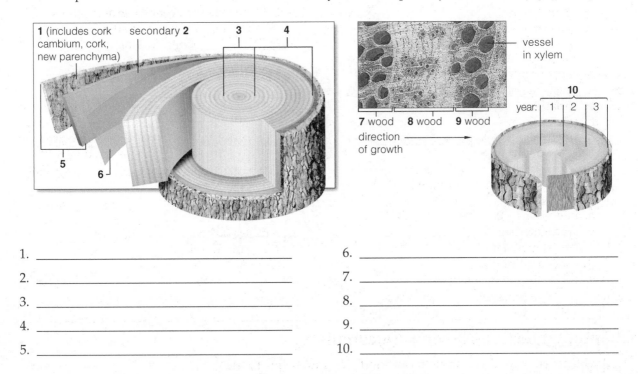

1. _____    6. _____

2. _____    7. _____

3. _____    8. _____

4. _____    9. _____

5. _____    10. _____

## Self-Quiz

___ 1. New plants grow and older plant parts lengthen through cell divisions at _____ meristems present at root and shoot tips; older roots and stems of woody plants increase in diameter through cell divisions at _____ meristems. [p.427]
   a. lateral; lateral
   b. lateral; apical
   c. apical; apical
   d. apical; lateral

___ 2. Which of the following is *not* characteristic of monocots? [p.427]
   a. scattered vascular bundles in the stem
   b. one cotyledon on the embryo
   c. vascular bundles in a ring in the stem
   d. parallel leaf venation
   e. Floral parts in threes, or multiples thereof

___ 3. Which of the following is *not* considered a type of simple tissue? [pp.428–429]
   a. xylem
   b. parenchyma
   c. collenchyma
   d. sclerenchyma
   e. a and d

___ 4. Of the following cell types, which one does not appear in vascular tissues? [p.429]
   a. vessel members
   b. cork cells
   c. tracheids
   d. sieve-tube members
   e. companion cells

___ 5. New stems and leaves form on plants by the activity of _____. [p.430]
   a. internodes
   b. apical meristems
   c. lateral meristems
   d. cork cambium
   e. cotyledons of the embryo

___ 6. The major photosynthetic cell layer of many leaves is found in the _____. [p.432]
   a. lower epidermis
   b. leaf vein
   c. palisade mesophyll
   d. upper epidermis
   e. spongy mesophyll

___ 7. A primary root and its lateral branchings represent a _____ system. [p.435]
   a. lateral root
   b. adventitious root
   c. taproot
   d. branch root
   e. fibrous root

___ 8. A root vascular cylinder is composed of _____ [p.435]
   a. epidermis, phloem, and xylem
   b. pericycle and epidermis
   c. pericycle and endodermis
   d. pericycle, phloem, and xylem
   e. epidermis, cortex, and pith

___ 9. Plants whose vegetative growth and seed formation continue year after year are _____ plants. [p.436]
   a. annual
   b. perennial
   c. nonwoody
   d. herbaceous

___10. "Wood that forms at the start of the growing season and has large-diameter, thin-walled xylem cells," describes _____. [p.437]
   a. hard wood
   b. late wood
   c. soft wood
   d. heartwood
   e. early wood

## Chapter Objectives/Review Questions

1. Be able to list ways that a drought can affect civilization. [pp.424–425]
2. Distinguish between the ground tissue system, the vascular tissue system, and the dermal tissue system. [p.426]
3. Distinguish between primary and secondary growth. [p.427]
4. Distinguish between monocots and dicots by listing their characteristics and citing examples of each group. [p.427]

5. Be able to visually identify and generally describe the simple tissues and cells called parenchyma, collenchyma, and sclerenchyma. [pp.427–428]
6. Name and describe the functions of the conducting cells in xylem and phloem. [p.429]
7. Understand the importance of the epidermis. [p.429]
8. Understand the difference between a terminal and lateral bud. [p.430]
9. Be able to visually distinguish between monocot stems and dicot stems, as seen in cross section; identify the cells present by name and function. [p.431]
10. Distinguish between deciduous and evergreen leaves. [p.432]
11. Describe the structure (cells and layers) and major functions of tissues composing a leaf. [p.433]
12. Distinguish a taproot system from a fibrous root system. [p.434]
13. Describe the origin and function of root hairs. [p.435]
14. Describe the formation of cork and bark by cork cambium activity. [p.436]
15. Distinguish early wood from late wood; heartwood from sapwood. [p.437]
16. Explain the origin of the annual growth rings (tree rings) seen in a cross section of a tree trunk. [p.437]

## Chapter Summary

Seed-bearing (1) _____ plants have a (2) _____ system, which includes stems, leaves, and reproductive parts.

Most also have a (3) _____ system. Such plants consist mostly of ground tissues. Their vascular tissues

distribute (4) _____, nutrients, and products of (5) _____. Their dermal tissues cover all surfaces exposed to

the (6) _____.

Plants lengthen or thicken only at active (7) _____: zones where undifferentiated cells are dividing rapidly.

Meristems near the tips of young shoots and roots drive (8) _____ growth, or the lengthening of plant parts.

Ground, vascular, and (9) _____ tissue systems of monocot and eudicot stems and leaves show patterns of

(10) _____. Patterns of leaf growth and internal leaf structure maximize (11) _____ interception, (12) _____

conservation, and (13) _____ exchange.

Ground vascular and dermal tissue systems of monocot and (14) _____ roots show different patterns of

organization. Root systems absorb water and (15) _____ ions, and may (16) _____ the plant.

In many plants, older branches or roots put on (17) _____ growth; they thicken during successive growing

seasons. Extensive secondary growth is known as (18) _____.

## Integrating and Applying Key Concepts

1. How do you think maple syrup is made from maple trees? Which specific systems of the plant are involved, and why are maple trees tapped only at certain times of the year?
2. Many plants can be asexually reproduced using stem cuttings. What tissues of the stem allow the stem to produce an entirely new plant?

# 27

# PLANT NUTRITION AND TRANSPORT

## INTRODUCTION

This chapter explains the processes by which plants move water and nutrients throughout the tissues of the plant. The chapter introduces several important principles of nutrient and water movement, including the cohesion-tension theory and the translocation of organic nutrients.

## FOCAL POINTS

- The importance of symbiotic relationships to plant nutrition. [pp.444–445]
- The principles of the cohesion-tension theory. [pp.446–447]
- Figure 27.16 [p.451] illustrates the translocation process.

## Interactive Exercises

*Leafy Clean-Up Crews* [pp.440–441]

## 27.1. PLANT NUTRIENTS AND AVAILABILITY IN SOIL [pp.442–443]

*Selected Words*

Phytoremediation [p.440], *macro*nutrients [p.442], *micro*nutrients [p.442]

*Boldfaced Terms*

[p.442] nutrient _____

_____

[p.442] soil _____

_____

[p.442] humus _____

_____

[p.442] loams _____

_____

[p.443] topsoil _____

_____

[p.443] leaching _____

_____

[p.443] soil erosion _____

_____

## Matching

Match each of the following terms with the correct description.

1. _____ micronutrients [p.442]
2. _____ leaching [p.443]
3. _____ soil erosion [p.443]
4. _____ macronutrients [p.442]
5. _____ soil [p.442]
6. _____ humus [p.442]
7. _____ loams [p.442]
8. _____ topsoil [p.443]
9. _____ nutrient [p.442]
10. _____ phytoremediation [p.440]

a. an element or molecule that has some essential role in growth and survival
b. nutrients that are required in amounts above 0.5 percent of the plant's dry weight
c. trace nutrients
d. mineral particles mixed with variable amounts of decomposing organic material
e. decomposing organic material
f. soil that is equal part sand, silt, and clay
g. the layer of soil most important for plant growth
h. the use of plants to remove toxic compounds from soil
i. loss of soil due to the action of wind or water
j. the removal of nutrients from soil by water

## 27.2. HOW DO ROOTS ABSORB WATER AND MINERAL IONS? [pp.444–445]

### Selected Words

hyphae [p.444], endodermal cells [p.445]

### Boldfaced Terms

[p.444] mycorrhiza _____

_____

[p.444] nitrogen fixation _____

_____

[p.444] root nodules _____

_____

[p.444] root hairs _____

_____

[p.445] vascular cylinder _____

_____

[p.445] Casparian strip _____

_____

[p.445] exodermis _____

_____

## Dichotomous Choice

Circle the word in parentheses that makes each statement correct.

1. Mycorrhizae and root nodules are two examples of (mutualism/parasitism). [p.444]
2. Plants lack the ability to use (ammonia/gaseous nitrogen) directly. [p.444]
3. Nitrogen-fixing bacteria reside in localized swellings on legume plants known as (root hairs/root nodules). [p.444]
4. In the mycorrhizae, the fungus receives (sugars/minerals) from the plant. [p.444]
5. In a mycorrhizal interaction between a young root and a fungus, the plant benefits by receiving (nitrogen/minerals) from the fungus. [p.444]

## Labeling

Identify each numbered part of the following illustration. [p.445]

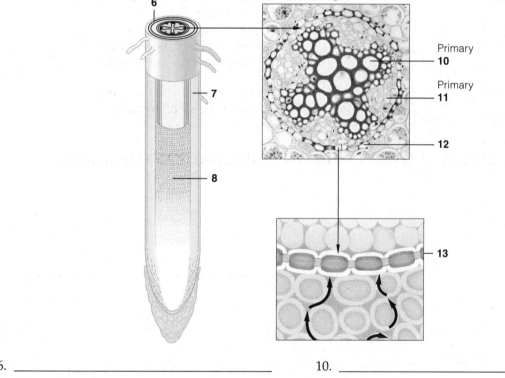

6. _____
7. _____
8. _____
9. _____

10. _____
11. _____
12. _____
13. _____

## Matching

Match each of the following structures of the plant root with its correct function. [p.445]

14. _____ Casparian strip
15. _____ vascular cylinder
16. _____ exodermis
17. _____ endodermal cells

a. separates the vascular tissue from the root cortex; secretes the Casparian strip
b. a waxy band that forces water to move through the endodermal cells
c. contains the xylem and phloem
d. located just below the surface, these cells have a Casparian strip that prevents water and solute flow

**27.3. HOW DOES WATER MOVE THROUGH PLANTS?** [pp.446–447]

**27.4. HOW DO STEMS AND LEAVES CONSERVE WATER?** [pp.448–449]

*Selected Words*

xylem [p.446], stomata [p.446]

*Boldfaced Terms*

[p.446] transpiration _____

_____

[p.446] tracheids _____

_____

[p.446] vessel members _____

_____

[p.446] cohesion-tension theory _____

_____

[p.448] cuticle _____

_____

[p.448] guard cells _____

_____

*Concept Map*

Provide the missing terms in the following concept map of the cohesion-tension theory. [pp.446–447]

1. _____
2. _____
3. _____
4. _____
5. _____

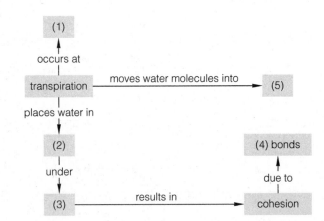

## Matching

Match each of the following with its function. [pp.448–449]

6. _____ stomata
7. _____ cuticle
8. _____ stoma
9. _____ abscisic acid (ABA)
10. _____ guard cells

a. a waxy covering that prevents water loss
b. the opening between two guard cells
c. the cells that are responsible for regulating the movement of gas into leaves
d. signals stomata to close during water shortages
e. the combination of guard cells and stoma

## 27.5. THE BAD OZONE [p.449]

## 27.6. HOW DO ORGANIC COMPOUNDS MOVE THROUGH PLANTS? [pp.450–451]

### Selected Words

Ozone [p.449], *source* [p.450], mesophylls [p.451], *sink* [p.451]

### Boldfaced Terms

[p.450] phloem _____

_____

[p.450] sieve tubes _____

_____

[p.450] companion cells _____

_____

[p.450] translocation _____

_____

[p.451] pressure flow theory _____

_____

## Matching

Choose the most appropriate definition for each term. [pp.450–451]

1. _____ translocation
2. _____ sieve tube
3. _____ companion cells
4. _____ source
5. _____ sink
6. _____ pressure flow theory

a. cells found in phloem that are responsible for the movement of organic compounds
b. process that moves sucrose and organic compounds through phloem
c. process by which internal pressure moves a solute-rich solution from a source to a sink
d. a region of a plant where organic compounds are being unloaded from the sieve-tube system
e. a region of a plant where organic compounds are being loaded into the sieve-tube system
f. cells adjacent to the sieve-tube system whose purpose is to supply energy for active transport

## Labeling

Provide the indicated terms in the diagram of the translocation process below. [p.451]

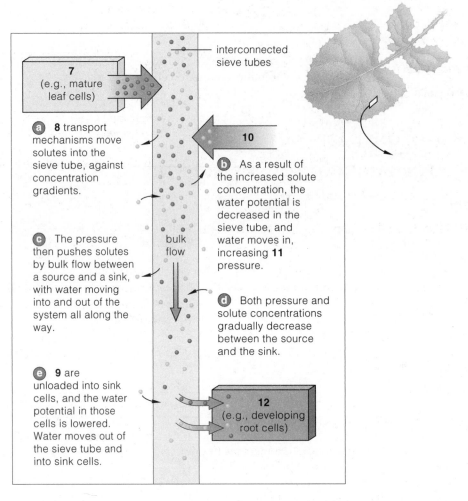

7. _____

8. _____

9. _____

10. _____

11. _____

12. _____

# Self-Quiz

1. Which of the following is not a macronutrient? [p.442]
   a. carbon
   b. manganese
   c. calcium
   d. nitrogen
   e. potassium

2. Gaseous nitrogen is converted to a plant usable form in the _____ of the plant. [p.444]
   a. root nodules
   b. root hairs
   c. mycorrhizae
   d. leaves
   e. stems

3. The removal of toxic compounds by plants is called _____. [p.440]
   a. photosynthesis
   b. transmutation
   c. translocation
   d. phytoremediation
   e. nitrification

4. The Casparian strip is located within the _____ of the plant. [p.445]
   a. roots
   b. leaves
   c. stems
   d. stomata

5. The cohesion-tension theory explains the movement of _____ in plants. [pp.446–447]
   a. minerals
   b. carbohydrates
   c. water
   d. bacteria

6. The stomata of a plant are involved in the exchange of _____. [p.448]
   a. water
   b. ammonia
   c. gases
   d. sugars

7. Sieve tubes are located in the _____ of the plant. [p.450]
   a. xylem
   b. phloem
   c. mycorrhizae
   d. root hairs

8. In the pressure flow theory, solutes are moved into the system by _____. [p.451]
   a. turgor pressure
   b. translocation
   c. transpiration
   d. active transport

9. A soil that has equal parts sand, silt, and clay is called _____. [pp.442–443]
   a. topsoil
   b. loam
   c. horizon
   d. humus

10. Which of the following processes represent the loss of nutrients in a soil? [p.443]
    a. translocation
    b. transpiration
    c. erosion
    d. leaching

## Chapter Objectives/Review Questions

1. Understand the concept of phytoremediation. [pp.440–441]
2. Distinguish between a macro- and micronutrient. [p.442]
3. Understand the components of soil. [pp.442–443]
4. Distinguish between soil erosion and leaching. [p.443]
5. Understand the importance of root nodules, mycorrhizae, and root hairs in the processing of water and minerals by a plant. [pp.444–445]
6. Understand the structure of a root hair and how the different components are involved in water and solute absorption. [p.445]
7. Define *transpiration*. [p.446]
8. Outline the steps of the water-cohesion theory. [pp.446–447]
9. Understand the importance of the cuticle and stomata in water conservation. [pp.448–449]
10. Understand how organic materials move in a plant. [pp.450–451]
11. Define *translocation*. [p.450]
12. Understand the pressure flow theory. [p.451]

## Chapter Summary

Fill in the blanks of the following items from the Key Concepts section of the chapter introduction. [p.441]

Many plant structures and functions are adaptations to limited amounts of (1) _____ and dissolved

(2) _____ ions. The (3) _____ systems of vascular plants absorb water and mine the soil for nutrients, and

many have symbionts that help them do so.

The amount of water and minerals available to plants to take up depends on the composition of the (4) _____. Soil is vulnerable to leaching and (5) _____.

Xylem distributes absorbed water and (6) _____ from roots to leaves. (7) _____ from leaves pulls up water molecules that are hydrogen-bonded to one another in long columns inside xylem. New molecules entering leaves replace the ones evaporating away.

A cuticle and (8) _____ help plants conserve water, a limited resource in most land habitats. Closed stomata stop water loss but also stop (9) _____ exchange. Some plant adaptations are trade-offs between water conservation and (10) _____.

(11) _____ distributes sucrose and other organic compounds from photosynthetic cells in leaves to living cells throughout the plant. (12) _____ compounds are actively loaded into conducting cells, then unloaded in growing tissues or storage tissues.

---

## Integrating and Applying Key Concepts

1. Under stressful conditions, plant roots may expel mutualistic bacteria and fungi. Explain the possible effects of this event and how the process can become cyclical.
2. Fungi are often pathogens of plants. How would you propose that by the process of natural selection that mycorrhizae may have evolved?

# 28

# PLANT REPRODUCTION AND DEVELOPMENT

## INTRODUCTION

This chapter introduces a typical plant life cycle, the structure of a flower, the formation of the fruit and seed, and the use of tropisms and hormones to control aspects of growth and development.

## FOCAL POINTS

- Figure 28.2a [p.456] outlines the life cycle of a typical flowering plant.
- Figure 28.2b [p.456] illustrates the structures of a flower.
- Table 28.3 [p.456] lists the major classes of plant hormones.
- Figure 28.23 [p.470] illustrates the phytochrome activation pathway.

## Interactive Exercises

*Imperiled Sexual Partners* [pp.454–455]

## 28.1. REPRODUCTIVE STRUCTURES OF FLOWERING PLANTS [pp.456–457]

*Selected Words*

Anther [p.456], stigma [p.456], style [p.456]

*Boldfaced Terms*

[p.456] sporophytes _____

_____

[p.456] flowers _____

_____

[p.456] gametophytes _____

_____

[p.456] stamens _____

_____

[p.456] pollen grains _____

_____

[p.456] carpels _____

_____

[p.456] ovule _____

_____

[p.456] seed _____

_____

[p.456] pollination vectors _____

_____

[p.457] nectar _____

_____

## Identification

Identify each of the numbered components in the diagram below. [p.456]

1. _____

2. _____

3. _____

4. _____

5. _____

6. _____

7. _____

8. _____

9. _____

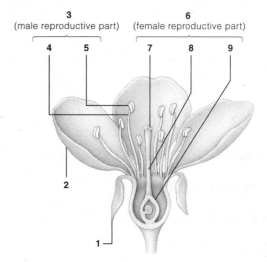

## Identification

Identify each of the numbered components in the diagram below. [p.456]

10. _____

11. _____

12. _____

13. _____

14. _____

15. _____

16. _____

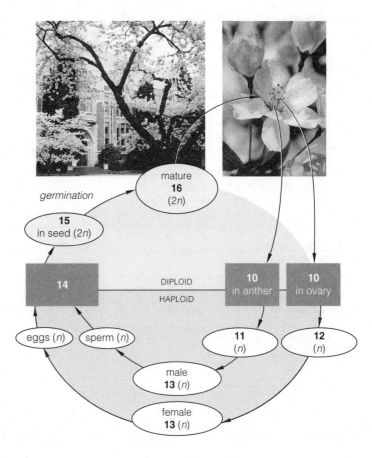

## 28.2. A NEW GENERATION BEGINS [pp.458–459]

## 28.3. FROM ZYGOTES TO SEEDS AND FRUITS [pp.460–461]

## 28.4. SEED DISPERSAL—THE FUNCTION OF FRUITS [p.462]

### Selected Words

True fruits [p.460], accessory fruits [p.460], dehiscent [p.460], indehiscent [p.460], drupes [pp.460–461], berry [p.461], pome [p.461], "parachute" [p.462]

## Boldfaced Terms

[p.458] microspores _____

_____

[p.458] megaspores _____

_____

[p.458] endosperm mother cell _____

_____

[p.458] pollination _____

_____

[p.458] double fertilization _____

_____

[p.459] endosperm _____

_____

[p.460] cotyledons _____

_____

[p.460] fruit _____

_____

## Identification

Identify the numbered parts of the eudicot life cycle. [pp.458–459]

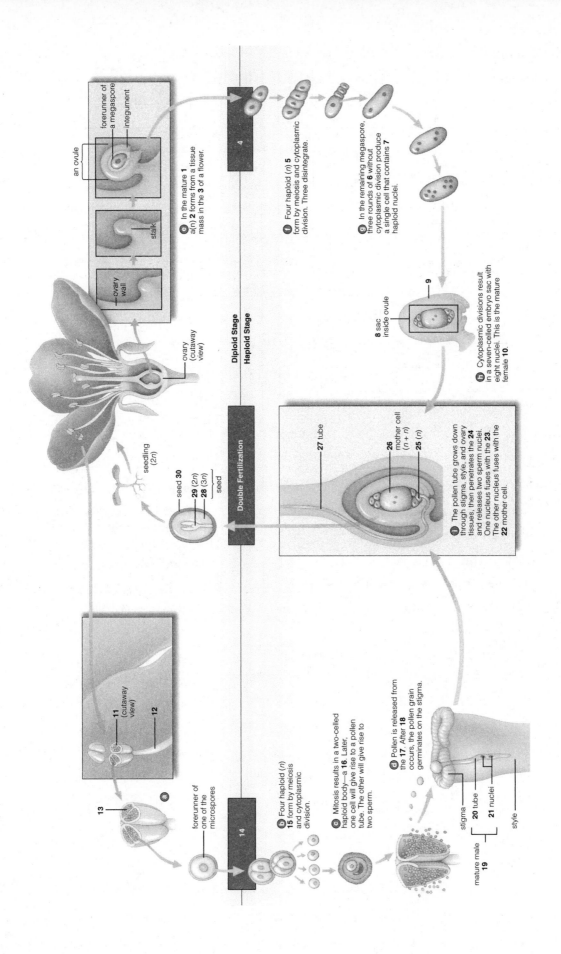

an ovule

forerunner of a megaspore

integument

**e** In the mature **1** a(n) **2** forms from a tissue mass in the **3** of a flower.

**4**

ovary wall

stalk

ovary (cutaway view)

**f** Four haploid (*n*) **5** form by meiosis and cytoplasmic division. Three disintegrate.

**g** In the remaining megaspore, three rounds of **6** without cytoplasmic division produce a single cell that contains **7** haploid nuclei.

**8** sac inside ovule

**9**

**h** Cytoplasmic divisions result in a seven-celled embryo sac with eight nuclei. This is the mature female **10**.

**Diploid Stage**

**Haploid Stage**

**Double Fertilization**

**27** tube

**26** mother cell (*n* + *n*)

**25** (*n*)

**i** The pollen tube grows down through stigma, style, and ovary tissues, then penetrates the **24** and releases two sperm nuclei. One nucleus fuses with the **23**. The other nucleus fuses with the **22** mother cell.

seedling (2*n*)

seed **30**

**29** (2*n*)

**28** (3*n*)

seed

**11** (cutaway view)

**12**

**a**

**13**

forerunner of one of the microspores

**14**

**b** Four haploid (*n*) **15** form by meiosis and cytoplasmic division.

**c** Mitosis results in a two-celled haploid body—a **16**. Later, one cell will give rise to a pollen tube. The other will give rise to two sperm.

**d** Pollen is released from the **17**. After **18** occurs, the pollen grain germinates on the stigma.

stigma

**20** tube

**21** nuclei

mature male **19**

style

1. _____
2. _____
3. _____
4. _____
5. _____
6. _____
7. _____
8. _____
9. _____
10. _____
11. _____
12. _____
13. _____
14. _____
15. _____

16. _____
17. _____
18. _____
19. _____
20. _____
21. _____
22. _____
23. _____
24. _____
25. _____
26. _____
27. _____
28. _____
29. _____
30. _____

## Matching

Match each of the following descriptions to the correct term. [pp.460–461]

31. _____ fruits that contain floral parts

32. _____ seeds are dispersed inside intact fruits

33. _____ contain a pit

34. _____ has one to many seeds, no pit, and a fleshy fruit

35. _____ fruits that contain only the ovary wall and its contents

36. _____ has seeds in an elastic core

37. _____ fruit that splits along seams to release seeds

a. dehiscent
b. indehiscent
c. drupes
d. berry
e. pome
f. true fruit
g. accessory fruit

## 28.5. ASEXUAL REPRODUCTION OF FLOWERING PLANTS [pp.462–463]

### Selected Words

Runner [p.463], rhizome [p.463], corm [p.463], tuber [p.463], bulb [p.463]

### Boldfaced Terms

[p.462] vegetative growth _____

_____

[p.463] tissue culture propagation _____

_____

### Matching

Match each of the following forms of asexual reproduction to the correct description. [p.463]

1. _____ plants arise from buds on short underground stems

2. _____ plants arise from aboveground horizontal stem

3. _____ plants arise from auxiliary buds on underground carbohydrate-storing stems

4. _____ plants arise from auxiliary buds on underground stems

5. _____ plant derived from a parent plant cell that is not reversibly differentiated

6. _____ plants arise from underground horizontal stem

7. _____ new plant arises from a structure dropped from the parent plant

a. rhizome
b. tissue culture propagation
c. corm
d. runner
e. tuber
f. bulb
g. vegetative growth

## 28.6. OVERVIEW OF PLANT DEVELOPMENT [pp.464–465]

### Selected Words

Meristem [p.464], cotyledon [p.464], "hypocotyl" [p.465]

### Boldfaced Terms

[p.464] germination _____

_____

[p.464] growth _____

_____

## Identification

Identify each of the numbered parts of the plant diagrams below. [p.465]

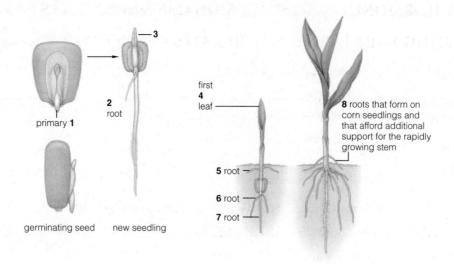

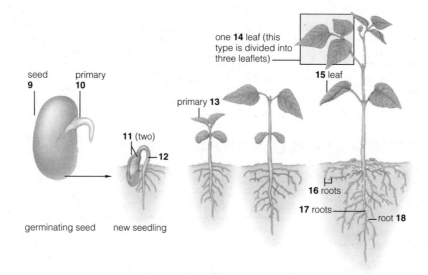

1. _____    10. _____

2. _____    11. _____

3. _____    12. _____

4. _____    13. _____

5. _____    14. _____

6. _____    15. _____

7. _____    16. _____

8. _____    17. _____

9. _____    18. _____

*TNOW:* For an animated version of this exercise, refer to Chapter 28: Art Labeling; Pattern of Growth for a Eudicot.

## 28.7. PLANT HORMONES AND OTHER SIGNALING MOLECULES [pp.466–467]

## 28.8. ADJUSTING THE DIRECTION AND RATES OF GROWTH [pp.468–469]

### Selected Words

Hormones [p.466], *jasmonates* [p.467], *FT protein* [p.467], *salicylic acid* [p.467], *nitric oxide* [p.467], *systemin* [p.467], tropisms [p.468], *flavoproteins* [p.468], thigmotropism [p.469]

### Boldfaced Terms

[p.466] gibberellins _____

_____

[p.466] auxins _____

_____

[p.466] apical dominance _____

_____

[p.466] abscission _____

_____

[p.466] cytokinins _____

_____

[p.467] ethylene _____

_____

[p.467] abscisic acid (ABA) _____

_____

[p.468] gravitropism _____

_____

[p.468] statoliths _____

_____

[p.468] phototropism _____

_____

*Choice*

For each of the following, choose the appropriate plant hormone from the list below.

> a. gibberellins [p.466]    b. cytokinins [pp.466–467]    c. auxins [p.466]
> d. ethylene [p.467]    e. abscisic acid [p.467]

1. _____ compound that triggers stem elongation

2. _____ stimulates rapid cell division in root and shoot meristems and in maturing fruit

3. _____ prevents the growth of lateral buds along the lengthening stem

4. _____ the only gaseous plant hormone

5. _____ helps seeds germinate by stimulating cell division and elongation

6. _____ causes stomata to close in water-stressed plants

7. _____ helps control the growth of most tissues, fruit ripening, leaf dropping, and other aging responses

8. _____ applied to nursery stock before shipping to induce dormancy

9. _____ induces and maintains dormancy in buds and seeds

10. _____ opposes the effects of auxins; makes lateral buds grow

11. _____ stimulates transport of some photosynthetic products to seeds; in seeds it triggers protein synthesis and embryo formation

12. _____ keeps leaves from aging; used to prolong shelf life in cut flowers

13. _____ used to brighten the rinds of citrus fruits

*Concept Map*

Provide the missing word or words in the concept map of tropism below. [pp.468–469]

14. _____

15. _____

16. _____

17. _____

18. _____

19. _____

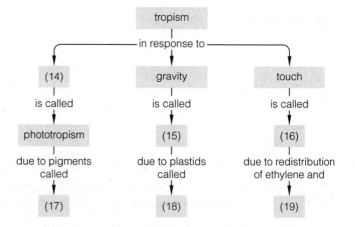

## 28.9. SENSING RECURRING ENVIRONMENTAL CHANGES [pp.470–471]

## 28.10. ENTERING AND BREAKING DORMANCY [p.472]

*Selected Words*

*Circa* [p.470], *dies* [p.470], *short-day* plants [p.470], *long-day* plants [p.470], *day-neutral* plants [p.470], *vernalis* [p.471]

## Boldfaced Terms

[p.470] biological clock _____

_____

[p.470] circadian rhythm _____

_____

[p.470] solar tracking _____

_____

[p.470] phytochromes _____

_____

[p.470] photoperiodism _____

_____

[p.471] vernalization _____

_____

[p.472] abscission _____

_____

[p.472] senescence _____

_____

[p.472] dormancy _____

_____

## Dichotomous Choice

Circle the words in parentheses that make each statement correct. [pp.470–471]

1. Summer blooming irises are (short-day/long-day) plants.
2. Chrysanthemums are examples of (short-day/long-day) plants.
3. Sunflowers are (short-day/day-neutral) plants that form flowers when they mature, without regard to the season.
4. "Short-day" plants flower only when the nights are (shorter/longer) than the critical value.
5. "Long-day" plants flower only when nights are (shorter/longer) than a critical value.
6. Phytochrome is activated by (red/far-red) wavelengths.
7. When a dark interval is interrupted by a pulse of (red/far-red) light, phytochrome becomes inactivated.
8. If red light is followed by a pulse of far-red light, phytochrome is (activated/inactivated) and plants behave as though the dark period was uninterrupted.

## Matching

Match each of the following terms its correct description.

9. _____ abscission [p.472]

10. _____ senescence [p.472]

11. _____ dormancy [p.472]

12. _____ vernalization [p.471]

a. the period from full maturity to death of the plant
b. the shedding of plant parts
c. a period of arrested growth
d. flowering only after exposure to low winter temperatures

---

# Self-Quiz

1. The portion of the carpel that contains the ovule is the _____. [p.456]
   a. stigma
   b. anther
   c. style
   d. ovary
   e. filament

2. The phase of the life cycle of a plant that gives rise to spores in the _____. [p.456]
   a. female gametophyte
   b. male gametophyte
   c. seed
   d. sporophyte
   e. vegetative growth

3. An immature fruit is a(n) _____ and an immature seed is a(n) _____. [p.460]
   a. ovary; megaspore
   b. ovary; ovule
   c. megaspore; ovule
   d. ovule; ovary

4. In flowering plants, one sperm nucleus fuses with that of an egg, while the sperm fuses with the endosperm. This is called _____. [p.458]
   a. vernalization
   b. double fertilization
   c. vegetative propagation
   d. meiosis

5. Which of the following is not a fruit? [pp.460–461]
   a. drupes
   b. pomes
   c. berry
   d. seed

6. An example of asexual reproduction in flowering plants is _____. [pp.462–463]
   a. a clone of quaking aspen plants
   b. runners
   c. bulbs
   d. tissue culture propagation
   e. all of the above

7. In dry climates, germination depends on _____. [p.464]
   a. number of daylight hours
   b. temperature
   c. water
   d. soil oxygen level
   e. all of the above

8. Which of the following controls the growth of plant tissues, fruit ripening, leaf drop, and other aging responses? [pp.466–467]
   a. auxins
   b. cytkinins
   c. gibberellins
   d. ethylene
   e. abscisic acid

9. The sum of all processes that lead to the death of a plant or some of its parts is called _____. [p.472]
   a. dormancy
   b. vernalization
   c. abscission
   d. senescence
   e. none of the above

10. Phytochrome is activated by _____. [pp.470–471]
    a. red wavelengths of light
    b. water
    c. auxins
    d. far-red wavelengths of light
    e. none of the above

# Chapter Objectives/Review Questions

1. Understand the life cycle of a flowering plant. [p.456]
2. Be able to label the parts of a flower. [p.456]
3. Distinguish between microspore and megaspores. [p.458]
4. Distinguish between pollination and fertilization. [p.458]
5. Understand the process of double fertilization, the structure involved, and the chromosome state (ploidy) of the resulting tissues. [pp.458–459]
6. Understand the terminology associated with seeds and fruits. [pp.460–461]
7. Understand the basic mechanisms of seed dispersal. [p.462]
8. Understand the major methods of asexual reproduction in plants. [pp.462–463]
9. Understand the conditions under which germination occurs. [p.464]
10. For each of the plant hormones, recognize its influence on plant growth and development. [pp.466–467]
11. Understand the major forces of tropism in plants. [pp.468–469]
12. Understand the concept of a biological clock. [p.470]
13. Understand photoperiodism. Be able to distinguish between short-day, long-day, and day-neutral plants. [p.470]
14. Understand the activation/deactivation pathway of phytochrome. [pp.470–471]
15. Understand the differences among dormancy, senescence, and abscission. [p.472]

# Chapter Summary

(1)_____ reproduction is the dominant reproductive mode of flowering plant life cycles, which typically depend on (2) _____. Spores and gametes form in the specialized reproductive shoots called (3) _____.

Sperm-bearing male gametophytes (pollen grains) and female (4) _____ that bear eggs inside ovules form in reproductive parts of flowers. After (5) _____, ovules mature into seeds, each an embryo (6) _____ and tissues that nourish and protect it.

As seeds develop, tissues of the (7) _____ and often other floral parts mature into fruits, which function in seed (8) _____. Air currents and (9) _____ are the main dispersal agents.

Many species of flowering plants also reproduce (10) _____ by vegetative growth and other mechanisms.

Interactions among (11) _____ and other signaling molecules control plant growth and development. Hormone synthesis starts at seed (12) _____ and guides all events of the life cycle, such as (13) _____ and shoot development, flowering, fruit formation, and dormancy.

Plant cell receptors that govern hormone (14) _____ respond to (15) _____ cues, such as gravity, sunlight, and seasonal shifts in (16) _____ length and temperature.

# Integrating and Applying Key Concepts

1. A lumber company removes all but a single oak tree from a forest. What changes have occurred in the environment of the oak tree? Which hormones and tropisms may be involved in the response of the oak to the changing environment?

2. You have been hired by a company to bring Costa Rican plants to the United States for sale. The plants require 12 hours of continuous darkness per night in order to flower and produce seeds. Explain what is happening within this plant.

3. As an agricultural consultant you have been assigned the task of maximizing profits by growing plants as quickly as possible and reducing the amount of loss due to damage or rotting on the way to market. Describe the hormones that you would most likely use in your work and when they would be applied to the plant.

# 29

# ANIMAL TISSUES AND ORGAN SYSTEMS

## INTRODUCTION

All animals are comprised of a set of common tissue types. These tissue types are then organized into specialized organ systems that perform specific tasks for the organism. In this chapter you will be introduced to the major tissue types and organ systems found in the animals.

## FOCAL POINTS

- Figure 29.5 [p.479] illustrates the major types of cell junctions.
- Figure 29.12b-c [p.484] illustrates the directional terms necessary to understand the organization and location of organ systems.
- Figure 29.13 [p.485] illustrates the location and function of the human organ systems.
- Figure 29.14a [p.486] diagrams the structure of human skin.

## Interactive Exercises

*Open or Close the Stem Cell Factories* [pp.476–477]

### 29.1. EPITHELIAL TISSUE [pp.478–479]

### 29.2. CONNECTIVE TISSUE [pp.480–481]

*Selected Words*

*Simple* epithelium [p.478], *stratified* epithelium [p.478], glandular epithelium [pp.478–479]

*Boldfaced Terms*

[p.478] epithelium _____

_____

[p.478] microvilli _____

_____

[p.479] glands_____

_____

[p.479] exocrine glands_____

_____

[p.479] endocrine glands _____

_____

[p.479] tight junctions _____

_____

[p.479] adhering junctions _____

_____

[p.479] gap junctions _____

_____

[p.480] connective tissues _____

_____

[p.480] loose connective tissue _____

_____

[p.480] dense, irregular connective tissue _____

_____

[p.480] dense, regular connective tissue _____

_____

[p.480] cartilage _____

_____

[p.481] adipose tissue _____

_____

[p.481] bone tissue _____

_____

[p.481] blood _____

_____

## Matching

Match each of the following to its correct definition.

1. _____ endocrine glands [p.479]

2. _____ loose connective tissue [p.480]

3. _____ glands [p.479]

4. _____ adhering junctions [p.479]

5. _____ dense, irregular connective tissue [p.480]

6. _____ exocrine glands [p.479]

7. _____ cartilage [p.480]

8. _____ dense, regular connective tissue [p.480]

a. glands that secrete products into interstitial fluid; do not possess ducts
b. contains fibroblasts and fibers dispersed widely throughout the matrix
c. contains fibroblasts and fibers in an orderly configuration
d. living cells within calcium hardened secretions
e. cells are derived from stem cells in bone
f. contain the body's main energy reservoir
g. contains ducts or tubes that deliver secretions to a free epithelial surface
h. a mixture of collagen fibers and glycoproteins

9. _____ tight junctions [p.479]

10. _____ adipose tissue [p.481]

11. _____ blood [p.481]

12. _____ bone tissue [p.481]

13. _____ gap junctions [p.479]

i. fibroblasts and fibers are packed into the matrix, but not orientated in one direction

j. hold cells together at distinct spots

k. permit ions and small molecules to pass freely from one cell to another

l. these seal epithelial membranes together

m. the general term for organs that secrete substances onto the skin, or into a body cavity or interstitial fluid

## Choice

Choose the appropriate class of tissue for each statement.

        a. epithelial tissue [pp.478–479]     b. connective tissue [pp.480–481]

14. _____ fibroblasts are usually the main cell type

15. _____ one free surface is exposed to the environment

16. _____ cells within an extracellular matrix

17. _____ arranged in a sheet-like configuration

18. _____ cells are usually squamous, cuboidal, or columnar in shape

## 29.3. MUSCLE TISSUES [pp.482–483]

## 29.4. NERVOUS TISSUE [p.483]

### Selected Words

*Contract* [p.482], *muscle fibers* [p.482], *"voluntary muscle"* [p.483], *"involuntary"* muscle [p.483]

### Boldfaced Terms

[p.482] skeletal muscle tissue_____

_____

[p.482] cardiac muscle tissue _____

_____

[p.483] smooth muscle tissue_____

_____

[p.483] nervous tissue _____

_____

[p.483] neurons _____

_____

## True–False

If the statement is correct, place a "T" in the blank. If the statement is incorrect, correct the underlined word to make the statement true. [pp.482–483]

1. _____ <u>Neurons</u> are the primary cells of the nervous system.

2. _____ The <u>skeletal</u> muscle tissues contain the highest number of mitochondria and the lowest level of glycogen of all muscle tissues.

3. _____ Cells of the <u>nervous</u> system function by contraction.

4. _____ Smooth muscle tissue controls the operation of <u>voluntary</u> muscles.

5. _____ Cells of the <u>nervous</u> system respond to stimuli using electrical and chemical signals.

6. _____ <u>Skeletal</u> muscle tissue lines the walls of most soft internal organs.


## 29.5. OVERVIEW OF MAJOR ORGAN SYSTEMS [pp.484–485]

## 29.6. VERTEBRATE SKIN—EXAMPLE OF AN ORGAN SYSTEM [pp.486–487]

### Selected Words

*Germ* layers [p.484], *dorsal* [p.484], *ventral* [p.484], *anterior* [p.484], *posterior* [p.484], dermis [p.486], epidermis [p.486], keratin [p.486], *Herpes simplex* [p.486]

### Boldfaced Terms

[p.484] ectoderm _____

_____

[p.484] mesoderm _____

_____

[p.484] endoderm _____

_____

[p.484] division of labor _____

_____


### Matching

Match each of the primary germ layers with the adult tissues it forms. [p.484]

1. _____ ectoderm

2. _____ endoderm

3. _____ mesoderm

a. forms internal skeleton and muscle, circulatory, reproductive, and urinary systems
b. forms inner lining of gut and linings of major organs from the embryonic gut
c. forms outer layer of skins and tissues of the nervous system

## Labeling

Identify each numbered part of the accompanying illustration that reviews the directional terms and place of symmetry of the human body. [p.484]

4. _____

5. _____

6. _____

7. _____

8. _____

9. _____

10. _____

11. _____

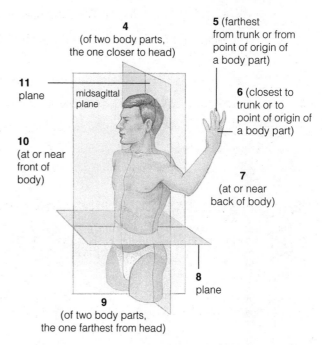

**4**
(of two body parts,
the one closer to head)

**5** (farthest
from trunk or from
point of origin of
a body part)

**11**
plane

midsagittal
plane

**6** (closest to
trunk or to
point of origin of
a body part)

**10**
(at or near
front of
body)

**7**
(at or near
back of body)

**8**
plane

**9**
(of two body parts,
the one farthest from head)

*TNOW:* For an animation of this material, refer to Chapter 29: Art Labeling: Directional Terms and Planes of Symmetry.

## Labeling and Matching

Write the name of each organ system described by the statement. Then, match the letter from the illustration to the correct statement. [p.485]

12. _____ ( ) Rapidly transports many nutrients to and from cells; helps stabilize internal pH and temperature.

13. _____ ( ) Rapidly delivers oxygen to the tissue that bathes all living cells; removes carbon dioxide waste of the cells; helps regulate pH.

14. _____ ( ) Maintains the volume and composition of the internal environment; excretes excess fluid and blood-borne wastes.

15. _____ ( ) Supports and protects body parts; provides muscle attachment sites; produces red blood cells; stores calcium, phosphorous.

16. _____ ( ) Hormonally controls body function; works with nervous system to integrate short- and long-term activities.

17. _____ ( ) *Female:* produces eggs; after fertilization, affords a protected, nutritive environment for the development of new individuals. *Male:* produces and transfers sperm to the female. Hormones of both systems also influence other organ systems.

18. _____ ( ) Ingests food and water; mechanically and chemically breaks down food and absorbs small molecules into the internal environment; eliminates food residues.

19. _____ ( ) Moves body and its internal parts; maintains posture; generates heat.

20. _____ (   ) Detects internal and external stimuli; controls and coordinates responses to stimuli; integrates all organ system activities.

21. _____ (   ) Protects body from injury, dehydration, and some pathogens; controls its temperature; excretes some wastes; receives some external stimuli.

22. _____ (   ) Collects and returns some tissue fluid to the bloodstream; defends the body against infection and tissue damage.

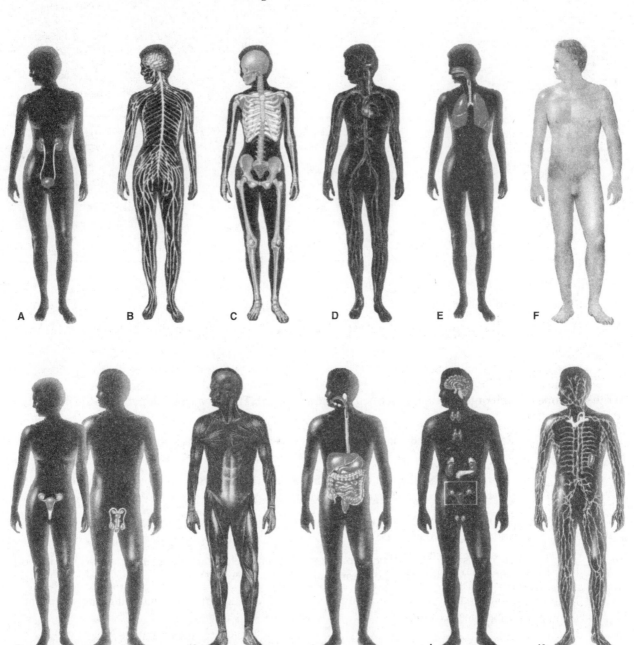

## Labeling

Label each of the numbered components of the diagram below. [p.486]

23. _____

24. _____

25. _____

26. _____

27. _____

28. _____

oil gland

27 blood vessels

sensory neuron

sweat gland

28

---

# Self-Quiz

_____ 1. The mesoderm gives rise to which of the following? [p.484]
   a. the lining of the intestinal tract
   b. the skin
   c. muscles and connective tissue
   d. the nervous system
   e. none of the above

_____ 2. Which body system maintains the pH and temperature of the internal environment? [p.485]
   a. nervous
   b. urinary
   c. reproductive
   d. circulatory

_____ 3. Which body system protects the body from dehydration and injury? [p.485]
   a. respiratory
   b. circulatory
   c. integumentary
   d. urinary
   e. skeletal

_____ 4. Which of the following types of junctions seal epithelial cells together? [p.479]
   a. adhering junctions
   b. tight junctions
   c. gap junctions
   d. all of the above

_____ 5. UV radiation has the biggest negative effect on the _____. [p.486]
   a. nervous system
   b. skin
   c. adipose tissue
   d. reproductive system

_____ 6. A gland that excretes compounds via a duct or tube is called a _____ gland. [p.479]
   a. secretory
   b. digestive
   c. endocrine
   d. exocrine

_____ 7. In an upright human, the back of the body is called the _____. [p.484]
   a. posterior
   b. anterior
   c. dorsal
   d. ventral

_____ 8. Which of the following forms of connective tissue is the most common in the human body? [p.480]
   a. loose
   b. dense, regular
   c. dense, irregular
   d. none of the above

9. Which of the following is under voluntary control? [p.482]
   a. skeletal muscle
   b. smooth muscle
   c. cardiac muscle
   d. all of the above

10. Which of the following is not a form of connective tissue? [pp.480–481]
   a. blood
   b. cartilage
   c. adipose tissue
   d. bone
   e. epithelium

## Chapter Objectives/Review Questions

1. Understand the basic forms and functions of epithelium tissue. [p.478]
2. Distinguish between an endocrine and exocrine gland. [p.479]
3. Understand the function of the major types of cell junctions. [p.479]
4. Understand the basic forms and functions of connective tissue. [p.480]
5. Recognize the importance of cartilage, adipose tissue, blood, and bone as a connective tissue. [pp.480–481]
6. Understand the role of muscle tissue and the three major forms of muscle tissue. [pp.482–483]
7. Recognize the role of nervous tissue in the body. [p.483]
8. Understand the body systems that form from each of the three germ layers in the embryo. [p.484]
9. Recognize the directional terms associated with human anatomy. [p.484]
10. For each body system, understand its general role in the human body. [pp.485–486]
11. Understand the structure and function of human skin. [p.486]

## Chapter Summary

Epithelial, (1) _____, muscle, and (2) _____ tissues are the basic categories of tissues in nearly all animals.

(3) _____ line the body surface and its internal cavities and tubes. They have protective and (4) _____ functions.

Connective tissues bind, support, strengthen, protect, and (5) _____ other tissues. They include soft connective tissues, (6) _____, bone, blood, and (7) _____ tissue.

(8) _____ tissues help move the body and its parts. The three kinds are skeletal, (9)_____, and smooth muscle tissue.

Nervous tissue provides local and long-distance lines of (10) _____ among cells. Its cellular components are (11) _____ and neuroglia.

Vertebrate organ systems (12) _____ the tasks of survival and reproduction for the body as a whole; they show a division of labor. Different systems arise from (13) _____, mesoderm, and endoderm, the primary tissue layers that form in the early (14) _____.

Human (15) _____ is an example of an organ system. It has epithelial layers, connective tissue, adipose tissue, glands, blood vessels, and (16) _____ receptors. It helps protect the body from injury and some pathogens, conserve (17) _____ and control body temperature, excrete (18) _____, and detect some external (19) _____.

# Integrating and Applying Key Concepts

1. As a stem cell researcher, you have a line of cells that you wish to turn into nervous tissue. What type of characteristics would those cells have to develop? How about if you decided to make connective tissues from these cells? What would be the major differences in how the nervous cells and connective cells developed?
2. What are some of the uses of adult and embryonic stem cells?

# 30

# NEURAL CONTROL

## INTRODUCTION

While we share many characteristics with other animals, it is our ability to feel emotions that sets the human apart from other species. The human brain is the organ responsible for this characteristic, and for many other traits that we possess. This chapter describes the vertebrate nervous system, with special attention paid to the organization and structure of the nervous system of humans. The chapter briefly discusses the evolution of the nervous system, and then takes an in-depth look at how nerve cells actually transmit messages to other cells; this focus includes much information about the chemicals involved in these messages. One main example is how the nerve cell "talks" to a muscle cell to initiate movement. The chapter then discusses drugs that can interfere with the function of the nervous system. Finally, the chapter concludes with a detailed look at the human brain, and to the organization of the human nervous system. The various parts of the brain are discussed in detail, including the most recently evolved cerebral cortex and the limbic system, the parts of the brain that allow humans to experience emotion.

## FOCAL POINTS

- Figure 30.2 [p.492] introduces the lines of neural communication.
- Figure 30.4 [p.493] shows the functional divisions of the vertebrate nervous system.
- Figure 30.5 [p.493] illustrates some of the major nerves of the human nervous system.
- Figure 30.6 [p.494] animates a neuron.
- Figure 30.8 [p.496] illustrates the propagation of an action potential.
- Figure 30.10 [p.498] illustrates a chemical synapse.
- Figure 30.16 [p.503] diagrams the stretch reflex.
- Figure 30.17 [p.504] shows the sympathetic and parasympathetic nerves.
- Figure 30.20 [p.507] illustrates the human brain.

## Interactive Exercises

*In Pursuit of Ecstasy* [pp.490–491]

### 30.1. EVOLUTION OF NERVOUS SYSTEMS [pp.492–493]

### 30.2. NEURONS—THE GREAT COMMUNICATORS [pp.494–495]

### 30.3. A LOOK AT ACTION POTENTIALS [pp.496–497]

*Selected Terms*

*Efferent axons* [p.493], *efferent axons* [p.493], *input* zones [p.494], *trigger* zone [p.494], *conducting* zone [p.494], *output* zone [p.494], *ion gradients* [p.495], graded [p.496]

## Boldfaced Terms

[p.492] nervous systems _____

_____

[p.492] neurons _____

_____

[p.492] neuroglia _____

_____

[p.492] sensory neurons _____

_____

[p.492] interneurons _____

_____

[p.492] motor neurons _____

_____

[p.492] nerve net _____

_____

[p.492] cephalization _____

_____

[p.492] nerves _____

_____

[p.492] ganglion _____

_____

[p.493] central nervous system _____

_____

[p.493] peripheral nervous system _____

_____

[p.494] dendrites _____

_____

[p.494] axons _____

_____

[p.494] resting membrane potential _____

_____

[p.494] action potential _____

_____

[p.496] threshold level _____

_____

[p.496] positive feedback _____

_____

[p.496] all or nothing event _____

_____

## Complete the Table

Use the table below to compare dendrites and axons. [p.494]

| Dendrites | Axons |
| --- | --- |
| 1. | Carry information away from the cell body |
| One per cell in most sensory neurons; many per cell in most other neurons | 2. |
| Branch close to cell body | 3. |
| 4. | Insulating sheath for most |

## Labeling

Use the following diagram to label the parts of a neuron involved in the generation of an action potential.

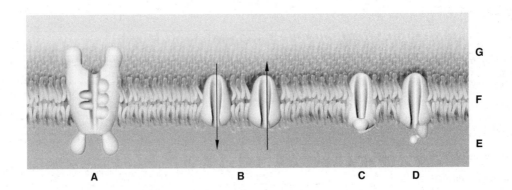

5. _____ cytoplasm of a neuron

6. _____ plasma membrane

7. _____ interstitial fluid

8. _____ gated channels that are open during an action potential; Na$^+$ and K$^+$ can pass through

9. _____ Sodium-potassium pump actively transports Na$^+$ and K$^+$ against the concentration gradient.

10. _____ Passive transporters with open channels allow K$^+$ to leak back in the cell, down the concentration gradient.

11. _____ gated channels that are shut while the cell is at rest

## True–False

12. Without ion gradients, an action potential cannot be generated.

13. In a resting cell, for every three Na$^+$ inside the cell, there are thirty outside. Likewise, for every K$^+$ inside, there is one outside the cell.

## Labeling

Label the indicated components of a neuron in the diagram below. [p.484]

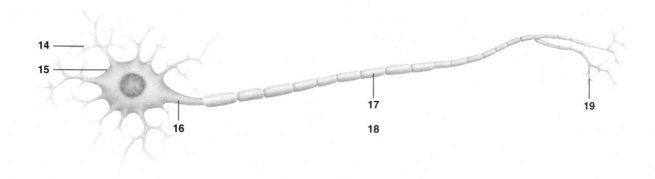

14. _____
15. _____
16. _____
17. _____
18. _____
19. _____

## Matching [pp.494–497]

Match each of the following terms with its correct description.

_____ 20. sodium-potassium pumps

_____ 21. resting membrane potential

_____ 22. neurons

_____ 23. excitable cells

_____ 24. neuroglia

_____ 25. action potential

_____ 26. axons

_____ 27. dendrites

a. The mechanism that maintains the correct ion concentration across the membrane.
b. The conducting and output zones of the neuron are located here.
c. cells that support the metabolic activity of a neuron
d. Along with neurons, these types of cells may generate an action potential.
e. the communication units of the nervous system
f. This is normally –70 millivolts.
g. a brief, abrupt reversal of the electrical charge across a membrane
h. input zones for information

## True–False

The following statements pertain to the propagation of an action potential. If the statement is true, write "T" in the space provided. If the statement is false, correct the underlined word or words to make the statement true. [pp.496–497]

_____ 28. Once a threshold level is reached there is an <u>all-or-nothing</u> membrane response.

_____ 29. In positive feedback, the flow of sodium ions <u>decreases</u> as an outcome of its own occurrence.

_____ 30. <u>Calcium</u> pumps are responsible for restoring the resting membrane potential.

_____ 31. Action potentials move along the axon as a result of <u>self-propagating</u> events.

_____ 32. Action potentials spread <u>away</u> from the membrane area where it started.

## 30.4. HOW NEURONS SEND MESSAGES TO OTHER CELLS [pp.498–499]

## 30.5. A SMORGASBORD OF SIGNALS [pp.500–501]

### Selected Terms

Presynaptic cell [p.498], postsynaptic cell [p.498]

### Boldfaced Terms

[p.498] chemical synapse _____

_____

[p.498] neurotransmitters _____

_____

[p.498] neuromuscular junction _____

_____

[p.498] acetylcholine (ACh) _____

_____

[p.499] synaptic integration _____

_____

[p.500] neuromodulators _____

_____

[p.500] substance P _____

_____

[p.500] enkephalins _____

_____

[p.500] endorphins _____

_____

### Matching

Match each of the following definitions with the most appropriate term. [pp.498–499]

_____ 1. The name given to the region between the output zone of one neuron and the input zone of a second neuron.

_____ 2. The summing of the incoming signals at the postsynaptic neuron.

_____ 3. Receptors on this cell respond to neurotransmitters released into the chemical synapse.

_____ 4. The result of signals that nudge a cell membrane in the direction of an action potential.

_____ 5. This cell releases a neurotransmitter in response to an action potential.

_____ 6. The result of signals that move the cell membrane away from an action potential.

a. postsynaptic neuron
b. synaptic integration
c. chemical synapse
d. excitatory effect
e. inhibitory effect
f. presynaptic neuron

## Matching

For each of the following, indicate the type of neurotransmitter that may be responsible for the response indicated. Some answers may be used more than once. [p.500]

_____ 7. influences cells involved with sleeping

_____ 8. influences emotions, dreaming, and alertness

_____ 9. initiates muscle contraction

_____ 10. a common inhibitory neurotransmitter

_____ 11. inhibits the response of cells involved in memory

_____ 12. is involved in award-based learning

_____ 13. influences sensory perception

a. serotonin
b. norepinephrine
c. acetylcholine
d. GABA
e. dopamine

## Labeling [p.498]

Label the indicated components of the chemical synapse in the diagram below:

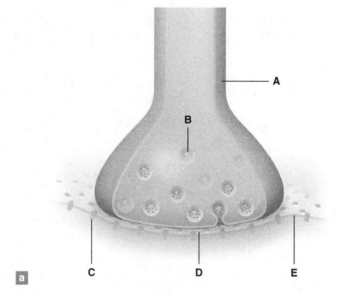

14. _____ plasma membrane of the postsynaptic cell

15. _____ plasma membrane of an axon ending in a presynaptic cell

16. _____ synaptic vesicle

17. _____ membrane receptor for neurotransmitter

18. _____ synaptic cleft

*Complete the Table* [p.500]

Fill in the following table.

| Disease | Neurotransmitter | Pathology Associated with This Neurotransmitter |
|---|---|---|
| Myasthenia gravis | 19. | 20. |
| 21. | Acetylcholine | Low levels in the brain |
| 22. | Dopamine | Dopamine secreting neurons in the brain are destroyed |

## 30.6. DRUGS AND DISRUPTED SIGNALING [p.501]

## 30.7. ORGANIZATION OF NEURONS IN NERVOUS SYSTEMS [pp.502–503]

## 30.8. WHAT ARE THE MAJOR EXPRESSWAYS? [pp.504–505]

### Selected Terms

*Cannabis* [p.501], THC [p.501], diverging circuits [p.502], converging circuits [p.502], reverberating circuits [p.502], multiple sclerosis [p.502], spinal nerves [p.504], cranial nerves [p.502], viscera [p.504], roots [p.505]

### Boldfaced Terms

[p.501] drug addiction _____

_____

[p.502] myelin sheath _____

_____

[p.503] reflex _____

_____

[p.503] muscle spindles _____

_____

[p.504] somatic nerves _____

_____

[p.504] autonomic nerves _____

_____

[p.504] sympathetic neurons _____

_____

[p.504] parasympathetic neurons _____

_____

[p.505] fight-flight response _____

_____

[p.505] spinal cord _____

_____

[p.505] gray matter _____

_____

[p.505] white matter _____

_____

## *List* [p.501]

List the eight warning signs of drug addiction:

1. _____     5. _____
2. * _____     6. _____
3. _____     7. _____
4. _____     8. _____

## *Labeling*

Match each of the following descriptions to its correct label in the diagram of a stretch reflex. [p.491]

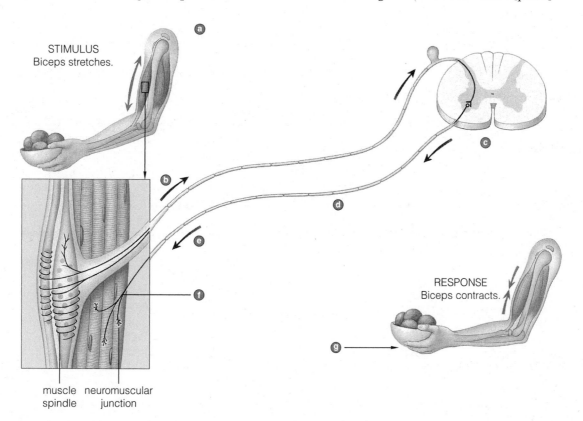

9. _____ Weight on an arm muscle stretches muscle spindles in the muscle sheath.

10. _____ Axon endings of the sensory neuron release a neurotransmitter that stimulates a motor neuron.

11. _____ Axon endings of the motor neuron synapse with muscle fibers in the stretched muscle.

12. _____ Stimulations make the stretched muscle contract.

13. _____ ACh released from the motor neuron's axon endings stimulates cells making up the muscle fibers.

14. _____ Action potentials self-propagate along the motor neuron axon.

15. _____ Stretching stimulates sensory receptor endings in the muscle spindle.

## True–False

If the statement is true, write "T" in the space provided. If false, correct the underlined word to make the statement true.

_____ 16. All interneurons are in the <u>central</u> nervous system. [p.493]

_____ 17. Rapid signal transmission occurs in the <u>gray</u> matter. [p.505]

_____ 18. The brain and spinal cord are part of the <u>autonomic</u> nervous system. [p.504]

_____ 19. The <u>peripheral</u> nervous system carries signals in and out of the central nervous system. [p.493]

_____ 20. Mylenated axons are present in the white matter. [p.505]

## Matching

Match each of the following terms to its correct definition. [pp.504–505]

_____ 21. cranial nerves

_____ 22. parasympathetic nerves

_____ 23. autonomic nerves

_____ 24. fight-flight response

_____ 25. spinal nerves

_____ 26. somatic nerves

_____ 27. sympathetic nerves

a. These nerves tend to slow down activity and begin housekeeping tasks.
b. There are 31 pairs of these in the body.
c. These carry signals about moving the head, trunk, and limbs.
d. These deal with internal organs and structures.
e. This is caused by release of the neurotransmitter epinephrine (adrenaline).
f. These nerves are activated in times of stress or danger.
g. Twelve pairs of these connect directly to the brain.

## Labeling

Provide the correct label for each of the indicated nerves in the diagram below. [p.494]

28. _____

29. _____

30. _____

31. _____

32. _____

33. _____

34. _____

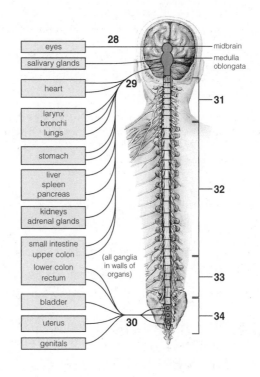

## Labeling [p.505]

Complete the following using the diagram of the spinal cord and vertebra.

35. _____ spinal cord

36. _____ location of intervertebral disk

37. _____ meninges

38. _____ spinal nerve

39. _____ vertebra

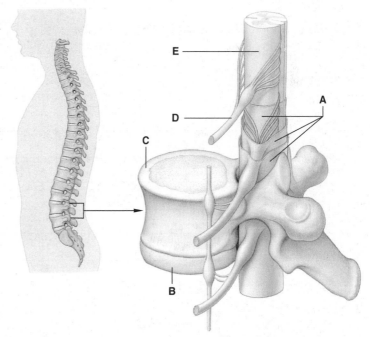

Match the class of drug with the effects it has on the nervous system or an example of each.

          a. stimulants     b. depressants     c. analgesics     d. hallucinogens

40. _____ Marijuana is an example.

41. _____ distort sensory perception

42. _____ LSD

43. _____ make users feel alert; can interfere with fine motor skills

44. _____ slow motor responses by inhibiting ACh output

45. _____ Heroin is an example.

46. _____ Cocaine is an example.

47. _____ mimic the body's natural painkillers

48. _____ Ecstasy is an example.

49. _____ Overdoses can cause strokes and heart attacks.

50. _____ can be strongly addictive

## 30.9. THE VERTEBRATE BRAIN [pp.506–507]

## 30.10. THE HUMAN CEREBRUM [pp.508–509]

## 30.11. NEUROGLIA [p.510]

### Selected Terms

*Motor* areas [p.508], *sensory* areas [p.508], *association* areas [p.508], Broca's area [p.508], "gut reactions" [p.509], *short-term* memory [p.509], *long-term* memory [p.509], *declarative* memory [p.509]

### Boldfaced Terms

[p.506] brain stem _____

_____

[p.506] medulla oblongata _____

_____

[p.506] cerebellum _____

_____

[p.506] pons _____

_____

[p.506] cerebrum _____

_____

[p.506] thalamus _____

_____

[p.506] hypothalamus _____

_____

[p.506] cerebrospinal fluid _____

_____

[p.506] blood brain barrier _____

_____

[p.508] cerebral cortex _____

_____

[p.509] limbic system _____

_____

[p.510] neuroglial cells _____

_____

[p.510] growth factors _____

_____

## Matching

Match each of the following subdivisions of the brain to its correct function. [p.506]

_____ 1. cerebellum

_____ 2. hypothalamus

_____ 3. medulla oblongata

_____ 4. pons

_____ 5. brain stem

_____ 6. cerebrum

_____ 7. thalamus

a. This area helps to control motor skills and posture.
b. This is a coordinating center for sensory input.
c. This exists as two large hemispheres and is where most signal integration occurs.
d. This provides homeostatic control for the body.
e. This ancient tissue is found in the midbrain, forebrain, and hindbrain.
f. This acts as a traffic center between the cerebellum and forebrain.
g. This contains reflex centers for respiration and circulation.

## Fill in the Blanks [pp.496–497]

A blood-brain barrier protects the brain and (8) _____ of vertebrates from harmful substances. It exerts some control over which solutes enter (9) _____ fluid.

The barrier works at the wall of (10) _____ that service the brain. In most brain regions, (11) _____ junctions fuse all abutting cells that make up the wall of capillaries, so water-soluble substances must pass through the cells to reach the brain. Transport (12) _____ in the plasma membrane allow (13) _____, other vital nutrients, and some (14) _____ across. They bar many toxins and wastes, such as (15) _____. The barrier is not perfect; some toxins such as (16) _____, alcohol, caffeine, or mercury can pass through. Inflammation, toxins, and traumatic blows can destroy this barrier and compromise (17) _____ function.

## Labeling

Label each of the indicated components of the human brain. [p.497]

18. _____

19. _____

20. _____

21. _____

22. _____

23. _____

24. _____

25. _____

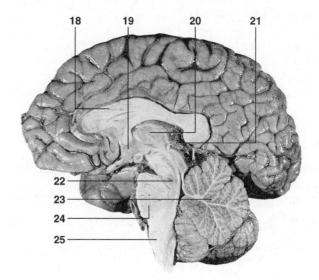

## Labeling

Indicate whether each of the following statements is associated with the motor area (M), sensory area (S), or association area (A) of the cerebral cortex. [p.507]

_____ 26. This contains the Broca's area.

_____ 27. Input from the skins and joints are processed here.

_____ 28. The integrations of many inputs occur in this area.

_____ 29. This controls the coordinated movement of skeletal muscles.

_____ 30. Intellect and personality are the result of activity in this area.

_____ 31. Sound and color are perceived here.

## True–False

If the statement is true, write "T" in the space provided. If false, correct the underlined word so that the statement reads is correct.

_____ 32. The control of <u>emotion</u> is provided by the limbic system. [p.509]

_____ 33. The limbic system is evolutionarily related to the <u>vision</u> system. [p.509]

_____ 34. <u>Short-term</u> memory can store a tremendous amount of information regarding sensory input. [p.509]

_____ 35. Alzheimer's disease specifically affects the operation of the <u>hypothalamus</u>. [p.509]

_____ 36. The memory system processes facts and skills <u>together</u>. [p.509]

# Self-Quiz

1. Which of the following group of animals is the first to display a simple nervous system? [p.492]
   a. vertebrates
   b. flatworms
   c. earthworms
   d. Cnidarians

2. Which of the following is active in a fight-flight response? [pp.494–495]
   a. somatic nerves
   b. autonomic nerves
   c. sympathetic nerves
   d. parasympathetic nerves

3. The input zones of a neuron are located in which of the following structures? [p.494]
   a. axon
   b. dendrite
   c. cell body
   d. nucleus

4. The brain and spinal cord comprise what part of the nervous system? [p.493]
   a. central nervous system
   b. peripheral nervous system
   c. limbic system
   d. autonomic system

5. Which of the following is responsible for control of the motor skills? [p.506]
   a. brain stem
   b. pons
   c. hypothalamus
   d. cerebellum

6. Caffeine and nicotine belong to what group of drugs? [p.501]
   a. depressants
   b. hallucinogens
   c. analgesics
   d. psychedelics
   e. stimulants

7. Which of the following process sensory input? [p.485]
   a. sensory neurons
   b. motor neurons
   c. interneurons
   d. transneurons

8. Which of the following is responsible for establishing the resting membrane potential in a neuron? [p.495]
   a. calcium pump
   b. sodium-potassium pump
   c. osmosis
   d. positive feedback

9. Homeostatic control of the internal environment occurs in the _____. [p.506]
   a. thalamus
   b. hypothalamus
   c. cerebrum
   d. medulla oblongata

10. Emotions are controlled by which of the following areas of the brain? [p.506]
    a. hypothalamus
    b. limbic system
    c. motor areas
    d. association areas

# Chapter Objectives/Review Questions

This section lists general and detailed chapter objectives that can be used as review questions. You can make maximum use of these items by writing answers on a separate sheet of paper. Fill in answers where blanks are provided. To check for accuracy, compare your answers with information given in the chapter or glossary.

1. Describe the evolution of nervous systems from Cnidarians to earthworms. [pp.492–493]
2. Describe the basic divisions of the human nervous system. [p.493]
3. Label the structure and zones of a neuron. [p.494]
4. Understand the concept of membrane potentials, including action potentials and resting membrane potentials. [pp.494–497]
5. How does a neuron restore a membrane potential following the generation of an action potential? [pp.496–497]

6. List the steps in the propagation of an action potential. [pp.496–497]
7. Indicate the role of the presynaptic and postsynaptic neurons. [p.498]
8. How does synaptic integration process information from the neurons? [p.499]
9. What is the role of the Schwann cell in a nerve? [p.502]
10. Understand the mechanism by which various drug classes interact with the nervous system. [p.501]
11. How does multiple sclerosis affect the structure of a nerve? [p.502]
12. What is a reflex? [p.503]
13. What is the difference between gray matter and white matter? [p.505]
14. What is the difference between a somatic nerve and an autonomic nerve? [p.504]
15. Describe the role of the sympathetic and parasympathetic nervous systems. [p.504]
16. What is the fight-flight response and how is it controlled? [p.505]
17. For each of the major areas of the brain, be able to identify its location and general function. [pp.496–497]
18. What substances can freely pass through the blood-brain barrier? [pp.506–507]
19. Describe the general function of the motor, sensory, and association areas of the brain. [p.506]
20. What is the purpose of the limbic system? [p.509]
21. Describe the differences between short- and long-term memory. [p.509]

## Chapter Summary

Excitable cells called (1) _____ interconnect and form communication lines of animal nervous systems.

In (2) _____ symmetrical animals, the neurons connect as a nerve net. In (3) _____ symmetrical

animals, the concentration of neurons is at one end and one or more cords run the length of the body.

Messages flow along a neuron's (4) _____ _____. These messages are brief (5) _____ propagating

reversals in the distribution of (6) _____ _____ across the membrane. At an output zone, they are

transduced to a (7) _____ signal that may stimulate or (8) _____ activity in another cell. (9) _____

drugs interfere with the information flow between cells. (10) _____ _____ are the simplest routes of

information flow.

The central nervous system of vertebrates consists of the (11) _____ and the (12) _____ _____. The

(13) _____ nervous system consists of many pairs of nerves that connect the brain and spinal cord to

the rest of the body. The (14) _____ ____ is the most recently evolved part of the brain. Supporting cells in

the human brain are called (15) _____—the brain cannot function without them.

## Integrating and Applying Key Concepts

1. Suppose that anger is eventually determined to be caused by excessive amounts of specific transmitter substances in the brains of angry people. Also suppose that an inexpensive antidote to anger that neutralizes these anger-producing transmitter substances is readily available. Can violent murderers now argue that they have been wrongfully punished because they were victimized by their brain's transmitter substances and could not have acted in any other way? Suppose an antidote is prescribed to curb violent tempers in an easily angered person. Suppose also that the person forgets to take the pill and subsequently murders a family member. Can the murderer still claim to be victimized by transmitter substances?

# 31

# SENSORY PERCEPTION

## INTRODUCTION

In this chapter you will examine the different types of receptors that are used to detect both external and internal stimuli. The chapter then discusses some of the special senses (sight, hearing, taste, vision, etc.).

## FOCAL POINTS

- The five classes of sensory receptors are presented on p.514.
- Figure 31.9 [p.518] diagrams the structure of the human ear.
- Figure 31.12 [p.521] diagrams the structure of the human eye.

## Interactive Exercises

*A Whale of a Dilemma* [pp.512–513]

## 31.1. OVERVIEW OF SENSORY PATHWAYS [p.514]

## 31.2. SOMATIC SENSATIONS [p.515]

### Selected Words

*Frequency* of action potentials [p.514], *number* of axons [p.514], Meissner's corpuscles [p.515], Pacinian corpuscles [p.515], Ruffini endings [p.515], *somatic* pain [p.515], *visceral* pain [p.515]

### Boldfaced Terms

[p.514] sensory systems _____

_____

[p.514] stimulus _____

_____

[p.514] sensation _____

_____

[p.514] mechanoreceptors _____

_____

[p.514] pain receptors _____

_____

[p.514] thermoreceptors _____

_____

[p.514] chemoreceptors _____

_____

[p.514] osmoreceptors _____

_____

[p.514] photoreceptors _____

_____

[p.514] sensory adaptation _____

_____

[p.515] somatic sensations _____

_____

[p.515] pain _____

_____

## Matching

Match each of the following stimuli to the correct receptor. [p.514]

1. _____ temperature changes
2. _____ changes in pressure, position, or acceleration
3. _____ tissue damage
4. _____ change in the concentration of solutes in a fluid
5. _____ specific substances dissolved in fluid
6. _____ light energy

a. pain receptors
b. photoreceptors
c. mechanoreceptors
d. thermoreceptors
e. chemoreceptors
f. osmoreceptors

## True–False

If the statement is true place a "T" in the blank. If the statement is incorrect, correct the underlined word so that the statement is true.

7. _____ Activation of a pain receptor in the skin would be an example of <u>visceral</u> pain. [p.515]

8. _____ The Meissner's corpuscles of the skin are responsible for detecting <u>temperature</u> changes. [p.515]

9. _____ <u>Sensory adaptation</u> occurs when a neuron stops paying attention to a constant stimulus. [p.514]

10. _____ The brain assesses a stimulus based on the <u>amplitude</u> and number of action potentials. [p.514]

11. _____ A <u>stimulus</u> is a form of energy that activates receptor endings of a sensory neuron. [p.514]

## 31.3. SAMPLING THE CHEMICAL WORLD [p.516]

## 31.4. KEEPING THE BODY BALANCED [p.517]

## 31.5. COLLECTING, AMPLIFYING, AND SORTING OUT SOUNDS [pp.518–519]

### Selected Words

*Sweet* [p.516], *sour* [p.516], *salty* [p.516], *bitter* [p.516], *umami* [p.516], *dynamic* equilibrium [p.517], *static* equilibrium [p.517], vertigo [p.517], *amplitude* [p.518], *frequency* [p.518], *outer* ear [p.518], *middle* ear [p.518], *inner* ear [p.518], *koklias* [p.518], organ of Corti [p.519]

### Boldfaced Terms

[p.516] olfaction _____

[p.516] olfactory receptors _____

[p.516] pheromones _____

[p.516] vomeronasal organ _____

[p.516] taste receptors _____

[p.517] organs of equilibrium _____

[p.517] vestibular apparatus _____

[p.517] hair cells _____

[p.518] hearing _____

[p.518] cochlea _____

## Choice

Choose the appropriate sense for each statement. Some statements may have more than one answer.

a. balance [p.517]    b. taste [p.516]    c. hearing [pp.518–519]    d. smell [p.516]

1. _____ involves the organ of Corti

2. _____ involves the vestibular apparatus

3. _____ involves the use of chemoreceptors

4. _____ conflicting sensory inputs may cause vertigo

5. _____ detects the frequency and amplitude of pressure variations

6. _____ involves structures found in the inner ear

7. _____ detects pheromones

8. _____ detects umami

## True–False

If the statement is true place a "T" in the blank. If the statement is incorrect, correct the underlined word so that the statement is true.

9. _____ Vomeronasal organs are involved in the detection of <u>taste</u> in mammals and reptiles. [p.516]

10. _____ <u>Taste</u> receptors are primarily involved in the detection of volatile chemicals. [p.516]

11. _____ Umami and plant toxins are detected by <u>taste</u> receptors. [p.516]

12. _____ A <u>pheromone</u> is a chemical secreted by one individual to influence the social behavior of other members of the species. [p.516]

13. _____ The utricle and saccule are involved in the process of <u>static</u> equilibrium. [p.517]

14. _____ <u>Dynamic</u> equilibrium involves the use of a jellylike mass to sense changes in orientation of the head. [p.517]

15. _____ The semicircular canals are part of the organs of equilibrium in mammals. [p.517]

## Labeling

Label each component of the human ear in the diagram below. [p.518]

16. _____

17. _____

18. _____

19. _____

20. _____

21. _____

22. _____

23. _____

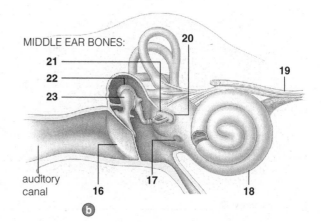

MIDDLE EAR BONES:

21

22

23

20

19

auditory canal

16

17

18

b

## Matching

Match each of the following terms with the correct definition or function. [pp.518–519]

24. _____ inner ear

25. _____ frequency

26. _____ middle ear

27. _____ amplitude

28. _____ outer ear

29. _____ basilar membrane

30. _____ tectorial membrane

a. the mechanoreceptors of the ear are attached to this
b. the cochlea is located here
c. this membrane responds to different frequencies
d. this contains the eardrum, anvil, stirrup, and hammer
e. this area is adapted for gathering sounds in the air
f. the number of wave cycles per second
g. the intensity, or loudness, of a sound

## 31.6. DO YOU SEE WHAT I SEE? [pp.520–521]

## 31.7. FROM THE RETINA TO THE VISUAL CORTEX [p.522]

## 31.8. VISUAL DISORDERS [p.523]

### Selected Words

*Compound* eyes [p.520], *camera* eyes [p.520], *sclera* [p.521], *cornea* [p.521], *pupil* [p.521], *iris* [p.521], *cillary* body [p.521], *choroid* [p.521], *optic* nerve [p.521], *retina* [p.521]

### Boldfaced Terms

[p.520] vision _____

_____

[p.520] eyes _____

_____

[p.520] photoreceptors _____

_____

[p.520] visual field _____

_____

[p.520] lens _____

_____

[p.520] retina _____

_____

[p.520] visual accommodation _____

_____

[p.522] rod cells _____

_____

[p.522] cone cells _____

_____

## Labeling

First, label the indicated structures of the human eye. Then, match each structure to the correct function or description from the list provided. [pp.520–521]

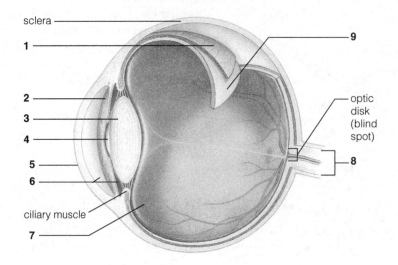

1. _____ ( )
2. _____ ( )
3. _____ ( )
4. _____ ( )
5. _____ ( )
6. _____ ( )
7. _____ ( )
8. _____ ( )
9. _____ ( )

a. the opening that lets light into the eye
b. keeps the lens moist
c. relays information from the eye to the brain for interpretation
d. the tissue that is densely packed with photoreceptors
e. made of transparent crystalline proteins
f. darkly pigmented tissue that absorbs wavelengths missed by the photoreceptors
g. contracts to adjust the size of the pupil
h. bends light in what is called visual accommodation
i. fills the area in the eye behind the lens

*TNOW:* For an animation of this exercise, refer to Chapter 31: Art Labeling: Anatomy of the Human Eye.

## True–False

If the statement is true place a "T" in the blank. If the statement is incorrect, correct the underlined word so that the statement is true.

10. _____ Information from the right visual field is delivered to the <u>right</u> cerebral hemisphere. [p.522]

11. _____ <u>Cone</u> cells are responsible for detecting dim light. [p.522]

12. _____ There are <u>four</u> types of cone cells. [p.522]

13. _____ Rod cells and cone cells are derived from <u>cilia</u>. [p.522]

14. _____ The visual pigment of <u>cone</u> cells is rhodopsin. [p.522]

## Matching

Match each of the following disorders to the correct description. [p.523]

15. _____ glaucoma

16. _____ farsightedness

17. _____ cataracts

18. _____ color blindness

19. _____ macular degeneration

20. _____ nearsightedness

21. _____ astigmatism

a. problems focusing on close objects
b. destruction of photoreceptors clouds straight-ahead vision
c. due to problems with one of the three types of cone cells
d. problems focusing on objects far away
e. excess accumulation of aqueous humor in the eye
f. clouding of the lens
g. unevenly curved cornea

---

# Self-Quiz

_____ 1. Which of the following does the brain not use to assess a stimulus? [p.514]
   a. the number of axons that the stimulus recruits
   b. which nerve pathways are carrying the signal
   c. the frequency of the action potentials
   d. the amplitude of the action potentials
   e. all of the above are utilized.

_____ 2. Amplification of sound waves occurs in the _____ [p.518]
   a. inner ear
   b. outer ear
   c. middle ear
   d. vestibular apparatus

_____ 3. Which of the following is associated with the sense of balance? [p.517]
   a. vestibular apparatus
   b. cochlea
   c. basilar membrane
   d. eardrum

_____ 4. Which of the following is the location of the photoreceptors of the eye? [p.521]
   a. retina
   b. cornea
   c. pupil
   d. lens
   e. none of the above

_____ 5. The accumulation of excess aqueous humor behind the eye is called _____. [p.523]
   a. cataracts
   b. glaucoma
   c. cancer
   d. astigmatism
   e. none of the above

_____ 6. Mechanoreceptors are involved in which of the following senses? [p.517]
   a. taste
   b. smell
   c. balance
   d. vision

_____ 7. Which of the following is not involved in a somatic sensation? [p.515]
   a. Meissner's corpuscles
   b. Pacinian corpuscle
   c. olfactory receptors
   d. Ruffini endings

_____ 8. Accommodation involves the ability to _____. [p.521]
   a. change the sensitivity of the rods and cones by means of neurotransmitters
   b. change the shape of the lens by using special muscles
   c. change the shape of the cornea
   d. adapt to rapid changes in light intensity
   e. all of the above

_____ 9. Rods and cones are _____. [p.522]
   a. mechanoreceptors
   b. osmoreceptors
   c. chemoreceptors
   d. photoreceptors

_____10. Which of the following is due to the action of a chemoreceptor? [p.516]
   a. light
   b. smell
   c. taste
   d. all of the above
   e. only b or c

# Chapter Objectives/Review Questions

1. List the six kinds of receptors and identify the type of stimulus energy that each type detects. [p.514]
2. List the three mechanisms by which the brain assesses a stimulus. [p.514]
3. Understand the processing of somatic sensations by the brain. [p.515]
4. Explain the difference between somatic pain and visceral pain. [p.515]
5. Explain the differences and similarities between olfactory receptors and taste receptors. [p.516]
6. Explain the function of the vomeronasal organ. [p.516]
7. Explain the process by which the body assesses balance. [p.517]
8. Explain the difference between static and dynamic equilibrium. [p.517]
9. Explain what occurs during vertigo and motion sickness. [p.517]
10. Distinguish between frequency and amplitude in the context of sound. [p.518]
11. Describe the functions of each part of the human ear. [pp.518–519]
12. Identify the structures of the human eye and give their function. [pp.520–521]
13. Define the term *visual accommodation*. [pp.520–521]
14. List the differences between rod cells and cone cells. [p.522]
15. Trace the movement of a visual signal from the eye to the corresponding cerebral hemisphere. [p.522]
16. Know the major vision disorders and their physiological causes. [p.523]

# Chapter Summary

Sensory systems are front doors of the (1) _____ system. Each has sensory receptors, (2) _____ pathways to the brain, and brain regions that receive and (3) _____ the sensory input. A (4) _____ is a form of energy that activates a specific type of sensory receptor. Information becomes encoded in the (5) _____ and (6) _____ of action potentials sent to the brain along particular nerve pathways.

Touch, pressure, pain, temperature, and muscle sense are (7) _____ sensations. These sensations start at (8) _____ in skin, muscles, and the wall of internal organs.

The senses of smell and taste require (9) _____, which bind molecules of specific substances that have become dissolved in the fluid bathing them.

The sense of (10) _____ starts at mechanoreceptors, which detect (11) _____, velocity, acceleration, and other forces that influence the position and motion of the body or specific parts of it.

In vertebrates, the sense of (12) _____ starts with structures that collect, amplify, and sort out pressure variations caused by sound waves. The variations trigger (13) _____ in mechanoreceptors that send signals to the brain.

Most organisms have (14) _____ pigments, but vision requires eyes with a dense array of (15) _____ and image formation in the brain. Paired, camera-like eyes of cephalopoda and vertebrates collect and process information on distance, brightness, shape, position, and movement of visual (16) _____. A sensory pathway starts at the (17) _____ and ends in the (18) _____.

# Integrating and Applying Key Concepts

1. Echolocation is common in some mammals (such as bats), but not in primates. How would our special senses have to change in order for us to utilize echolocation? What types of receptors would we use?
2. The eyes of humans and squid are very similar. Explain why this may be an example of convergent evolution.

# 32

# ENDOCRINE CONTROL

## INTRODUCTION

This chapter explores the mechanism by which the body uses hormones to control the activity of glands and organs. The chapter introduces the different types of hormones and then examines the function of the major endocrine glands in the human body.

## FOCAL POINTS

- Figure 32.2 [p.529] illustrates the major endocrine glands of the human body.
- Figure 32.2 [p.531] illustrates the differences between steroid and peptide hormones.
- Figure 32.7 [p.534] diagrams a negative feedback loop.
- Table 32.3 [p.540] and Table 32.4 [p.541] list the source and function of the major human hormones.

## Interactive Exercises

*Hormones in the Balance* [pp.526–527]

## 32.1. INTRODUCING THE VERTEBRATE ENDOCRINE SYSTEM [pp.528–529]

*Selected Words*

Atrazine [p.526], "hormone" [p.528], *hormon* [p.528]

*Boldfaced Terms*

[p.528] animal hormones _____

_____

[p.528] neurotransmitters _____

_____

[p.528] local signaling molecules _____

_____

[p.528] pheromones _____

_____

[p.528] endocrine system _____

_____

## Matching

Choose the most appropriate definition for each term. [p.528]

1. _____ neurotransmitters
2. _____ animal hormones
3. _____ local signaling molecules
4. _____ pheromones

a. integrate social behavior
b. secretory products of endocrine glands
c. change the chemical conditions of nearby tissues
d. released from the axons of nerve endings

## Labeling

For each of the following, first provide the name of the endocrine gland. Second, select the hormone(s) secreted by the gland from the list provided. [p.529]

5. _____( )
6. _____( )
7. _____( )
8. _____( )
9. _____( )
10. _____( )
11. _____( )
12. _____( )
13. _____( )
14. _____( )

a. cortisol, aldosterone, epinephrine and norepinephrine
b. ACTH, TSH, FSH, and LH
c. Insulin
d. sex-releasing and inhibiting hormones, ADH and oxytocin
e. melatonin
f. estrogens and progesterone
g. parathyroid hormone
h. thymosins
i. testosterone
j. calcitonin and thyroid hormone

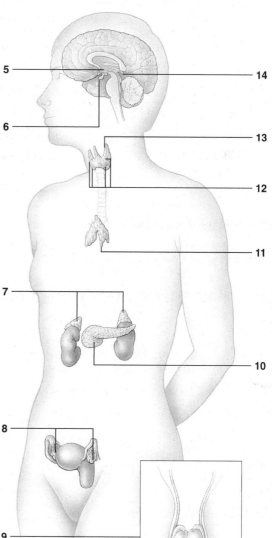

**TNOW:** For an animation of this material, refer to Chapter 32: Art Labeling: Human Endocrine Organs.

## 32.2. NATURE OF HORMONE ACTION [pp.530–531]

*Selected Words*

*Steroid* hormones [pp.530–531], *amine* hormones [p.530], *peptide* hormones [pp.530–531], *protein* hormones [p.530]

*Short Answer*

1. List the four categories of hormones. [p.530] _____

   _____

   _____

   _____

2. List the factors that influence a cell's sensitivity to a hormone. [pp.530–531] _____

   _____

   _____

   _____

   _____

   _____

*Choice*

Indicate which class of hormones each of the following statements describes. [pp.531–532]

        a. peptide     b. steroid

3. _____ Binds to receptors on the plasma membrane.

4. _____ Directly stimulates transcription into mRNA.

5. _____ Lipid-soluble molecules that are derived from cholesterol.

6. _____ May involve the use of second messenger systems.

7. _____ Binds to receptors in the cytoplasm or nucleus.

## Labeling

Provide the missing terms in the following diagrams of hormone action. [p.531]

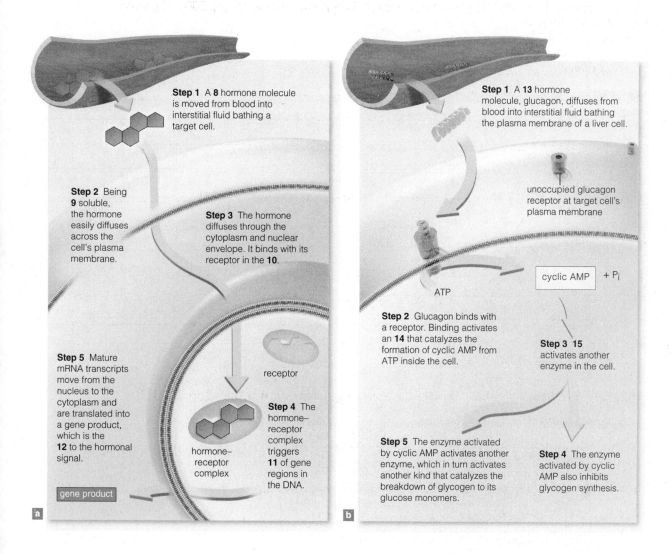

**Step 1** A **8** hormone molecule is moved from blood into interstitial fluid bathing a target cell.

**Step 2** Being **9** soluble, the hormone easily diffuses across the cell's plasma membrane.

**Step 3** The hormone diffuses through the cytoplasm and nuclear envelope. It binds with its receptor in the **10**.

**Step 5** Mature mRNA transcripts move from the nucleus to the cytoplasm and are translated into a gene product, which is the **12** to the hormonal signal.

gene product

receptor

hormone–receptor complex

**Step 4** The hormone–receptor complex triggers **11** of gene regions in the DNA.

a

**Step 1** A **13** hormone molecule, glucagon, diffuses from blood into interstitial fluid bathing the plasma membrane of a liver cell.

unoccupied glucagon receptor at target cell's plasma membrane

cyclic AMP  + P$_i$

ATP

**Step 2** Glucagon binds with a receptor. Binding activates an **14** that catalyzes the formation of cyclic AMP from ATP inside the cell.

**Step 3** **15** activates another enzyme in the cell.

**Step 5** The enzyme activated by cyclic AMP activates another enzyme, which in turn activates another kind that catalyzes the breakdown of glycogen to its glucose monomers.

**Step 4** The enzyme activated by cyclic AMP also inhibits glycogen synthesis.

b

8. _____

9. _____

10. _____

11. _____

12. _____

13. _____

14. _____

15. _____

## 32.3.  THE HYPOTHALAMUS AND PITUITARY GLAND [pp.532–533]

## 32.4.  THYROID AND PARATHYROID GLANDS [p.534]

### Selected Words

*Posterior* lobe [p.532], *anterior* lobe [p.532]

### Boldfaced Terms

[p.532] hypothalamus _____

_____

[p.532] pituitary gland _____

_____

[p.532] releasers _____

_____

[p.532] inhibitors _____

_____

[p.534] thyroid gland _____

_____

[p.534] parathyroid glands _____

_____

### Choice

For each statement, choose the appropriate endocrine gland from the list below.

> a. anterior pituitary [p.532]     b. posterior pituitary [p.532]     c. hypothalamus [p.532]
> d. thyroid [p.534]                    e. parathyroid [p.534]

1. _____ Hormones from this gland have the major control over blood calcium levels.

2. _____ Secretes calcitonin

3. _____ Secretes oxytocin and ADH

4. _____ Secretes growth hormones

5. _____ Abnormal secretions can lead to diabetes insipidus.

6. _____ Secretes ACTH and LH

7. _____ Contains neurons that secrete hormones

8. _____ Goiter and Graves, disease are associated with this gland.

9. _____ Requires iodine

10. _____ Directly connected to the pituitary glands

## 32.5. THE ADRENAL GLANDS [p.535]

## 32.6. PANCREATIC HORMONES [p.536]

## 32.7. BLOOD SUGAR DISORDERS [p.537]

### Selected Words

Pancreatic islets [p.536], *alpha* cells [p.536], *beta* cells [p.536], *delta* cells [p.536], diabetes mellitus [p.537], hypoglycemia [p.537]

### Boldfaced Terms

[p.535] adrenal cortex _____

_____

[p.535] cortisol stress response _____

_____

[p.535] adrenal medulla _____

_____

### Choice

For each of the following, choose the appropriate endocrine organ from the following list.

        a. pancreatic islets [pp.536–537]      b. adrenal gland [p.535]

1. _____ is involved in the stress response

2. _____ Diabetes mellitus is associated with this endocrine gland.

3. _____ secretes somatostatin

4. _____ has both endocrine and exocrine functions

5. _____ secretes glucagons and insulin

6. _____ Cushing Syndrome is associated with this endocrine gland.

7. _____ The fight-flight response is controlled from here.

### True–False

If the statement is true, place a "T" in the blank. If the statement is false, correct the underlined word so that the statement is correct.

8. _____ The adrenal <u>cortex</u> releases aldosterone and cortisol. [p.535]

9. _____ Epinephrine and norepinephrine are involved in the <u>stress</u> response. [p.535]

10. _____ The beta cells of the pancreatic islets secrete the <u>glucagon</u> hormone. [p.536]

11. _____ Directly after a meal, when blood glucose levels are high, <u>alpha</u> cells are active. [p.536]

12. _____ The <u>somatostatin</u> hormone is involved in regulating digestion and absorption. [p.536]

13. _____ Somatostatin is secreted by the delta cells of the <u>adrenal medulla</u>. [p.536]

14. _____ <u>Type I diabetes</u> is the result of an autoimmune response. [p.537]

15. _____ In hypoglycemia, <u>low</u> blood glucose levels disrupt body functions. [p.537]

16. _____ <u>Type II</u> diabetes may result in ketoacidosis. [p.537]


## 32.8. OTHER ENDOCRINE GLANDS [p.538]

## 32.9. A COMPARATIVE LOOK AT A FEW INVERTEBRATES [p.539]

### Selected Words

Testes [p.538], ovaries [p.538], "winter blues" [p.538], thymus [p.538]

### Boldfaced Terms

[p.538] gonads _____

_____

[p.538] puberty _____

_____

[p.538] pineal gland _____

_____

[p.538] biological clock _____

_____

[p.539] molting _____

_____

[p.539] ecdysone _____

_____


### Matching

Match each of the following terms to the correct statement.

1. _____ pineal gland

2. _____ ecdysone

3. _____ melatonin

4. _____ biological clock

5. _____ gonads

6. _____ puberty

7. _____ thymus

a. internal timing mechanism
b. hormones from this gland are associated with T-cell maturity
c. endocrine gland that senses changes in light levels
d. invertebrate hormone that controls molting
e. primary reproductive glands
f. hormone associated with seasonal affective disorder
g. developmental stage when reproductive glands mature

# Self-Quiz

___ 1. Goiter is associated with a problem in the
_____ gland. [p.534]
   a. pineal
   b. adrenal
   c. pituitary
   d. thyroid
   e. parathyroid

___ 2. Hormones from the parathyroid gland
regulate the balance of _____ in the
body. [p.534]
   a. iron
   b. calcium
   c. sex hormones
   d. potassium
   e. magnesium

___ 3. The fight-flight response is regulated by the
_____. [p.535]
   a. adrenal gland
   b. pineal gland
   c. testes
   d. thyroid
   e. kidneys

___ 4. This class of chemicals helps to integrate
social behavior. [p.528]
   a. neurotransmitters
   b. pheromones
   c. hormones
   d. local signaling molecules
   e. none of the above

___ 5. This endocrine gland is responsible
for the secretion of melatonin. [p.538]
   a. hypothalamus
   b. pituitary gland
   c. ovaries
   d. pineal gland
   e. pancreatic islets

___ 6. The neurons in this gland secrete hormones
instead of neurotransmitters. [p.532]
   a. pituitary
   b. adrenal
   c. pineal
   d. hypothalamus
   e. thyroid

___ 7. Which of the following conditions can
be due to a problem with the pituitary?
[pp.532–533]
   a. Grave's disease
   b. diabetes mellitus
   c. hypoglycemia
   d. seasonal affective disorder
   e. none of the above

___ 8. Which of the following statements is true
regarding the steroid hormones?
[pp.530–531]
   a. They can diffuse directly across the
   plasma membrane.
   b. They bind to receptors in the cytoplasm
   or nucleus.
   c. The stimulate gene transcription
   d. They are made from cholesterol.
   e. All of the above are correct.

___ 9. The _____ cells of the pancreas
produce insulin. [p.536]
   a. alpha
   b. beta
   c. gamma
   d. delta
   e. omega

___ 10. Which of the following is associated
with the activity of T cells in the immune
system? [p.538]
   a. thymus
   b. thyroid
   c. adrenal gland
   d. pineal gland
   e. hypothalamus

# Chapter Objectives/Review Questions

1. Understand the categories of signaling molecules and the function of each in the body. [p.528]
2. Know the location and role of each gland in the endocrine system. [p.539]
3. Know the four types of hormones. [p.530]
4. Understand the five conditions that affect a cell's sensitivity to a hormone. [p.530]
5. Compare the mode of action of a steroid and peptide hormone. [pp.530–531]
6. Understand the interaction of the hypothalamus and the pituitary gland. [p.532]
7. List the major secretions of the anterior and posterior pituitary glands. [pp.532–533]
8. Understand the concept of negative feedback. [p.534]
9. Understand the role and secretions of the thyroid and parathyroid glands. [p.534]
10. Understand the role and secretions of the adrenal gland. [p.535]
11. Understand the role of the pancreatic islets, its secretions, and the diseases associated with abnormal function of this gland. [pp.536–537]
12. Understand the role of the pineal gland and melatonin. [p.538]
13. Understand the relationship between ecdysone and invertebrate molting. [p.539]

# Chapter Summary

(1) _____ and other signaling molecules regulate the pathways that control metabolism, growth, (2) _____, and reproduction. Nearly all vertebrates have an (3) _____ system composed of the same hormone sources.

A hormone signal acts on any cell that has (4) _____ for it. Receptor activation leads to (5) _____ of the signals and a response in the targeted cell.

Invertebrates, the (6) _____ and pituitary gland are connected structurally and functionally. Together, they coordinate the activities of many other (7) _____.

(8) _____ feedback loops to the hypothalamus and pituitary control secretions from many glands. Other glands secrete hormones in response to local changes in the internal (9) _____.

Hormones control (10) _____ and other events in invertebrate life cycles. Vertebrate hormones and receptors for them first evolved in ancestral lineages of (11) _____.

(12) _____ organ systems are a focus of many chapters in this unit. This secretion is an overview of tissue-specific responses to hormones including the impact of variations in (13) _____.

# Integrating and Applying Key Concepts

1. Many people are concerned about the presence of growth hormones in milk and other animal products. These hormones are given to animals to stimulate the production of milk or muscle mass in order to increase agricultural yield. To be harmful, these would have to mimic the action of a human growth-related hormone. Compare this to the material presented on ecdysone and molting in invertebrates and state why you think, or don't think, that this may be a problem.
2. What would be the possible effects of a tumor on the anterior pituitary that limited its function? The posterior?

# 33

# STRUCTURAL SUPPORT AND MOVEMENT

## INTRODUCTION

This chapter examines how muscles and bones interact in the process of movement. The chapter introduces the basic principles of how skeletons and muscles function, as well as some of the diseases that are common with these systems.

## FOCAL POINTS

- Figure 33.4 [p.547] illustrates some of the major bones of the human body.
- Figure 33.10 [p.551] illustrates some of the major muscles of the human body.
- Figure 33.12 [p.553] illustrates the action of a sarcomere.

## Interactive Exercises

*Pumping up Muscles* [pp.544–545]

## 33.1. ANIMAL SKELETONS [pp.546–547]

## 33.2. ZOOMING IN ON BONES AND JOINTS [pp.548–549]

*Selected Words*

"Andro" [p.544], creatine [p.545], *axial* [p.546], *appendicular* [p.546], *compact* bone [p.548], *spongy* bone [p.548], cartilage [p.548], osteoporosis [p.549], *fibrous* joint [p.549], *cartilaginous* joint [p.549], *synovial* joint [p.549], sprain [p.549], arthritis [p.549], osteoarthritis [p.549], rheumatoid arthritis [p.549]

*Boldfaced Terms*

[p.546] hydrostatic skeleton _____

_____

[p.546] exoskeleton _____

_____

[p.546] endoskeleton _____

_____

[p.546] vertebrae _____

_____

[p.546] intervertebral disks _____

_____

[p.548] osteoblasts _____

_____

[p.548] osteocytes _____

_____

[p.548] osteoclasts _____

_____

[p.548] red marrow _____

_____

[p.548] yellow marrow _____

_____

[p.549] joints _____

_____

[p.549] ligaments _____

_____

## True–False

If the statement is true, place a "T" in the space provided. If false, correct the underline word so that the statement is true. [pp.544–545]

_____ 1. Creatine is an intermediate of the sex hormones testosterone and estrogen.

_____ 2. The use of androstenedione has been proven to increase muscle mass.

_____ 3. Creatine is produced by the muscles of the body as an energy source.

_____ 4. Excess creatine has been shown to increase stress on the kidneys.

## Choice

For each of the following, choose the form of skeleton that is best described by the statement. [p.546]

        a. endoskeleton    b. exoskeleton    c. hydrostatic skeleton

_____ 5. The vertebrates possess this form of skeleton.

_____ 6. Arthropods possess this form of skeleton.

_____ 7. Internal body parts accept the applied force of muscle contraction.

_____ 8. Muscle contractions redistribute a volume of fluid in the body cavity.

_____ 9. Earthworms and soft-bodied invertebrates have this form of skeleton.

_____ 10. External body parts receive the applied force of muscle contraction.

## Labeling

Match the statement with the appropriate label from the figure. [p.547]

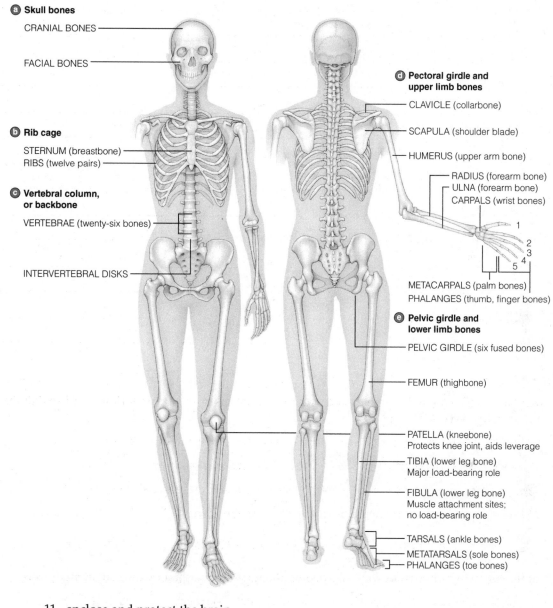

**a** Skull bones

CRANIAL BONES

FACIAL BONES

**b** Rib cage

STERNUM (breastbone)
RIBS (twelve pairs)

**c** Vertebral column, or backbone

VERTEBRAE (twenty-six bones)

INTERVERTEBRAL DISKS

**d** Pectoral girdle and upper limb bones

CLAVICLE (collarbone)

SCAPULA (shoulder blade)

HUMERUS (upper arm bone)

RADIUS (forearm bone)
ULNA (forearm bone)
CARPALS (wrist bones)

1
2
3
4
5

METACARPALS (palm bones)
PHALANGES (thumb, finger bones)

**e** Pelvic girdle and lower limb bones

PELVIC GIRDLE (six fused bones)

FEMUR (thighbone)

PATELLA (kneebone)
Protects knee joint, aids leverage

TIBIA (lower leg bone)
Major load-bearing role

FIBULA (lower leg bone)
Muscle attachment sites;
no load-bearing role

TARSALS (ankle bones)
METATARSALS (sole bones)
PHALANGES (toe bones)

_____ 11. enclose and protect the brain

_____ 12. These bones are involved in locomotion.

_____ 13. support the skull and upper extremities

_____ 14. contains bones with extensive muscle attachments

_____ 15. contains fibrous, cartilaginous structures that absorb movement-induced stresses

_____ 16. supports the weight of the backbone

_____ 17. protects the heart and lungs

***TNOW:*** For an interactive exercise on the bones of the human body, refer to Chapter 33: Art Labeling: Major Bones of the Human Body.

## Matching

Match each of the following terms with its correct definition. [pp.548–549]

_____ 18. yellow marrow

_____ 19. cartilage

_____ 20. osteocytes

_____ 21. red marrow

_____ 22. osteoblasts

_____ 23. osteoclasts

_____ 24. joints

_____ 25. ligaments

a. cells that break down bone
b. this consists primarily of fat
c. the first skeleton to form in vertebrate embryos
d. the major site of blood cell formation
e. areas of contact, or near-contact, between bones
f. bone-forming cells
g. the most common bone cells in adults
h. straps of dense connective tissue

## Labeling

Label each of the indicated structures of the below diagram. [p.548]

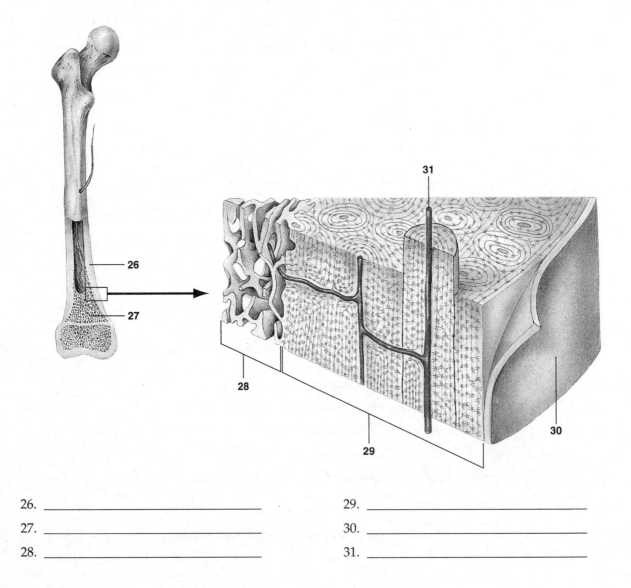

26. _____

27. _____

28. _____

29. _____

30. _____

31. _____

## Concept Map

Provide the missing terms for the numbered items in the accompanying concept map of bone formation and remodeling. [pp.548–549]

32. _____

33. _____

34. _____

35. _____

36. _____

37. _____

38. _____

39. _____

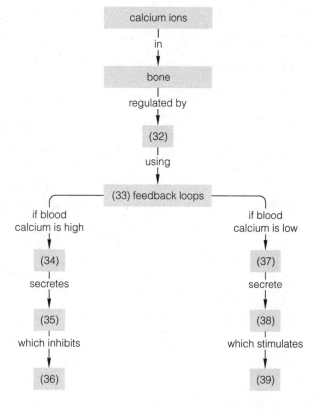

## Matching

Match each of the following ailments to its correct description. [p.549]

_____ 40. osteoarthritis

_____ 41. strain

_____ 42. sprain

_____ 43. osteoporosis

_____ 44. rheumatoid arthritis

a. when cartilage in movable joints wears off
b. tearing of the ligaments or tendons in a joint
c. a decrease in bone density
d. membranes in joints become inflamed due to an autoimmune response
e. stretching or twisting a joint too far

## 33.3. SKELETAL-MUSCULAR SYSTEMS [pp.550–551]

## 33.4. HOW DOES SKELETAL MUSCLE CONTRACT? [pp.552–553]

### Selected Words

*Skeletal* muscle [p.550], smooth muscle [p.550], cardiac muscle [p.550]

### Boldfaced Terms

[p.550] muscle fiber _____

_____

[p.550] tendon _____

_____

[p.552] myofibrils _____

_____

[p.552] sarcomeres _____

_____

[p.552] actin _____

_____

[p.553] myosin _____

_____

[p.553] sliding filament model _____

_____


### True–False

If the statement is true, place a "T" in the space provided. If the statement is false, correct the underlined term so that the statement is correct. [p.550]

_____ 1. The functional partner of bone is <u>smooth</u> muscle.

_____ 2. Skeletal muscles are composed of bundles of <u>muscle fibers</u>.

_____ 3. <u>Ligaments</u> attach skeletal muscle to bone.

_____ 4. In the human body, most skeletal muscles act as a <u>lever</u> system.

## Labeling

Label the muscles in the following diagram. Then match named muscle to its correct function. [p.551]

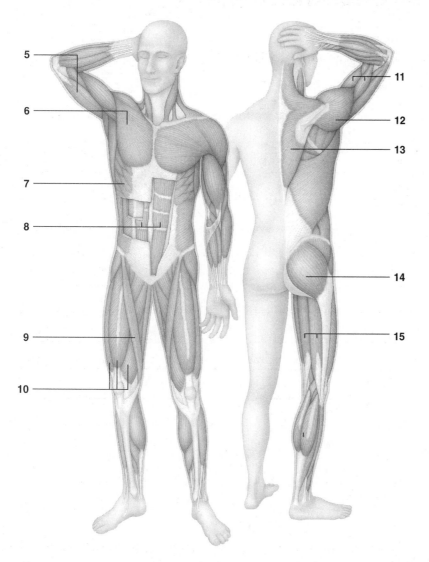

5. _____ (  )

6. _____ (  )

7. _____ (  )

8. _____ (  )

9. _____ (  )

10. _____ (  )

11. _____ (  )

12. _____ (  )

13. _____ (  )

14. _____ (  )

15. _____ (  )

a. extends and rotates the thigh outward when walking and running
b. bends the thigh at the hip, bends lower leg at the knee
c. bends the forearm at the elbow
d. raises the arm
e. straightens the forearm at the elbow
f. lifts the shoulder blade and draws the head back
g. draws thigh backward and bends the knee
h. draws the arm forward and in toward the body
i. flexes the thigh at hips and extends the leg at the knee
j. compresses the abdomen
k. depresses the thoracic cavity and bends the backbone

*Matching*

Match each of the following definitions to its correct term. [pp.552–553]

_____ 16. the motor protein for muscle contraction

_____ 17. threadlike, cross-banded structures within a muscle fiber

_____ 18. These run parallel to the muscle's long axis.

_____ 19. short, ATP-driven movements of the myosin heads that pull actin filaments towards the sarcomere center

_____ 20. the basic unit of muscle contraction

a. myosin
b. sarcomeres
c. myofibrils
d. muscle fibers
e. sliding-filament model

*Sequence*

Place each of the following items in its correct sequence. Indicate the first event with "1," the second event with "2," and so on. [p.553]

_____ 21. New ATP binds to the myosin heads and they detach from actin. Hydrolysis of the ATP will return the heads to their original orientation, ready to act again.

_____ 22. Binding makes each myosin head tilt toward the center of the sarcomere and slide the bound actin filaments along with it.

_____ 23. The myosin filaments are at rest. They were energized earlier by ATP.

_____ 24. Release of calcium from a cellular storage system allows formation of cross bridges.

_____ 25. ADP and phosphate are released as the myosin heads drag the actin filaments inward, pulling the Z lines closer together.

## 33.5. FROM SIGNAL TO RESPONSES [pp.554–555]

## 33.6. OH *CLOSTRIDIUM*! [p.556]

*Selected Words*

*Isotonically* [p.554], *isometrically* [p.554], creatine phosphate [p.555], muscular dystrophies [p.555], *Clostridium botulinium* [p.556], botulism [p.556], *C. tetani* [p.556], tetanus [p.556]

*Boldfaced Terms*

[p.554] sarcoplasmic reticulum _____

_____

[p.554] motor unit _____

_____

[p.554] muscle twitch _____

_____

[p.554] tetanus _____

_____

[p.554] muscle tension _____

_____

[p.554] muscle fatigue _____

_____

## Matching

Match each of the following terms to its correct definition. [pp.554–555]

_____ 1. muscle fatigue

_____ 2. muscle tension

_____ 3. isometrically

_____ 4. motor unit

_____ 5. isotonically

_____ 6. tetanus

_____ 7. muscular dystrophies

_____ 8. muscle twitch

a. A brief generation of contractile force
b. The mechanical force exerted by a muscle on an object
c. Muscles that develop tension but can not shorten
d. The progressive weakening and degeneration of muscles
e. When contracting muscles shorten and move a load
f. A sustained series of muscle twitches
g. A motor neuron and all of the muscle fibers that it controls
h. A muscle that is in an ongoing state of muscle contraction

## Complete the Table

Indicate the disease that is caused by each of the organisms below. [p.556]

| Clostridium tetani | 9. |
| --- | --- |
| Clostridium botulinum | 10. |

# Self-Quiz

1. The basic unit of muscle contraction is the _____. [p.552]
   a. myofibril
   b. muscle fiber
   c. myosin
   d. sarcomere

2. Which of the following is not involved in the regulation of calcium levels in bone? [p.549]
   a. PTH
   b. vitamin D
   c. myosin
   d. calcitonin

3. In which of the following do internal body parts accept the applied force of contraction? [p.546]
   a. exoskeleton
   b. hydrostatic skeleton
   c. endoskeleton
   d. all of the above

4. Which of the following is the functional partner of bone? [p.550]
   a. cardiac muscle
   b. smooth muscle
   c. skeletal muscle
   d. all of the above

5. Which of the following components of bone is responsible for the formation of red blood cells? [p.548]
   a. yellow marrow
   b. osteoclasts
   c. osteoblasts
   d. red marrow
   e. osteocytes

6. The _____ attaches the bundles of muscle fiber to the bone. [p.549]
   a. myofibril
   b. tendon
   c. ligament
   d. joint

7. A brief generation of contractile force in a muscle is called _____. [p.554]
   a. tetanus
   b. muscle twitch
   c. muscle fatigue
   d. muscle tension

8. Which of the following correctly describes aerobic exercise? [p.555]
   a. long duration, intense activity
   b. long duration, not intense activity
   c. short duration, intense activity
   d. short duration, not intense activity

9. The sliding-filament model describes which of the following? [p.553]
   a. the action of the sarcomere
   b. the function of the joints
   c. the skeletal model of an exoskeleton
   d. a form of exercise for muscle

10. Which of the following is not usually used as an energy source for muscle? [p.554]
   a. creatine phosphate
   b. glucose
   c. fatty acids
   d. amino acids

---

## Chapter Objectives/Review Questions

1. Describe the general characteristics of the three primary types of skeletons. [p.546]
2. List some of the challenges of life on land that an endoskeleton helps to solve. [p.546]
3. Identify the functions of the primary bone groups in humans. [p.547]
4. Understand the structure of bone and the role of osteocytes, osteoclasts, osteoblasts, red marrow, and yellow marrow. [p.548]
5. Identify the roles of calcitonin, vitamin D, and PTH in the process of bone remodeling. [pp.548–549]
6. Understand how bones connect with one another and the major diseases/ailments of these locations. [p.549]
7. Recognize the role and structure of a tendon. [p.550]
8. Know the three types of muscle. [p.550]
9. Understand the structure of a sarcomere. [p.552]
9. Understand the process by which a muscle contracts. [pp.552–553]
10. List the principles of the sliding-filament model. [p.553]
11. Define the terms *muscle tension*, *muscle twitch*, *tetanus*, and *muscle fatigue* as they relate to muscle contraction. [pp.554–555]
12. Understand the causes of the diseases tetanus and botulism. [p.554]

---

## Chapter Summary

(1) _____ force exerted against some type of (2) _____ moves the animal body. Many invertebrates have a (3) _____ skeleton, which is a fluid-filled cavity. Others have an (4) _____ of hardened structures at the body surface. Vertebrates have an (5) _____, an internal skeleton of cartilage, bone, or both.

Bones are (6) _____-rich organs that help the body move. They also protect and support soft organs, and store (7) _____. (8) _____ cells form in some bones. Cartilage or ligaments connect bones at (9) _____.

(10) _____ muscles are bundles of muscle fibers that interact with bones and with one another. Some cause movements by working as pairs or groups. Others oppose or (11) _____ the action of a partner muscle. (12) _____ attach skeletal muscles to bones.

A muscle fiber contains many (13) _____, each divided crosswise into (14) _____, the basic unit of contraction. Sarcomeres contain many parallel arrays of (15) _____ and myosin filaments. (16) _____-driven interactions between the arrays shorten sarcomeres, which collectively accounts for (17) _____ of a whole muscle.

Muscle fibers in a muscle are organized in (18) _____ that contract in response to signals from one motor (19) _____. Cross-bridges form in all sarcomeres and collectively exert (20) _____. A muscle (21) _____ only when the tensile force exceeds other, opposing forces. (22) _____ enhances the properties of whole muscle, and aging and disease diminish them.

## Integrating and Applying Key Concepts

1. Despite the risks of using anabolic steroids, many professional athletes continue to use steroids in an attempt to enhance their performance. From this chapter, speculate on a theoretical hormone that mimics one of the chemicals presented in the chapter. In addition, describe some problems that overuse of this chemical may present to the athlete over time.
2. While hormone-replacement therapy in women remains controversial, one of the benefits of an increase in estrogen is an increase in bone density. Why would this occur? What cells would it be stimulating?

# 34

# CIRCULATION

## INTRODUCTION

The human cardiovascular system is the result of millions of years of evolution and is one of the most important systems in the body. This system has evolved from an open circulatory system in arthropods and mollusks to a very sophisticated closed circulatory system in humans. This chapter discusses the evolution of the human cardiovascular system, and then goes on to point out its critical function: the movement of substances in and out of cellular environments. Human blood is composed not only of red blood cells, which carry oxygen, but also of white blood cells, for fighting pathogens and platelets for clotting blood. This chapter shows how these cells form and what their specific functions are. Then, the chapter goes on to discuss, in detail, the structure of the human heart and vascular system, as well as the lymphatic system. Finally, the importance of this system to human health is illustrated by discussions of diseases caused by abnormalities in the blood, or cardiovascular system. Some of these diseases include leukemia, anemia, hypertension, and cardiovascular disease. An understanding of these pathologies enables us to better understand not only what causes disease, but also how to protect our own health by preventing these diseases.

## FOCAL POINTS

- Figure 34.2 [p.560] compares open and closed circulatory systems.
- Figure 34.4 [p.562] tabulates the typical components of human blood.
- Figure 34.8 [p.565] illustrates the responses of humans with various blood types.
- Figure 34.10 [p.566] animates the systemic and pulmonary circuits of the human cardiovascular system.
- Figure 34.11 [p.567] diagrams the major vessels of the human cardiovascular system.
- Figure 34.12 [p.568] illustrates the human heart.
- Figure 34.21 [p.572] shows fluid movement at the capillary bed.
- Figure 34.27 [p.576] diagrams the components of the human lymphatic system.

---

## Interactive Exercises

*And Then My Heart Stood Still* [pp.558–559]

### 34.1. THE NATURE OF BLOOD CIRCULATION [pp.560–561]

*Selected Terms*

Open circulatory system [p.560], closed circulatory system [p.560], internally moistened sacs [p.561]

## Boldfaced Terms

[p.560] circulatory system _____

_____

[p.560] blood _____

_____

[p.560] heart _____

_____

[p.560] interstitial fluid _____

_____

[p.560] internal environment _____

_____

[p.560] blood capillaries _____

_____

[p.561] pulmonary circuit _____

_____

[p.561] systemic circuit _____

_____

## Matching

Match each of the following statements to the correct term. [pp.560–561]

_____ 1. carries oxygen-poor, carbon dioxide-rich blood to the lungs

_____ 2. Blood flow is slowest here, to allow the exchange of substances by diffusion.

_____ 3. may be either open or closed, depending on the type of organism

_____ 4. This is responsible for directly exchanging substances with cells.

_____ 5. carries oxygen rich blood to the tissues

a. pulmonary circuit
b. systemic circuit
c. capillaries
d. interstitial fluid
e. circulatory system

## Choice

Choose which of the following circulatory systems each statement applies to. [pp.560–561]

      a. fish     b. amphibians     c. birds and mammals

_____ 6. consists of two, separate circuits for the blood

_____ 7. utilize a two-chambered heart

_____ 8. utilize a three-chambered heart

_____ 9. utilize a four-chambered heart

_____ 10. Blood pressure is lowest in this group.

_____ 11. The development of a pulmonary circuit occurred in this group.

_____ 12. Blood mixes in a common ventricle.

## 34.2. CHARACTERISTICS OF BLOOD [pp.562–563]

## 34.3. BLOOD DISORDERS [p.564]

## 34.4. BLOOD TYPING [pp.564–565]

### Boldfaced Terms

[p.562] plasma _____

_____

[p.562] stem cell _____

_____

[p.562] red blood cells _____

_____

[p.563] cell count _____

_____

[p.563] white blood cells _____

_____

[p.563] platelets _____

_____

[p.564] anemias _____

_____

[p.564] agglutination _____

_____

[p.564] ABO blood typing _____

_____

[p.565] Rh blood typing _____

_____

### Choice

Choose one of the following blood components for each statement. Answers may be used more than once. [pp.562–563]

a. red blood cells     b. plasma     c. white blood cells     d. platelets

_____ 1. also called erythrocytes

_____ 2. also called leukocytes

_____ 3. the transport medium for blood cells and platelets

_____ 4. These initiate blood clotting.

_____ 5. Neutrophils, basophils, and macrophages are examples.

_____ 6. acts as a solvent for hundreds of plasma proteins

_____ 7. transport oxygen to aerobically transpiring cells

_____ 8. contains hemoglobin

_____ 9. identify "nonself" objects in the body

_____ 10. These have a 120-day lifespan.

## Complete the Table

Complete the following table, which describes the components of blood. [p.562]

| Components | Relative Amounts | Functions |
|---|---|---|
| **Plasma Portion (50%–60% of total volume)** | | |
| 11. | 91%–92% of plasma volume | Solvent |
| 12. | 7%–8% | Defense, clotting, lipid transport, roles in extracellular fluid volume, etc. |
| Ions, sugars, lipids, amino acids, hormones, vitamins, dissolved gases | 13. | Roles in extracellular fluid volume, pH, etc. |
| **Cellular Portion (40%–50% of total volume)** | | |
| 14. | 4,800,000–5,400,000 per microliter | $O_2$, $CO_2$ transport |
| 15. | 3,000–6,750 | Phagocytosis |
| 16. | 1,000–2,700 | Immunity |
| Monocytes (macrophages) | 150–720 | 17. |
| Eosinophils | 100–360 | 18. |
| 19. | 25–90 | Roles in inflammatory response, anticlotting |
| 20. | 250,000–300,000 | Roles in clotting |

## Choice

Indicate whether each of the following conditions is the result of a red blood cell disorder or a white blood cell disorder. [p.564]

      a. red blood cell disorder    b. white blood cell disorder

_____ 21. leukemia

_____ 22. sickle cell anemia

_____ 23. infectious mononucleosis

_____ 24. blood doping

_____ 25. thalassemias

_____ 26. hemolytic anemias

## Fill in the Blanks [pp.564–565]

Molecular variations in one kind of (27) _____ marker on red blood cells are analyzed with (28) _____ blood typing. If you are type A, your immune system will treat blood with type (29) _____ markers as foreign. If you are type B, your body will react against type (30) _____ markers. If your blood is type (31) _____, your body is familiar with both markers and won't react against either type; you are able to (32) _____ blood from anyone. If you are type O, both A and B markers are perceived as (33) _____. You can only receive blood from others who are also type (34) _____, but you can donate blood to anyone.

(35) _____ blood typing is based on the presence or absence of the Rh (36) _____. If you are type (37) _____, your body cells bear this marker. If you are type (38) _____ they do not. Usually people do not have (39) _____ against Rh markers, because their body has never been exposed to them. But an Rh– recipient of transfused (40) _____ blood will make antibodies, and they will remain in the blood.

41. Humans have four possible blood types (A, B, AB, and O). A court case has been filed by a mother with type O blood who has a son with type O blood. There are two fathers being accused; Edward has type AB blood, and Charles has type A blood. Which one of the men could be the father of the child?

## 34.5. THE HUMAN CARDIOVASCULAR SYSTEM [pp.566–567]

## 34.6. THE HEART IS A LONELY PUMPER [pp.568–569]

### Selected Terms

"Cardiovascular" [p.566], kardia [p.566], vasculum [p.566], pericardium [p.568], myocardium [p.568], diastole [p.568], systole [p.568]

### Boldfaced Terms

[p.566] aorta _____

_____

[p.566] cardiac cycle _____

_____

[p.566] cardiac muscle _____

_____

[p.569] cardiac conduction system _____

_____

[p.569] cardiac pacemaker _____

_____

## Labeling

First, name each of the components of the human circulatory system in the diagram below. Then, choose the correct function for each component from the list provided. [p.567]

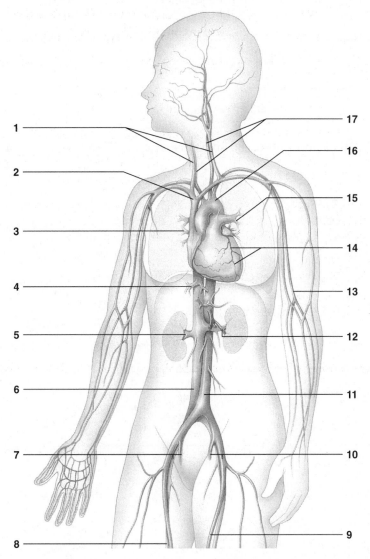

1. _____ ( )

2. _____ ( )

3. _____ ( )

4. _____ ( )

5. _____ ( )

6. _____ ( )

7. _____ ( )

8. _____ ( )

9. _____ ( )

10. _____ ( )

a. delivers nutrient rich blood from small intestine to liver for processing
b. carries blood away from thigh and inner knee
c. delivers blood to thigh and inner knee
d. delivers oxygenated blood from lungs to the heart
e. services the active cardiac muscles of heart
f. carry blood away from pelvic organs and lower abdominal wall
g. delivers blood to kidneys
h. receives blood from brain and tissues of the head
i. delivers blood to pelvic organs and lower abdominal wall
j. receives blood from veins of upper body
k. deliver oxygen-poor blood from heart to lungs
l. deliver blood to neck, head, and brain

11. _____ ( )
12. _____ ( )
13. _____ ( )
14. _____ ( )
15. _____ ( )
16. _____ ( )
17. _____ ( )

m. carries oxygenated blood away from heart; the largest artery
n. delivers blood to upper extremities
o. carries processed blood away from kidneys
p. delivers blood to arteries leading to the digestive tract, kidneys, pelvic organs, lower extremities
q. receives blood from all veins below diaphragm

## Labeling

Label each of the indicated components of the human heart. [p.568]

18. _____
19. _____
20. _____
21. _____
22. _____
23. _____
24. _____
25. _____
26. _____
27. _____
28. _____
29. _____
30. _____

## True–False

If the statement is correct, write "T" in the space provided. If it is incorrect, correct the underlined word so that the statement is true. [pp.568–569]

_____ 31. The contraction of the <u>right atrium</u> provides the driving force for blood circulation.

_____ 32. The <u>AV</u> node of the heart functions as the cardiac pacemaker.

_____ 33. The <u>AV</u> valves separate the atria from the ventricles.

_____ 34. Semilunar and AV valves are <u>one-way</u> valves.

_____ 35. The outermost connective tissue sac of the heart is called the <u>septum</u>.

## 34.7. PRESSURE, TRANSPORT, AND FLOW DISTRIBUTION [pp.570–571]

## 34.8. DIFFUSION AT CAPILLARIES, THEN BACK TO THE HEART [pp.572–573]

## 34.9. CARDIOVASCULAR DISORDERS [pp.574–575]

## 34.10. INTERACTIONS WITH THE LYMPHATIC SYSTEM [pp.576–577]

### Selected Terms

*Systolic* [p.571], *diastolic* [p.571]

### Boldfaced Terms

[p.570] arteries _____

_____

[p.570] arterioles _____

_____

[p.570] capillaries _____

_____

[p.570] venules _____

_____

[p.570] veins _____

_____

[p.570] blood pressure _____

_____

[p.573] ultrafiltration _____

_____

[p.573] capillary reabsorption _____

_____

[p.574] hemostasis _____

_____

[p.576] lymph vascular system _____

_____

[p.576] lymph _____

_____

[p.577] lymph nodes _____

_____

[p.577] spleen _____

_____

[p.577] thymus gland _____

_____

## Matching

Match each of the following words to the correct statement. [pp.570–571]

_____ 1. vasoconstriction

_____ 2. diastolic pressure

_____ 3. cardiac output

_____ 4. blood pressure

_____ 5. systolic pressure

_____ 6. vasodilation

_____ 7. arteries

_____ 8. venules

_____ 9. veins

_____10. arterioles

_____11. capillaries

a. carry blood rapidly away from the heart's ventricles
b. controls over blood flow operate here
c. These form diffusion zones.
d. the small vessels between capillaries and veins
e. act as blood reservoirs
f. This is the pressure caused by ventricular contractions.
g. an increase in the diameter of the blood vessels
h. a decrease in the diameter of the blood vessels
i. the peak pressure against the walls of the arteries
j. the lowest arterial pressure
k. how much blood the ventricles pump out

## Labeling [p.572]

Label the following picture with the appropriate letters:

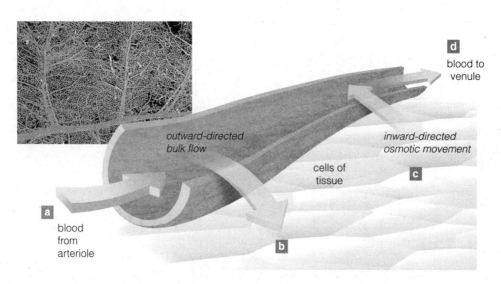

12. _____ ultrafiltration—fluid leaves capillaries for cells

13. _____ blood going to venule

14. _____ reabsorption—fluid entering capillaries from cells

15. _____ blood coming from arteriole

## Matching

Match each of the following cardiovascular disorders to its correct statement. [pp.574–575]

_____ 16. atherosclerosis

_____ 17. diabetes and hypertension

_____ 18. arrhythmias

a. chronically high blood pressure
b. The atria do not contract normally.
c. a serious condition in which the ventricle does not contract correctly

_____ 19. hemostasis

_____ 20. hypertension

_____ 21. arterial fibrillation

_____ 22. ventricular fibrillation

d. risk factors for coronary artery disease

e. abnormal rhythms in the heartbeat

f. the process that stops blood loss from damaged blood vessels

g. High LDL levels increase the risk of this condition.

## Fill in the Blanks [pp.576–577]

A portion of the circulatory system, called the (23) _____ system, consists of many tubes that collect and deliver (24) _____ and solutes from (25) _____ fluid to ducts of the circulatory system. Its main components are (26) _____ capillaries and vessels. Tissue fluid that moves into these vessels is called (27) _____.

The lymph vascular system serves three functions. First, its vessels are drainage channels for water and (28) _____ proteins that have leaked out from the (29) _____ at capillary beds and must be delivered back to the blood circulation. Second, the system takes up (30) _____ that the body has absorbed from the (31) _____ intestine and delivers them to the general circulation. Third, it delivers (32) _____, foreign cells, and cellular debris from tissues to the lymph vascular system's disposal centers, the lymph (33) _____.

# Self-Quiz

1. Which of the following is/are correct regarding a male with O+ blood? [pp.564–565]
   a. Both A and B markers are perceived as foreign by their body.
   b. They can donate blood to anyone.
   c. Their blood cells possess the Rh marker.
   d. If they marry an Rh– female, there is a risk of erythroblastosis fetalis.
   e. All of the above are correct.

2. The _____ act as blood reservoirs in the body. [p.570]
   a. veins
   b. capillaries
   c. ventricles
   d. arterioles
   e. arteries

3. The pulmonary circuit first arose in which group of organisms? [p.561]
   a. fish
   b. birds and mammals
   c. reptiles
   d. amphibians

4. Which of the following serves as the cardiac pacemaker? [p.569]
   a. the AV node
   b. the SA node
   c. the pericardium
   d. the left ventricle
   e. none of the above

5. The name of the major blood vessel that carries blood from the left ventricle of the heart is the _____. [p.566]
   a. femoral vein
   b. renal artery
   c. aorta
   d. jugular vein
   e. carotid arteries

6. Chronic elevated high blood pressure is called _____. [p.575]
   a. hypertension
   b. arteriosclerosis
   c. angina pectoris
   d. edema
   e. atrial fibrillation

7. Which of the following makes up the largest percent of blood volume? [p.562]
   a. platelets
   b. red blood cells
   c. plasma
   d. white blood cells

8. The _____ returns excess interstitial fluid to the circulatory system. [p.576]
   a. lymph vascular system
   b. pulmonary circuit
   c. systemic circuit
   d. renal vein
   e. vena cava

9. Neutrophils, basophils, and lymphocytes are all _____. [pp.562–563]
   a. red blood cells
   b. plasma proteins
   c. platelets
   d. white blood cells

10. Leukemia is a disease of the _____. [p.564]
    a. red blood cells
    b. platelets
    c. plasma proteins
    d. white blood cells

## Chapter Objectives/Review Questions

This section lists general and detailed chapter objectives that can be used as review questions. You can make maximum use of these items by writing answers on a separate sheet of paper. Fill in answers where blanks are provided. To check for accuracy, compare your answers with information given in the chapter or glossary.

1. Distinguish between open and closed circulatory systems. Provide examples of animals with one or the other. [p.560]
2. Describe how vertebrate circulatory systems have evolved from the fish model to the mammalian model. [pp.560–561]
3. Describe the composition of human blood, using percentages of total volume. [p.562]
4. State where erythrocytes, leukocytes, and platelets are produced. [pp.562–563]
5. Recognize the major forms of red and white blood cell disorders. [p.564]
6. Describe how blood is typed for the ABO blood group and for the Rh factor. [pp.564–565]
7. Know the major arteries and veins of the human body. [p.567]
8. Describe the functional links between the circulatory, respiratory, urinary, and digestive systems. [p.567]
9. Know the major structures of the human heart. [p.568]
10. Explain the process of cardiac muscle contraction. [pp.568–569]
11. Explain what causes high pressure and low pressure in the human circulatory system. [pp.570–571]
12. Describe how the structures of arteries, capillaries, and veins differ. [p.570]
13. Describe the exchanges that occur in the capillary bed regions and the mechanisms that cause the exchanges. [pp.572–573]
14. Explain hemostasis. [p.574]
15. Define hypertension [p.575]
16. State the significance of high- and low-density lipoproteins to cardiovascular disorders. [pp.574–575]
17. Define the major types of arrhythmias. [p.575]
18. Describe the composition and function of the lymphatic system. [pp.576–577]

## Chapter Summary

Many animals have either an open or a closed (1) _____ system that transports substances to and from all body (2) _____.

Vertebrate blood is fluid (3) _____ tissue. It consists of red blood cells, white blood cells, (4) _____, and diverse substances dissolved in (5) _____, the transport medium. Red blood cells function in (6) _____ exchange; white blood cells and platelets help (7) _____ tissues.

The human heart has (8) _____ chambers. Blood flows into its two atria and then into two (9) _____, which pump it into two separate circuits of blood vessels. One circuit extends through all body regions, the other through (10) _____ tissue only. Both circuits loop back to the (11) _____.

The heart pumps blood (12) _____, on its own. Blood pressure is highest in the heart's (13) _____, drops as it flows through arteries, and is lowest in the (14) _____. Adjustments at (15) _____ regulate how much blood volume is distributed to tissues. Exchange of gases, wastes, and nutrients between blood and tissues takes place at (16) _____.

Ruptured or clogged blood (17) _____ or abnormal heart (18) _____ cause problems. Some problems have a genetic basis, most are related to age or (19) _____.

A (20) _____ vascular system delivers the excess fluid that collects in tissues to the blood. Lymphoid organs cleanse blood of (21) _____ agents and other threats to health.

## Integrating and Applying Key Concepts

1. You observe that some people appear as though fluid had accumulated in their lower legs and feet. Their lower extremities resemble those of elephants. You inquire about what is wrong and are told that the condition is caused by the bite of a mosquito that is active at night. Construct a testable hypothesis that would explain (1) why the fluid was not being returned to the torso, as normal, and (2) what the mosquito did to its victims.

# 35

# IMMUNITY

## INTRODUCTION

The human body encounters millions of microbes—i.e., bacteria, viruses, fungi, and protozoa—each day, yet we rarely become sick. Why is that? How can we constantly be exposed to disease-causing organisms, yet remain relatively healthy? The answer is the immune system, an elaborate network of cells, tissue, and organs dispersed throughout the vertebrate body that works to combat many potential invading organisms. This chapter discusses the organization, structure, and function of the components of the mammalian immune system. From our exploration and study of the immune system, which really began in 1796 with the work of Edward Jenner, we have come to understand the exquisite design and impressive function of this system. Further, we also have come to understand autoimmune and allergic reactions; the consequences that occur when the immune system responds to inappropriate antigens. Last, this knowledge has allowed us to manipulate the immune system in order to prevent disease. Like Jenner in the 1700s, we have been able to prevent many diseases using vaccines, which are a way of "tricking" the immune system. This and other ways of immune system manipulation offer much promise in the treatment and prevention of many diseases.

## FOCAL POINTS

- Table 35.1 [p.583] delineates the vertebrate surface barriers present in innate immunity.
- Figure 35.5 [p.584] illustrates the ability of the complement system to destroy target cells.
- Figure 35.6 [p.585] depicts the steps that occur during an inflammatory response.
- Figure 35.7 [p.586] shows the process of antigen processing.
- Figure 35.8 [p.587] illustrates the interactions between the two arms of the adaptive immune system.
- Figure 35.11 [p.589] animates an antibody mediated immune response.
- Figure 35.13 [p.590] animates a cell mediated immune response.
- Figure 35.19 [p.595] diagrams the replication cycle of HIV.
- Table 35.5 [p.596] compares innate and adaptive immunity.
- Table 35.6 [p.596] summarizes the cells involved in an immune response.

# Interactive Exercises

*Viruses, Vulnerability, and Vaccines* [pp.580–581]

## 35.1. INTEGRATED RESPONSE TO THREATS [p.582]

## 35.2. SURFACE BARRIERS [p.583]

### Selected Terms

*Self* [p.582], *nonself* [p.582], *Clostridium tetani* [p.583], *Staphylococcus aureus* [p.583], *Lactobacillus* [p.583], *Staphylococcus epidermidis* [p.583], MRSA [p.583]

### Boldfaced Terms

[p.582] immunity _____

_____

[p.582] complement _____

_____

[p.582] innate immunity _____

_____

[p.582] adaptive immunity _____

_____

[p.582] antigen _____

_____

[p.582] cytokines _____

_____

[p.582] neutrophils _____

_____

[p.582] macrophages _____

_____

[p.582] dendritic cells _____

_____

[p.582] basophils _____

_____

[p.582] mast cells _____

_____

[p.582] eosinophils _____

_____

[p.582] B lymphocytes _____

_____

[p.582] T lymphocytes_____

_____

[p.582] natural killer cells_____

_____

[p.583] lysozyme _____

_____

## Matching

Match each of the following cell types to its correct function. [p.582]

_____ 1. macrophages

_____ 2. neutrophils

_____ 3. natural killer cells

_____ 4. B lymphocytes and T lymphocytes

_____ 5. eosinophils

_____ 6. basophils and mast cells

a. destroy parasitic worms
b. "big eaters" can destroy up to 100 bacterial cells
c. the most common immune system cell
d. secrete chemicals to assist inflammation
e. manufacture proteins that chemically recognize pathogens
f. bind to tumor cells, virus-infected cells, and bacterial cells

## Complete the Table

For each of the following, list the mechanism by which the structure provides a first line of defense. [p.583]

| Structure | Defense Mechanism |
|---|---|
| Eyes | 7. |
| Urinary tract | 8. |
| Vertebrate skin | 9. |
| Mouth | 10. |
| Stomach | 11. |

## 35.3. INNATE IMMUNE RESPONSES [pp.584–585]

## 35.4. OVERVIEW OF ADAPTIVE IMMUNITY [pp.586–587]

## 35.5. THE ANTIBODY-MEDITATED IMMUNE RESPONSE [pp.588–589]

### Selected Terms

Self vs. nonself recognition [p.586], specificity [p.586], diversity [p.586], memory [p.586], effector cells [p.586]

## Boldfaced Terms

[p.584] acute inflammation _____

_____

[p.585] fever _____

_____

[p.586] MHC markers _____

_____

[p.586] T cell receptors _____

_____

[p.587] antibody mediated immune response_____

_____

[p.587] antibodies _____

_____

[p.587] cell-mediated immune response _____

_____

[p.588] B cell receptor _____

_____

## Choice

Choose the form of innate response that is associated with the statement. Some answers may be used more than once. [pp.584–585]

        a. fever      b. antimicrobial proteins      c. inflammatory response

_____ 1. capillaries become "leaky"

_____ 2. proteins tag microbes for destruction

_____ 3. the hypothalamic thermostat is reset

_____ 4. endogenous pyrogens stimulate prostaglandin release

_____ 5. mast cells in connective tissue release histamine

_____ 6. complement proteins attack microbes or interfere with reproduction

_____ 7. Edema is caused by a rush of fluid to the site of an infection

## Sequence

Place the following events of the inflammatory response in their correct sequence. The letter of the first event is placed in #8, the second event in #9, etc. [pp.584–585]

_____ 8.
_____ 9.
_____ 10.
_____ 11.
_____ 12.

    a. bacteria invade cells and damage tissue
    b. complement proteins attack bacteria
    c. localized edema occurs
    d. neutrophils and macrophages engulf debris and invades
    e. mast cells in tissue release histamine

*Fill in the Blanks* [pp.586–587]

Four defining features characterize the (13) _____ immune system of all jawed vertebrates. These four

characteristics are: (14) _____, (15) _____, (16) _____, and (17)_____.

Self vs. nonself recognition starts with (18) _____ patterns that give each cell or virus its identity.

The cells of the immune system can recognize these patterns as foreign. Specificity means that each B or T

lymphocyte binds to only (19) _____ antigen.

Diversity refers to the billions of different antigen (20)_____. Memory refers to the ability of the

immune system to (21) _____ an antigen.

*Matching*

Match each of the following to its correct statement. [pp.586–589]

_____ 22. antigen-presenting cells

_____ 23. B cells

_____ 24. helper T cell

_____ 25. MHC markers

_____ 26. cytotoxic T cells

_____ 27. antibodies

a. induces T and B cells to start division and differentiation
b. are involved in the cell-mediated responses
c. specific recognition proteins on the plasma membrane
d. bind pieces of antigen to MHC markers, forming emergency flags
e. are involved in the antibody-mediated responses
f. a molecule that can bind to specific antigens

*Labeling*

Label the following parts of the antibody molecule:

28. antigen binding site _____

29. constant region of the heavy chain _____

30. variable region of the heavy chain _____

31. variable region of the light chain _____

32. constant region of the light chain _____

*Matching*

Match the class of antibody with its major function or characteristic:

a. IgD    b. IgA    c. IgE    d. IgM    e. IgG

33. _____ abundant in mucous and exocrine secretions

34. _____ signals the release of histamines and cytokines from the cell to which it is attached

35. _____ main antibody in the blood; activates complete, neutralizes toxins, etc.

36. _____ secreted as a pentamer, also functions as a B cell receptor

37. _____ always membrane bound, functions as B cell receptor

## 35.6. THE CELL-MEDIATED RESPONSE [pp.590–591]

## 35.7. REMEMBER TO FLOSS [p.591]

### Selected Terms

"Natural killers" [p.591], *Porphyromonas gingivalis* [p.591]

### Boldfaced Terms

[p.586] plaque _____

_____

### Labeling

Label each of the indicated molecules or cells in the diagram below. [p.590]

1. _____
2. _____
3. _____
4. _____
5. _____
6. _____
7. _____

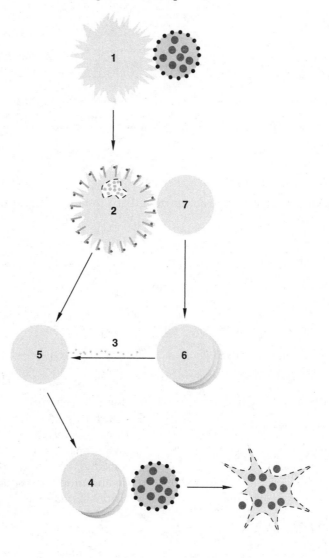

## 35.8.  DEFENSES ENHANCED OR COMPROMISED [pp.592–593]

## 35.9.  AIDS REVISITED—IMMUNITY LOST [pp.594–595]

### Selected Terms

Active immunization [p.592], passive immunization [p.592], primary immunodeficiency [p.593], secondary immunodeficiency [p.593], *Pneumocystis carinii* [p.594]

### Boldfaced Terms

[p.592] immunization _____

_____

[p.592] vaccine _____

_____

[p.592] allergen _____

_____

[p.592] allergy _____

_____

[p.593] autoimmune response _____

_____

### Choice

List three vaccines that should be administered at 2 months of age: [p.592]

1.  _____

2.  _____

3.  _____

### Matching

Match each of the terms with its correct definition. [pp.592–593]

_____  4. passive immunization

_____  5. anaphylactic shock

_____  6. asthma and hay fever

_____  7. multiple sclerosis

_____  8. AIDS

_____  9. active immunization

_____ 10. rheumatoid arthritis

_____ 11. allergen

_____ 12. Grave's disease

a. a life-threatening response to an allergen
b. an autoimmune response of the thyroid gland
c. antibodies that fight infection but do not activate the immune system
d. a harmless substance that invokes inflammation
e. forms of allergies
f. a vaccine is used to immunize the individual
g. a secondary immune deficiency
h. an autoimmune response in the nervous system
i. chronic inflammation of the joints and heart

## Labeling

Label the diagram in Figure 35.3 of the virus that causes AIDS. [p.595]

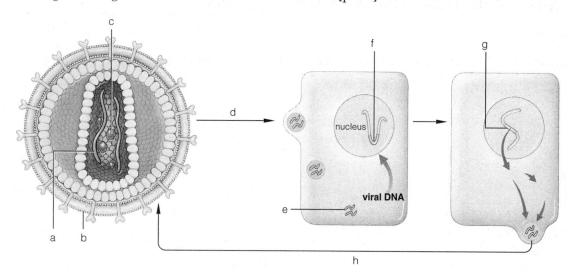

13. _____ viral DNA integrated into host DNA

14. _____ viral RNA enters a host cell

15. _____ viral DNA is transcribed

16. _____ lipid envelop of virus with proteins

17. _____ viral DNA forms by reverse transcription

18. _____ viral RNA

19. _____ viral enzyme reverse transcriptase

20. _____ new virus particles produced

# Self-Quiz

1. Which of the following is not an innate mechanism? [p.584]
   a. macrophages
   b. B lymphocytes
   c. neutrophils
   d. eosinophils
   e. all of the above are nonspecific

2. Which of the following is *not* associated with the immune system of vertebrates [p.586]?
   a. diversity
   b. memory
   c. self vs. nonself recognition
   d. migration

3. Which of the following is not a physical defense mechanism? [p.583]
   a. lysozyme in saliva
   b. pH of urinary tract and stomach
   c. complement proteins
   d. keratin-packed epithelial cells

4. _____ are divided into two groups: T cells and B cells. [p.582]
   a. macrophages
   b. lymphocytes
   c. platelets
   d. complement cells
   e. cancer cells

5. _____ immunoglobulins are secreted by B cells and comprise the main antibody in the blood. [p.588]
   a. IgA
   b. IgE
   c. IgM
   d. IgG

6. The markers for every cell in the human body are referred to by the letters _____. [p.586]
   a. HIV
   b. MBC
   c. RNA
   d. DNA
   e. MHC

7. HIV primarily infects _____. [p.594]
   a. neutrophils
   b. eosinophils
   c. macrophages, dendritic cells, and T helper cells
   d. mast cells

8. Histamine is associated with which of the following? [p.585]
   a. antibody production
   b. activation of helper T cells
   c. inflammatory response
   d. immunological memory

9. Which of the following is not an autoimmune disease? [p.593]
   a. hay fever
   b. Grave's disease
   c. rheumatoid arthritis
   d. multiple sclerosis

10. A virus-infected cell of the body would be targeted for destruction by _____ cells. [p.590]
    a. cytotoxic T
    b. macrophages
    c. eosinophils
    d. B

## Chapter Objectives/Review Questions

This section lists general and detailed chapter objectives that can be used as review questions. You can make maximum use of these items by writing answers on a separate sheet of paper. Fill in answers where blanks are provided. To check for accuracy, compare your answers with information given in the chapter or glossary.

1. Distinguish between the major cells of the immune system. [p.582]
2. Identify the major physical barriers to infection in the body. [p.583]
3. Outline the process of the inflammatory response. [pp.584–585]
4. Understand the role of antimicrobial proteins and fever in the immune response. [p.585]
5. Recognize the cells associated with adaptive immunity. [pp.586–587]
6. Understand the role of lymphocytes in the immune response. [p.586]
7. Understand the concept of memory. [p.586]
8. List the function of each class of immunoglobulins. [p.588]
9. List the steps in a cell-mediated response. [pp.584–585]
10. Understand the major processes in the study of immunotherapy. [p.590]
11. Distinguish between passive and active immunity. [p.592]
12. Understand the major autoimmune disorders. [p.593]
13. Recognize the significance of AIDS. [pp.594–595]

## Chapter Summary

The vertebrate body has three lines of defense against invading microorganisms. These are (1) _____, _____, and _____. The (2) _____ immune response rids the body of most invaders before infection becomes established. The (3) _____ immune response targets specific pathogens and cancer cells. (4) _____, _____, and secretions at the body's surfaces function as barriers that exclude most microbes. (5) _____ immunity involves a set of general, immediate defenses. (6) _____white blood cells, which engulf pathogens, are major components of the innate immune system. (7) B lymphocytes produce _____, which bind specifically to pathogens. (8) _____ lymphocytes directly destroy diseased body cells in a cell meditated response. The distinction between (9) _____ and nonself is critical in an appropriate immune response. (10) _____ and _____ can develop if the immune system responds inappropriately.

## Integrating and Applying Key Concepts

1. Why would you not necessarily choose to treat a fever of 101.3 degrees F?

# 36

# RESPIRATION

## INTRODUCTION

The respiratory system plays a number of important roles in the human body. Not only does it deliver oxygen to, and remove carbon dioxide from, our tissues, but it also plays an important role in maintaining internal homeostasis. This chapter explores the principles of a respiratory system and the major types of respiratory systems in both vertebrates and invertebrates.

## FOCAL POINTS

• Figure 36.9 [p.604] diagrams the major structures of the human respiratory system.

## Interactive Exercises

*Up in Smoke* [pp.598–599]

### 36.1. THE NATURE OF RESPIRATION [p.600]

### 36.2. INVERTEBRATE RESPIRATION [p.601]

*Selected Words*

*Internal* respiratory surface [p.601]

*Boldfaced Terms*

[p.600] respiration _____

_____

[p.600] respiratory surface _____

_____

[p.600] respiratory proteins _____

_____

[p.600] hemoglobin _____

_____

[p.600] myoglobin _____

_____

[p.601] integumentary exchange _____

_____

[p.601] gills _____

_____

[p.601] tracheal system _____

_____

## Choice

For each of the following, choose the most appropriate factors affecting diffusion rates from the list below. [p.600]

      a. respiratory proteins     b. ventilation     c. surface-to-volume ratio

_____ 1. this determines the overall size of the organism

_____ 2. bind and release oxygen

_____ 3. animals without respiratory organs are usually small and flat

_____ 4. hemoglobin and myoglobin are examples

_____ 5. the movement of air or water across a respiratory surface

## Matching

Match each of the following terms to the correct statement. [pp.600–601]

_____ 6. integumentary exchange

_____ 7. tracheal system

_____ 8. hemoglobin

_____ 9. gills

_____ 10. myoglobin

_____ 11. book lungs

_____ 12. hemocyanin

a. a copper-based respiratory pigment of invertebrates
b. gas diffusion directly across the body surface
c. a series of branching tubes inside invertebrates such as insects and spiders
d. an iron-based respiratory pigment that has a higher affinity for oxygen than hemoglobin
e. the primary respiratory surface of aquatic invertebrates
f. the main respiratory pigment of vertebrates
g. thin sheets of tissue that are the respiratory surfaces of spiders

## 36.3. VERTEBRATE RESPIRATION [pp.602–603]

## 36.4. HUMAN RESPIRATORY SYSTEM [pp.604–605]

### Selected Words

*External* gills [p.602], *internal* gills [p.602], "Adam's apple" [p.605], laryngitis [p.605], "bronchial tree" [p.605], *pulmo* [p.605]

## Boldfaced Terms

[p.600] countercurrent exchange _____

_____

[p.600] lung _____

_____

[p.605] pharynx _____

_____

[p.605] larynx _____

_____

[p.605] glottis _____

_____

[p.605] epiglottis _____

_____

[p.605] trachea _____

_____

[p.605] bronchus _____

_____

[p.605] bronchioles _____

_____

[p.605] alveoli _____

_____

[p.605] diaphragm _____

_____

[p.605] intercostal muscles _____

_____

## Choice

For each of the following statements, choose the correct group of organisms to which the statement applies. Some questions may have more than one answer. [pp.602–603]

        a. fish     b. birds     c. amphibians     d. mammals

_____ 1. Larvae of this group have gills, but most adults have lungs.

_____ 2. Use a countercurrent exchange system.

_____ 3. Air in these organisms continuously flows over the respiratory surface.

_____ 4. These organisms "gulp" air into the lungs.

_____ 5. Use paired lungs.

## Labeling

First, identify each of the indicated structures in the human respiratory system. Then, choose the correct function for each structure from the list provided. [pp.604–605]

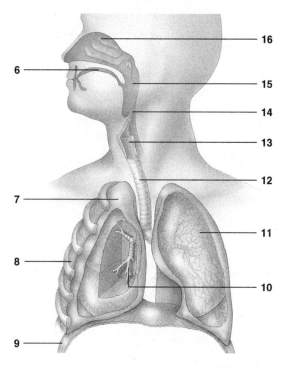

6. _____ ( )

7. _____ ( )

8. _____ ( )

9. _____ ( )

10. _____ ( )

11. _____ ( )

12. _____ ( )

13. _____ ( )

14. _____ ( )

15. _____ ( )

16. _____ ( )

a. a muscle located between the chest and abdominal cavities
b. supplemental airway
c. airway connecting the nasal cavity and larynx
d. connects larynx with bronchi in lungs
e. warms, moistens, and filters the incoming air
f. skeletal muscles with roles in breathing
g. branched structures of the lungs that end at the alveoli
h. lobed organ that enhances gas exchange between internal and external environments
i. closes off the larynx during swallowing
j. double-layer membrane that separates the lungs from other organs
k. voice box

*TNOW:* For an interactive exercise of this material, refer to Chapter 36: Art Labeling: Human Respiratory System.

# 36.5. GAS EXCHANGE AND TRANSPORT [pp.606–607]

# 36.3. CYCLIC REVERSALS IN AIR PRESSURE GRADIENTS [pp.608–609]

## Selected Words

Hemoglobin [p.606], myoglobin [p.606], carbaminohemoglobin [p.606], carbon monoxide [p.607], inhalation [p.608], exhalation [p.608]

## Boldfaced Terms

[p.606] respiratory membrane _____

_____

[p.606] heme group _____

_____

[p.606] oxyhemoglobin _____

_____

[p.607] carbonic anhydrase _____

_____

[p.608] respiratory cycle _____

_____

[p.608] vital capacity _____

_____

[p.608] tidal volume _____

_____

## Choice

Choose whether each statement is associated with oxygen or carbon dioxide transport. [pp.606–607]

      a. carbon dioxide transport    b. oxygen transport

_____ 1. Transport is enhanced by hemoglobin molecules.

_____ 2. The majority is transported as bicarbonate.

_____ 3. The transported gas is released when the blood is warm and at a low pH.

_____ 4. When bound, oxyhemoglobin is formed.

_____ 5. Carbaminohemoglobin is responsible for 30 percent of the transport.

_____ 6. The enzyme carbonic anhydrase removes it from the red blood cell.

## Labeling

For each numbered item in the diagram below, provide the missing term or terms. [p.609]

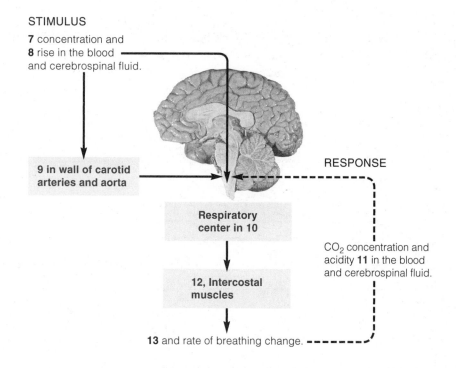

STIMULUS

**7** concentration and
**8** rise in the blood
and cerebrospinal fluid.

**9 in wall of carotid
arteries and aorta**

RESPONSE

**Respiratory
center in 10**

$CO_2$ concentration and
acidity **11** in the blood
and cerebrospinal fluid.

**12, Intercostal
muscles**

**13** and rate of breathing change.

7. _____

8. _____

9. _____

10. _____

11. _____

12. _____

13. _____

## 36.7. RESPIRATORY DISEASES AND DISORDERS [pp.610–611]

## 36.8. HIGH CLIMBERS AND DEEP DIVERS [pp.612–613]

### Selected Words

Apnea [p.610], sudden infant death syndrome (SIDS) [p.610], tuberculosis [p.610], pneumonia [p.610], bronchitis [p.610], emphysema [p.610], "tar" [p.611], "thin air" [p.612], *hypoxia* [p.611], nitrogen narcosis [p.612], "raptures of the deep" [p.612], "the bends" [p.612]

### Boldfaced Terms

[p.612] acclimatization _____

_____

[p.612] erythropoietin _____

_____

## Matching

Match each of the following terms to its correct definition. [pp.610–612]

_____ 1. apnea

_____ 2. emphysema

_____ 3. tuberculosis

_____ 4. bronchitis

_____ 5. smoking

_____ 6. hypoxia

_____ 7. nitrogen narcosis

_____ 8. SIDS

_____ 9. pneumonia

a. chronic infection of the epithelium lining of the bronchioles
b. destruction of the alveoli
c. may be caused by defects in the respiratory control centers
d. a stoppage in the pattern of breathing
e. caused by a bacterial infection that causes problem breathing and bloody mucus
f. cellular oxygen deficiency
g. also know as "the bends"
h. one of the leading causes of respiratory-related deaths
i. lung inflammation fills tissues with fluid

---

# Self-Quiz

1. The primary respiratory pigment of vertebrate cardiac muscle is _____. [p.600]
   a. hemocyanin
   b. hemoglobin
   c. hemerythrin
   d. myoglobin

2. Which of the following is not involved in carbon dioxide transport? [pp.606–607]
   a. carbonic anhydrase
   b. carbaminohemoglobin
   c. bicarbonate
   d. oxyhemoglobin

3. If someone "forgets" to breathe, it is called _____. [p.610]
   a. emphysema
   b. hyperventilation
   c. bronchitis
   d. hypoxia
   e. apnea

4. Which of the following is used by invertebrates as a respiratory system? [p.601]
   a. tracheal system
   b. gills
   c. integumentary exchange
   d. "breathing tubes"
   e. all of the above are used

5. The atmosphere contains _____ percent oxygen. [p.600]
   a. 0.04
   b. 21
   c. 78
   d. has not been determined

6. A countercurrent exchange system is found in the respiratory system of _____. [p.601]
   a. fish
   b. amphibians
   c. reptiles
   d. birds
   e. mammals

7. Under normal conditions, exhalation is a _____ process. [p.608]
   a. active
   b. passive

8. Which structure serves as a passageway for both the respiratory and digestive systems? [pp.604–605]
   a. bronchial tubes
   b. esophagus
   c. nasal cavity
   d. pharynx
   e. larynx

9. This structure separates the abdominal and thoracic cavities. [pp.604–605]
   a. pharynx
   b. diaphragm
   c. intercostals muscles
   d. rib cage
   e. esophagus

10. The majority of gas exchange in humans occurs in the _____. [p.605]
    a. tracheal system
    b. alveoli
    c. nasal cavity
    d. oral cavity

# Chapter Objectives/Review Questions

1. Understand how the human respiratory system is related to the circulatory system, to cellular respiration, and to the nervous system. [p.599]
2. Understand the factors that affect diffusion rates. [p.600]
3. Understand the role of respiratory pigments in gas exchange and know the major human respiratory pigments. [p.600]
4. Recognize the major respiratory systems of invertebrates. [p.601]
5. List the forms of respiratory structures found in vertebrates. [pp.602–603]
6. List all the principal parts of the human respiratory system and explain how each structure contributes to transporting oxygen from the external world to the bloodstream. [pp.604–605]
7. List the molecules and factors associated with oxygen and carbon dioxide transport. [pp.606–607]
8. Explain the sequence of events in the process of inhalation and exhalation. [pp.608–609]
11. Identify the causes of the major respiratory disorders. [pp.610–611]

# Chapter Summary

Aerobic respiration uses free (1) _____ and its (2) _____ wastes are removed before the internal environment's (3) _____ shifts dangerously. (4) _____ is the sum of the processes that move oxygen from the outside environment to all metabolically active (5) _____ and move carbon dioxide from those tissues to the outside.

Gas exchange occurs across the body surface or (6) _____ of aquatic invertebrates. In large invertebrates on land, it occurs across a moist, (7) _____ respiratory surface or at fluid-filled tips of branching (8) _____ that extend from the surface to internal tissues.

(9) _____, skin, and paired (10) _____ are gas exchange organs. Breathing ventilates (11) _____. (12) _____ is a transport medium. It picks up oxygen and gives up carbon dioxide at the (13) _____ surface. It also exchanges gases with the (14) _____ fluid that bathes cells.

(15) _____ exchange is not efficient when the rates of air flow and blood flow at the respiratory surface match. (16) _____ centers adjust the rate and (17) _____ of breathing.

Respiration can be disrupted by damage to respiratory centers in the (18) _____, physical obstructions, (19) _____, and inhalation of (20) _____, including (21) _____ smoke.

At high (22) _____, the human body makes short-term and long-term adjustments to the thinner air. Built-in respiratory mechanisms and specialized behaviors allow sea turtles and diving (23) _____ mammals to stay under water, at great depths, for long periods.

# Integrating and Applying Key Concepts

1. Consider animals such as the amphibians that have aquatic larval forms (tadpoles) and terrestrial adults. Outline the respiratory changes that you think might occur as an aquatic tadpole metamorphoses into a land-going juvenile.
2. In the movie *Waterworld*, Kevin Costner's character develops gills to aid his underwater breathing. Explain what changes would be necessary in the human respiratory system in order to make this possible.
3. Blood doping, a process by which athletes attempt to increase their total red blood cell count, has become a major issue in professional athletics. Explain what benefit an increased red blood cell count might be for an athlete.

# 37

# DIGESTION AND HUMAN NUTRITION

## INTRODUCTION

How true is the saying "You are what you eat"? This chapter allows the student to explore this question by describing, in detail, the structures and functions associated with the human digestive system. Following an in-depth discussion of the digestive processes, the chapter describes some diseases that are associated with various parts of the digestive system. Finally, the chapter discusses the human diet and the requirements of a healthy diet. Attention is given to the food groups, vitamins, and minerals needed to maintain a healthy diet. Unfortunately, even with this knowledge readily available to almost everyone, the majority of Americans are obese. Thus, the chapter concludes with a discussion of obesity and the many lifestyle choices and genetic components that contribute to this disease. As always, this knowledge is presented to afford the student the knowledge with which to better his or her own health.

## FOCAL POINTS

- Figure 37.3 [p.618] displays all the organ systems important in digestion.
- Figure 37.6 [p.620] presents an overview of the human digestive tract.
- Table 37.1 [p.622] summarizes the chemicals used in digestion.
- Figure 37.10 [p.624] illustrates the structure of the lining or the intestinal mucosa and the villi at the surface.
- Figure 37.11 [p.625] animates the digestion and absorption that occurs in the small intestine.
- Figure 37.13 [p.627] illustrates the major pathways of organic metabolism.
- Figure 37.14 [p.628] summarizes the recommended nutritional guidelines.
- Tables 37.3 and 37.4 [pp.630–631] categorize the vitamins and minerals essential to a healthy diet.
- Table 37.5 [p.634] summarizes the human digestive system.

---

## Interactive Exercises

*Hominoids, Hips, and Hunger* [pp.616–617]

## 37.1. THE NATURE OF DIGESTIVE SYSTEMS [pp.618–619]

## 37.2. OVERVIEW OF THE HUMAN DIGESTIVE SYSTEM [pp.620–621]

*Selected Terms*

Obesity [p.616], leptin [p.616], ghrelin [p.616], "stomach stapling" [p.616], cholecystokin [p.617], mechanical processing and motility [p.618], secretion [p.618], digestion [p.618], absorption [p.619], elimination [p.619]

## Boldfaced Terms

[p.618] incomplete digestive system _____

_____

[p.618] complete digestive system _____

_____

[p.619] ruminants _____

_____

[p.621] gastrointestinal tract _____

_____

[p.621] mouth _____

_____

[p.621] pharynx _____

_____

[p.621] esophagus _____

_____

[p.621] stomach _____

_____

[p.621] sphincter _____

_____

[p.621] small intestine _____

_____

[p.621] colon _____

_____

[p.621] rectum _____

_____

[p.621] anus _____

_____

## Fill in the Blanks [p.618]

Recall from Chapter 23, that flatworms are among the simplest animals with (1) _____ systems. Their digestive system is (2) _____ as it is in a few other invertebrates. This means that food enters and (3) _____ leave through a single opening at the body's surface. In a flatworm, the sac-like, branching (4) _____ cavity opens at the start of a (5) _____, a muscular tube. Food enters the sac, is partially (6) _____, and circulates to cells even as wastes are sent out. This two-way traffic does not favor regional (7) _____.

Most species of animals have a (8) _____ digestive system. Basically, they have a tube with an opening at one end for taking in (9) _____ and an opening at the other end for eliminating unabsorbed residues and (10) _____. The tube is divided into (11) _____ regions for food breakdown, (12) _____, and waste concentration.

## Labeling

First, identify each numbered structure in the illustration. Then, match each structure with its function from the list provided. [p.662]

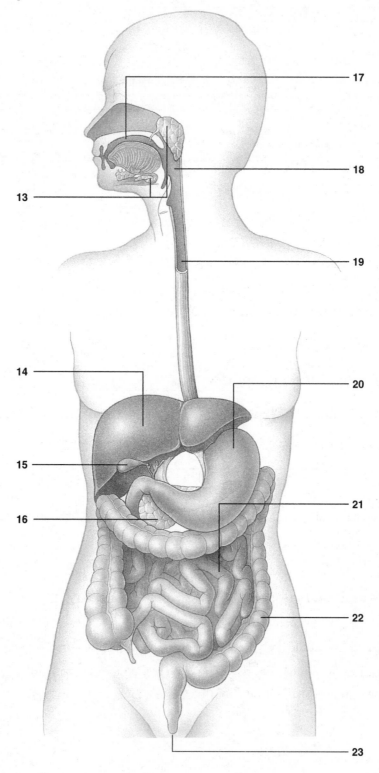

13. _____ ( )

14. _____ ( )

15. _____ ( )

16. _____ ( )

17. _____ ( )

18. _____ ( )

19. _____ ( )

20. _____ ( )

21. _____ ( )

22. _____ ( )

23. _____ ( )

a. secretes digestive enzymes, buffers, and insulin
b. tube that moves food from pharynx to stomach
c. stores and concentrates bile
d. secretes bile
e. entrance to system, polysaccharide digestion begins
f. major site of digestion and absorption
g. opening through which feces are expelled
h. entrance to respiratory system and tubular section of digestive system
i. muscular sac, protects against pathogens, protein digestion begins
j. secrete saliva, enzymes, buffers, and mucus
k. concentrates and stores undigested matter

## Matching

Match each of the following statements with the correct term. [p.620]

_____ 24. a ring of smooth muscles that regulate movement of food

_____ 25. a mixture of water, salivary amylase, and bicarbonate

_____ 26. the site where protein digestion begins

_____ 27. the entrance to the esophagus and the lungs

_____ 28. a muscular tube that moves food into the stomach

_____ 29. the site where carbohydrate digestion begins

_____ 30. the site where the digestion of most nutrients is completed

a. esophagus
b. small intestine
c. pharynx
d. mouth
e. stomach
f. sphincter
g. saliva

## 37.3. FOOD IN THE MOUTH [p.621]

## 37.4. FOOD BREAKDOWN IN THE STOMACH AND SMALL INTESTINE [pp.622–623]

## 37.5. ABSORPTION FROM THE SMALL INTESTINE [pp.624–625]

### Selected Terms

*Heliobacter pylori* [p.622]

### Boldfaced Terms

[p.621] tooth _____

_____

[p.622] mucosa _____

_____

[p.622] gastric fluid _____

_____

[p.622] chyme _____

_____

[p.623] bile _____

_____

[p.623] emulsification _____

_____

[p.623] segmentation _____

_____

[p.624]villi _____

_____

[p.624] brush border cells _____

_____

[p.624] microvilli _____

_____

[p.625] micelle formation _____

_____

## Complete the Table [p.622]

| Enzyme | Source | Where Active | Substrate |
|---|---|---|---|
| Salivary amylase | Salivary glands | Mouth stomach | (1) |
| Pancreatic amylase | Pancreas | (2) | Polysaccharides |
| (3) | Stomach lining | Stomach | Proteins |
| Trypsin and chymotrypsin | Pancreas | (4) | Proteins |
| Carboxypeptidase | (5) | Small intestine | Protein fragments |
| Aminopeptidase | (6) | Small intestine | Protein fragments |
| Lipase | Pancreas | Small intestine | (7) |
| Pancreatic nucleases | Pancreas | Small intestine | (8) |
| Intestinal nucleases | Intestinal lining | Small intestine | (9) |

## Matching

Match each of the following statements with the correct term. [pp.616–617]

_____ 10. This chemical causes the stomach to release acid.

_____ 11. The process by which bile salts break up fats for digestion

_____ 12. Secreted by the liver, this assists the process of fat digestion.

_____ 13. The three regions of the small intestine

_____ 14. A combination of hydrochloric acid, pepsinogens, and other compounds

_____ 15. The chemical that signals the pancreas to release digestive enzymes

_____ 16. An oscillating pattern of contractions in the small intestine

a. gastric fluid
b. gastrin
c. duodenum, jejunum, ileum
d. emulsification
e. bile
f. segmentation
g. CCK

## Labeling [p.625]

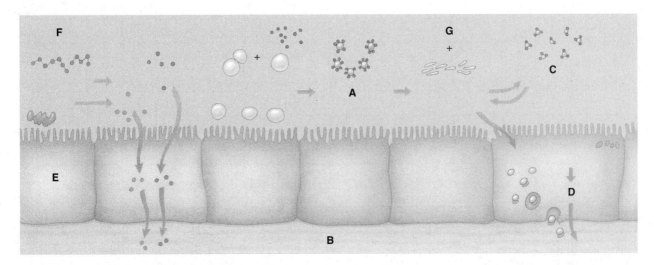

Please match the following structures and functions with the appropriate labels from the figure.

17. _____ interstitial fluid inside a villus

18. _____ micelles

19. _____ lumen of small intestine

20. _____ brush border cell

21. _____ lipoproteins

22. _____ emulsification of droplets

23. _____ bile salts and fat digestion products

## 37.6. THE LARGE INTESTINE [p.626]

## 37.7. WHAT HAPPENS TO ABSORBED ORGANIC COMPOUNDS? [p.627]

## 37.8. HUMAN NUTRITIONAL REQUIREMENTS [pp.628–629]

## 37.9. VITAMINS AND MINERALS [pp.630–631]

## 37.10. WEIGHTY QUESTIONS, TANTALIZING ANSWERS [pp.632–633]

### Selected Terms

*Escherichia coli* [p.626], trans-fatty acids [p.628], *complete* [p.629], *incomplete* [p.629], *Chenopium quinoa* [p.629], *Mediterranean diet* [p.629], *olive oil* [p.629], "low-carb" diets [p.629], "bad" cholesterol [p.629], *organic* [p.630], *inorganic* [p.630], *body mass index* [p.632], "beer belly" [p.632], *PYY* [p.633]

### Boldfaced Terms

[p.626] feces _____

_____

[p.626] appendix _____

_____

[p.627] organic metabolism _____

_____

[p.628] glycemic index (GI) _____

_____

[p.628] essential fatty acids _____

_____

[p.629] essential amino acids _____

_____

[p.630] vitamins _____

_____

[p.630] minerals _____

_____

[p.632] obesity _____

_____

### True–False

If the statement is true, write "T" in the blank. If the statement is false, correct it by changing the underlined word(s) and writing the correct word(s) in the answer blank.

_____ 1. The appendix has no known <u>digestive</u> functions. [p.626]

_____ 2. Water and <u>sodium ions</u> are absorbed into the bloodstream from the lumen of the large intestine. [p.621]

_____ 3. The risk of colon cancer increases with a <u>high</u>-fiber diet. [p.621]

_____ 4. Bacteria in the large intestine are <u>beneficial</u>. [p.621]

*Fill in the Blanks* [p.628]

List the recommended number of servings of each of the following foods:

5. dark green vegetables_____

6. orange vegetables_____

7. legumes _____

8. starchy vegetables _____

9. fruits_____

10. milk products_____

11. whole grains _____

12. fish, poultry, lean meat _____

*Complete the Table* [pp.630–631]

| Vitamin | Sources | Main Functions |
|---|---|---|
| **Fat-Soluble Vitamins** | | |
| A | 13. | Synthesis of visual pigments, bone, teeth |
| D | 14. | Promotes bone growth and mineralization; enhances calcium absorption |
| E | Whole grains, dark green vegetables, vegetable oils | 15. |
| 16. | Enterobacteria | Blood clotting, ATP formation |
| **Water-Soluble Vitamins** | | |
| Thiamin | Whole grains, legumes, lean meats | 17. |
| Riboflavin | Whole grains, poultry, fish | 18. |
| 19. | Green leafy vegetables, potatoes, peanuts, poultry | Coenzyme action |
| $B_6$ | 20. | Coenzyme in amino acid metabolism |
| Pantothenic acid | 21. | Coenzyme in glucose metabolism |
| 22. | Dark green vegetables, whole grains, lean meats | Coenzyme in nucleic acid and amino acid metabolism |
| $B_{12}$ | 23. | Coenzyme in nucleic acid metabolism |
| Biotin | Legumes, egg yolk | 24. |
| C | 25. | Collagen synthesis, structural role in bone, cartilage, teeth, used in carbohydrate metabolism |

## Matching

Match each of the following terms to the correct definition. [pp.626–627]

_____ 26. body mass index

_____ 27. obesity

_____ 28. leptin

_____ 29. ob gene

_____ 30. ghrelin

_____ 31. CCK

a. a measurement that is designed to determine the health risk of weight
b. the hormone encoded by the mouse ob gene
c. a hormone of the small intestine that makes you feel full
d. indicated by a BMI score >30
e. a hormone secreted by the stomach lining that makes you feel hungry
f. first discovered in mice; variations in this gene lead to an increase in weight

# Self-Quiz

1. The name of the enzyme(s) produced by the pancreas for protein digestion in the small intestine is _____. [p.622]
   a. salivary amylase
   b. lipase
   c. pancreatic nucleases
   d. trypsin and chymotrypsin
   e. aminopeptidase

2. The name of the hormone that signals the stomach to release acid is _____. [p.622]
   a. gastrin
   b. CCK
   c. leptin
   d. ghrelin
   e. insulin

3. This organ produces bile in the body. [p.623]
   a. liver
   b. gall bladder
   c. pancreas
   d. small intestine
   e. stomach

4. The _____ contains pepsinogens and hydrochloric acid. [p.622]
   a. saliva
   b. mucus
   c. gastric fluid
   d. feces

5. Which of the following is not truly part of the digestive system? [p.626]
   a. esophagus
   b. colon
   c. appendix
   d. jejunum, ileum, and duodenum
   e. oral cavity

6. Which of the following should be represented least in a healthy diet? [p.628]
   a. vegetables
   b. fruits
   c. grain products
   d. milk and milk products
   e. fats and refined sugars

7. Which of the following is not a fat-soluble vitamin? [p.630]
   a. D
   b. C
   c. E
   d. K
   e. A

8. A complete digestive system has both a mouth and an anus. [p.618]
   a. true
   b. false

9. The majority of digestion and absorption occurs in the _____. [pp.622–623]
   a. stomach
   b. liver
   c. small intestine
   d. large intestine
   e. pancreas

10. Which of the following is responsible for emulsifying fats? [p.625]
    a. CCK
    b. gastrin
    c. leptin
    d. bile
    e. saliva

# Chapter Objectives/Review Questions

This section lists general and detailed chapter objectives that can be used as review questions. You can make maximum use of these items by writing answers on a separate sheet of paper. Fill in answers where blanks are provided. To check for accuracy, compare your answers with information given in the chapter or glossary.

1. List the five tasks that a digestive system needs to perform for an organism. [p.618]
2. Distinguish between incomplete and complete digestive systems, and tell which is characterized by (a) specialized regions, and (b) two-way traffic. [p.618]
3. List all parts (in order) of the human digestive system through which food actually passes. Then, list the auxiliary organs that contribute one or more substances to the digestive process. [p.620]
4. List the enzyme(s) that act in (a) the oral cavity, (b) the stomach, and (c) the small intestine. Then, tell where each enzyme was originally made. [p.622]
5. Describe how the digestion and absorption of fats differ from the digestion and absorption of carbohydrates and proteins. [p.625]
6. Understand the role of gastrin. [p.622]
7. List the items that leave the digestive system and enter the circulatory system during the process of absorption. [p.625]
8. Describe the cross-sectional structure of the small intestine, and explain how its structure is related to its function. [p.624]
9. Explain how the human body manages to meet the energy and nutritional needs of the various body parts even though the person may be feasting sometimes and fasting at other times. [p.627]
10. Describe events that occur in the human colon. Describe three disorders of the large intestine. [p.626]
11. List the numerical range of servings permitted from each group, and also list some of the choices available. [p.628]
12. Define an essential fatty acid and an essential amino acid. [pp.628–629]
13. Know the difference between the fat-soluble and water-soluble vitamins. [p.630]
14. Understand the role of key minerals in human metabolism. [p.631]
15. Understand what the BMI represents. [p.632]
16. Recognize new research trends into understanding the genetics of obesity. [p.633]

# Chapter Summary

The digestive systems of some animals are (1) _____-like, without a separate entrance and exit, while the digestive systems of most animals consist of a (2) _____ that extends throughout the body. In complex animals, the digestive system interacts with other (3) _____ _____ to distribute (4) _____ and water, to dispose of (5) _____, and to maintain the internal environment.

In humans, digestion starts in the (6) _____, continues in the (7) _____, and is finished in the small intestine. Secretions from (8) _____ glands, the (9) _____ and the (10) _____ function in digestion. Most nutrients are absorbed in the (11) _____. The large intestine absorbs (12) _____ and concentrates and stores (13) _____. A healthy diet provides all (14) _____, (15) _____, and minerals necessary to support metabolism. Maintaining body weight requires balancing (16) _____ taken in with (17) _____ burned in metabolism and physical activity.

# Integrating and Applying Key Concepts

1. Suppose that you were a chemist for a pharmaceutical company. You have an interest in developing a drug to fight a specific disease. This drug needs to be activated by the enzymes of the stomach, and then absorbed through the stomach lining. What nutrient classes would you pattern your new drug on? What if you wanted it absorbed in the small intestine?

2. For one week keep a food diary of *everything* you eat. Calculate the amount of calories, fat, protein and fiber for each day. Also calculate the number of servings of vegetables (dark green, yellow, legumes, other) fruits, milk products, whole grains, and meats. Compare your food intake with the suggested guidelines and evaluate your personal diet.

# 38

# THE INTERNAL ENVIRONMENT

## INTRODUCTION

The previous chapter discussed what happens to the food that enters the vertebrate body. The process of digestion was discussed and the student learned how the body uses up the food it takes in to conduct various biological processes. As with any process, however, there is often waste produced. This chapter deals with how one form of the body's waste, urine, is produced in the kidneys and excreted from the body. This process is critical for the body to be able to maintain a balanced environment in which cells can survive and function. Also critical to the balanced environment inside the vertebrate body is temperature control. The mechanisms by which different organisms maintain their body temperature are discussed, with particular detail regarding the warm-blooded mammals.

## FOCAL POINTS

- Figure 38.3 [p.638] illustrates the integration of several body systems that work together to maintain homeostasis.
- Figure 38.6 [p.640] diagrams the human urinary system.
- Figure 38.8 [p.642] illustrates a nephron and the associated blood capillaries.
- Table 38.3 [p.648] summarizes the processes of urine formation.
- Table 38.4 [p.648] summarizes the hormonal effects on reabsorption.

---

## Interactive Exercises

*Truth in a Test Tube* [pp.636–637]

## 38.1. GAINS AND LOSSES IN WATER AND SOLUTES [pp.638–639]

## 38.2. DESERT RATS [p.639]

## 38.3. STRUCTURE OF THE URINARY SYSTEM [pp.640–641]

*Selected Terms*

*Diabetes mellitus* [p.630], *Dipodomys deserti* [p.639], "metabolic water" [p.633], renal [p.640]

*Boldfaced Terms*

[p.638] homeostasis _____

---

[p.638] urinary system _____

_____

[p.638] kidneys _____

_____

[p.638] urine _____

_____

[p.641] nephrons _____

_____

[p.641] loop of Henle _____

_____

[p.641] distal tubule _____

_____

[p.641] glomerulus _____

_____

[p.641] peritubular capillaries _____

_____

## Labeling

Below are two fishes. Label which one is a bony fish and which one is a marine fish, based on the flow of water.

1. A is a _____ fish.

2. B is a _____ fish.

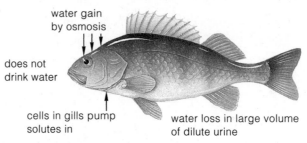

water gain
by osmosis

does not
drink water

cells in gills pump
solutes in

water loss in large volume
of dilute urine

A

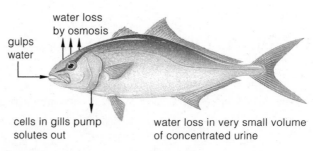

water loss
by osmosis

gulps
water

cells in gills pump
solutes out

water loss in very small volume
of concentrated urine

B

## Matching [pp.638–639]

Match the following terms with their best definition.

3. _____ homeostasis
4. _____ extracellular fluid
5. _____ urinary system
6. _____ urine
7. _____ kidneys
8. _____ ammonia
9. _____ urea
10. _____ intracellular fluid

a. the majority of the fluid in an animal; consists mainly of interstitial fluid
b. fluid inside the body cells
c. a pair of organs that filter blood
d. poison product of protein metabolism
e. most abundant metabolic waste product in human urine
f. the process that keeps the volume and composition of extracellular fluid within ranges that living cells can tolerate.
g. the collection of excess water and solutes in kidneys
h. present in most vertebrates to filter water and solutes from the blood

## Labeling [p.638]

Label the diagram with the correct letters, corresponding to the following systems:

11. _____ urinary system
12. _____ digestive system
13. _____ circulatory system
14. _____ respiratory system

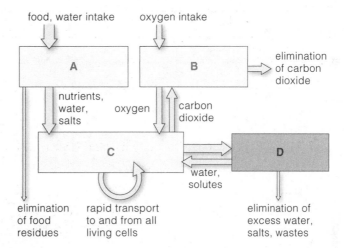

## Labeling
Label each part of the diagram below. [p.640]

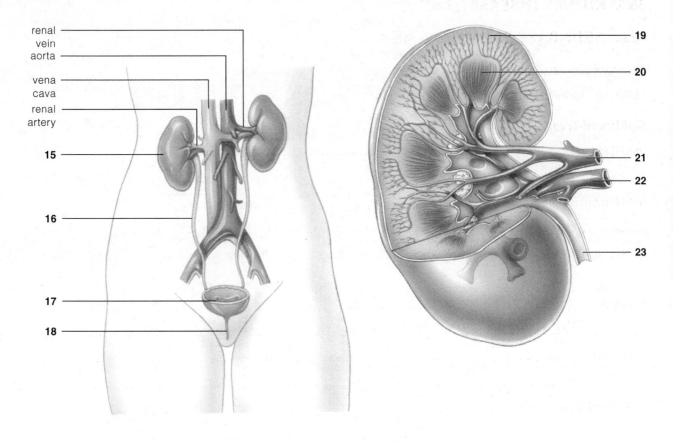

renal vein
aorta
vena cava
renal artery

15
16
17
18

19
20
21
22
23

15. _____

16. _____

17. _____

18. _____

19. _____

20. _____

21. _____

22. _____

23. _____

## 38.4. URINE FORMATION [pp.642–643]

## 38.5. KIDNEY DISEASE [p.644]

## 38.6. ACID-BASE BALANCE [p.644]

### Selected Terms

"Dialysis" [p.644], *hemodialysis* [p.644], *peritoneal dialysis* [p.644], *metabolic acidosis* [p.644]

### Boldfaced Terms

[p.642] glomerular filtration _____

_____

[p.642] tubular reabsorption _____

_____

[p.642] tubular secretion _____

_____

[p.643] ADH _____

_____

[p.643] aldosterone _____

_____

[p.644] acid-base balance _____

_____

[p.644] buffer system _____

_____

## Labeling

Choose the location in the nephron where each statement occurs. Answers may be used more than once. [p.642]

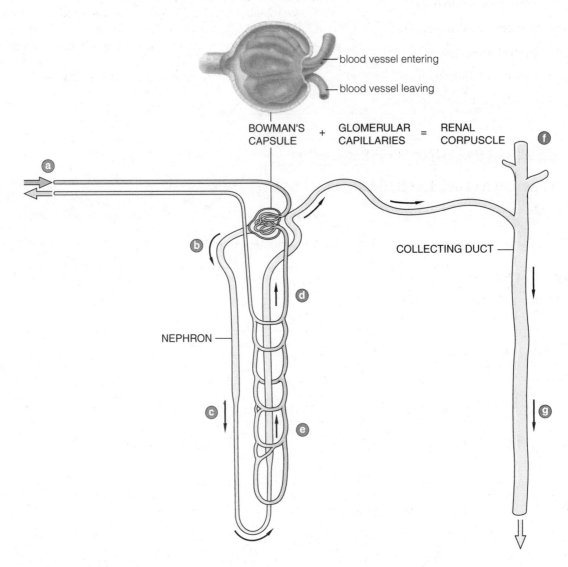

_____ 1. Tubular reabsorption

_____ 2. Tubular secretion

_____ 3. ADH promotes water reabsorption.

_____ 4. Filtration

_____ 5. Aldosterone increases sodium reabsorption.

_____ 6. Cells pumps chloride ions out of the loop of Henle.

_____ 7. Blood pumped from the heart travels to the renal artery, then into the kidneys.

_____ 8. Urinary excretion

## Matching

Match each of the following terms to the correct description.

_____ 9. uremic toxicity [p.638]

_____ 10. kidney stones [p.644]

_____ 11. diabetes mellitus [p.644]

_____ 12. hemodialysis [p.644]

_____ 13. peritoneal dialysis [p.644]

_____ 14. metabolic acidosis [p.644]

a. An external machine cleans the blood of toxins and wastes

b. Deposits of uric acid and calcium block the urethra or ureter

c. Waste metabolic products from proteins build in the blood

d. One of the major causes of kidney failure

e. A person's abdominal cavity is used to filter waste and toxins from the blood.

f. Kidneys can't secrete enough H+ ions to balance metabolism.


## 38.7. HEAT GAINS AND LOSSES [p.645]

## 38.8. TEMPERATURE REGULATION IN MAMMALS [pp.646–647]

### Selected Terms

Panting [p.646], *Ursus maritimus* [p.647], "brown" adipose tissue [p.647]

### Boldfaced Terms

[p.645] thermal radiation _____

_____

[p.645] conduction _____

_____

[p.645] convection _____

_____

[p.645] evaporation _____

_____

[p.645] ectotherms _____

_____

[p.645] endotherms _____

_____

[p.645] heterotherms _____

_____

[p.646] peripheral vasodilation _____

_____

[p.646] evaporative heat loss _____

_____

[p.646] fever _____

_____

[p.647] peripheral vasoconstriction _____

_____

[p.647] pilomotor response _____

_____

[p.647] shivering response _____

_____

[p.647] nonshivering heat production _____

_____

## Matching

Match each of the following terms to the correct statement.

_____ 1. convection [p.645]

_____ 2. heterotherms [p.645]

_____ 3. thermal radiation [p.645]

_____ 4. evaporation [p.645]

_____ 5. evaporative heat loss [p.646]

_____ 6. pilomotor response [p.647]

_____ 7. peripheral vasodilation [p.647]

_____ 8. endotherms [p.645]

_____ 9. conduction [p.645]

_____ 10. shivering response [p.647]

_____ 11. nonshivering heat production [p.647]

_____ 12. peripheral vasoconstriction [p.647]

_____ 13. ectotherms [p.645]

a. occurs as a result of sweating
b. the transfer of heat between objects in direct contact
c. creates a layer of still area next to the skin
d. rhythmic tremors of skeletal muscles to produce heat
e. organisms with high metabolic rates, such as mammals
f. mitochondria in brown adipose tissue generate heat
g. emission of heat from an object in the form of radiant energy
h. moving air or water moves heat
i. constriction of blood vessels to decrease blood flow
j. the conversion of a liquid to a gas moves heat
k. the blood vessels of the skin dilate, increasing blood flow
l. organisms that use the environment to gain heat
m. organisms that balance between being endothermic and ectothermic

## Self-Quiz

1. The functional unit of the kidney is the
   _____. [p.641]
   a. nephron
   b. nephridia
   c. renal artery
   d. kidney medulla

2. _____ is the hormone that promotes water reabsorption by the kidney. [p.643]
   a. insulin
   b. ADH
   c. aldosterone
   d. glycogen

3. Accumulations of uric acid and calcium in the kidneys may cause _____. [p.644]
   a. uremic toxicity
   b. metabolic acidosis
   c. kidney stones
   d. diabetes mellitus

4. The transfer of heat by the movement of air over a surface is called _____. [p.645]
   a. evaporation
   b. transduction
   c. convection
   d. thermal radiation

5. The _____ connects the kidneys and the urinary bladder. [p.640]
   a. urethra
   b. proximal tubule
   c. distal tubule
   d. ureter

6. The inability of the kidneys to maintain a proper acid-base balance may result in _____. [p.644]
   a. kidney stones
   b. diabetes insipidus
   c. diabetes mellitus
   d. uremic toxicity
   e. metabolic acidosis

7. _____ organisms use the external environment to gain body heat. [p.645]
   a. endothermic
   b. ectothermic
   c. heterothermic
   d. all of the above

8. A restriction of blood to the skin is caused by _____. [p.646]
   a. pilomotor response
   b. hyperthermia
   c. evaporative heat loss
   d. peripheral vasodilation
   e. peripheral vasoconstriction

9. Brown adipose tissue is involved in which of the following? [p.646]
   a. pilomotor response
   b. nonshivering heat production
   c. shivering response
   d. peripheral vasoconstriction

10. A reduction in the core temperature of an organism is called _____. [p.647]
   a. hyperthermia
   b. hypothermia
   c. isothermia
   d. heterothermia

## Chapter Objectives/Review Questions

This section lists general and detailed chapter objectives that can be used as review questions. You can make maximum use of these items by writing answers on a separate sheet of paper. To check for accuracy, compare your answers with information given in the chapter or glossary.

1. Explain how the urinary system allows animals to adapt to life on land. [p.639]
2. Give the location and function of the organs of the urinary system. [p.640]
3. Know the structure of a human kidney. [p.640]
4. Understand the process by which urine is formed. [pp.642–643]
5. Understand the role of ADH and aldosterone in maintaining the water balance. [p.643]
6. Understand the major diseases associated with the kidneys. [p.644]
7. Recognize the four methods of heat transfer. [p.645]
8. Distinguish between heterotherms, endotherms, and ectotherms. [p.645]
9. Understand how mammals respond to heat stress. [p.646]
10. Understand how mammals respond to cold stress. [pp.646–647]

# Chapter Summary

Animals continually produce metabolic (1) _____. They also continually gain and lose (2) _____ and solutes. Despite these continuous gains and losses, the overall composition and (3) _____ of (4) _____ fluid must be kept within a range that the cells can tolerate. This process is called (5) _____. In humans, the (6) _____ system interacts with other systems to accomplish these tasks. The human urinary system consists of two (7) _____, two ureters, a (8) _____, and a urethra. Inside each kidney, millions of (9) _____ filter water and solutes from the blood. Water and solutes not returned to the blood leave the body as (10) _____. Urine forms by three processes: (11) _____, (12) _____, and (13) _____. The hormones (14) _____ and (15) _____, as well as a thirst response, influence whether urine is concentrated or dilute. Heat losses and heat gains determine an animal's (16)_____.

# Integrating and Applying Key Concepts

1. The hemodialysis machine used in hospitals is expensive and time-consuming. So far, artificial kidneys capable of allowing people who have nonfunctional kidneys to purify their blood by themselves, without having to go to a hospital or clinic, have not been developed. Which aspects of the hemodialysis procedure do you think have presented the most problems in development of a method of home self-care? If you had an unlimited budget and were appointed head of a team to develop such a procedure and its instrumentation, what strategy would you pursue?

# 39

# ANIMAL REPRODUCTION AND DEVELOPMENT

## INTRODUCTION

Sexual reproduction is a defining characteristic of many species, including man. And while this process is costly from a biological standpoint, the rewards are great in that the products of sexual reproduction are genetically diverse and therefore able to respond to a changing environment. This chapter outlines the process involved in the animal life cycle, and then focuses, in detail, on the structure and function of the human reproductive system. The process of fertilization is outlined, as is the development of a human fetus. The student also will learn about the ability of humans to both prevent and promote pregnancy, using various physical and chemical methods. Continuing the focus on human health, the chapter then discusses many of the sexually transmitted diseases that may be spread through human sexual intercourse. A discussion of the birth process and the growth and maturation of the newborn conclude this chapter.

## FOCAL POINTS

- Figure 39.4 [p.654] diagrams the 6 stages of animal reproduction and development.
- Figure 39.12 [p.661] illustrates the components of the human male reproductive system and describes their functions.
- Figure 39.14 [p.663] shows the signaling pathways used in sperm formation.
- Figure 39.16 [p.664] illustrates the components of the human female reproductive system and describes their functions.
- Figure 39.18 [p.667] correlates the changes in the ovary and uterus in response to changing hormone levels.
- Figure 39.20 [p.669] illustrates the process of fertilization.
- Figure 39.21 [p.670] shows various methods of birth control and their effectiveness.
- Table 39.4 [p.672] estimates the number of various sexually transmitted diseases.
- Figure 39.26 [pp.674–675] illustrates the stages involved from fertilization to implantation.
- Figure 39.29 [pp.678–679] shows the human embryo at various stages of development.

## Interactive Exercises

*Mind-Boggling Births* [pp.650–651]

### 39.1. REFLECTIONS ON SEXUAL REPRODUCTION [pp.646–647]

### 39.2. STAGES OF REPRODUCTION AND DEVELOPMENT [pp.654–655]

## Selected Words

*In vitro* fertilization [p.651], *Rana pipiens* [p.654]

## Boldfaced Terms

[p.652] sexual reproduction _____

_____

[p.652] asexual reproduction _____

_____

[p.653] internal fertilization _____

_____

[p.653] yolk _____

_____

[p.654] fertilization _____

_____

[p.654] cleavage _____

_____

[p.654] blastula _____

_____

[p.654] gastrulation _____

_____

[p.654] germ layers _____

_____

[p.654] ectoderm _____

_____

[p.654] endoderm _____

_____

[p.654] mesoderm _____

_____

## Fill in the Blanks [p.652]

In earlier chapters, we considered the genetic basis of sexual reproduction. Again, (1) _____ and the formation of gametes typically occur in two prospective parents. At (2) _____ a gamete from one parent fuses with a gamete from the other and forms a (3) _____, the first cell of a new individual. We also looked at asexual reproduction, whereby a (4) _____ organism—one parent only—produces (5) _____. We now turn to examples of structural, behavioral, and ecological aspects of these two modes of animal reproduction.

Mutation aside, in cases of (6) _____, all offspring are (7) _____ the same as their individual parent. Phenotypically, they are also much the same. This can be advantageous if the (8) _____ does not vary over time. (9)_____ combinations that allowed the parent to (10)_____ will be expected to do the (11) _____ for the offspring.

Most (12) _____, however, live where opportunities, (13) _____ and danger are variable. They reproduce (14) and offspring inherit different mixes of (15) _____ from female and male parents. The resulting (16) _____ in traits can improve the likelihood that at least some of the offspring will (17) _____ and (18) _____ [p.646] if prevailing conditions change.

## Sequence

Arrange the following events in correct chronological sequence. Write the letter of the first event next to 18, the letter of the second event next to 19, and so on. [All from p.654]

19. _____
20. _____
21. _____
22. _____
23. _____
24. _____

a. gastrulation
b. fertilization
c. cleavage
d. growth, tissue specialization
e. organ formation
f. gamete formation

## Complete the Table

25. [p.654] Complete the table below by entering the correct germ layer (ectoderm, mesoderm, or endoderm) that forms the tissues and organs listed.

| Tissues/Organs | Germ Layer |
| --- | --- |
| Muscle, circulatory organs | a. |
| Nervous system tissues | b. |
| Inner lining of the gut | c. |
| Circulatory organs (blood vessels, heart) | d. |
| Outer layer of the integument | e. |
| Reproductive and excretory organs | f. |
| Organs derived from the gut | g. |
| Most of the skeleton | h. |
| Connective tissues of the gut and integument | i. |

## 39.3. EARLY MARCHING ORDERS [pp.656–657]

### Selected Words

Animal pole [p.656], vegetal pole [p.656]

### Boldfaced Terms

[p.656] sperm _____

_____

[p.656] oocytes _____

_____

[p.656] cytoplasmic localization _____

_____

### Matching

Choose the most appropriate answer for each term.

1. ___ oocyte [p.656]
2. ___ sperm [p.656]
3. ___ gray crescent [p.656]
4. ___ cytoplasmic localization [p.656]
5. ___ cleavage [p.656]
6. ___ blastula [p.654]
7. ___ vegetal pole [p.656]
8. ___ animal pole [p.656]
9. ___ gastrulation [p.657]
10. ___ Hilde mangold [p.657]
11. ___ rotation [p.656]

a. process of organization that gives rise to three germ layers; initiated by signals from the dorsal lip
b. first researcher to discover the cues for gastrulation
c. embryonic stage characterized by blastomeres and a fluid-filled cavity, the blastocoel
d. consists of paternal DNA and cellular equipment enabling this cell to reach and penetrate an egg
e. an area of intermediate pigmentation and maternal messages in some animal eggs
f. contains little yolk, pigment most concentrated here
g. by virtue of where they form, blastomeres end up with different maternal messages; helps seal the developmental fate of each cell's descendants
h. process that puts different parts of the egg cytoplasm into different blastomeres—not random
i. term for an immature egg; contains regionally localized aspects known as "maternal messages"
j. in a yolk rich egg, this contains most of the yolk
k. occurs following fertilization of the egg with sperm; the cortex of the egg moves to reveal the gray crescent

## 39.4. SPECIALIZED CELLS, TISSUES, AND ORGANS [pp.658–659]

### Selected Words

Apical ectodermal ridge [p.659], *physical* constraints [p.659], *architectural* constraints [p.659], *phyletic* constraints [p.659]

### Boldfaced Terms

[p.658] selective gene expression _____

_____

[p.658] cell differentiation _____

_____

[p.658] morphogens _____

_____

[p.658] embryonic induction _____

_____

[p.658] morphogenesis _____

_____

[p.659] apoptosis _____

_____

[p.659] pattern formation _____

_____

[p.655] homeotic genes _____

_____

### Matching [pp.658–659]

1. _____ selective gene expression
2. _____ morphogens
3. _____ embryonic induction
4. _____ homeotic genes
5. _____ apoptosis
6. _____ AER
7. _____ pattern formation
8. _____ cell differentiation
9. _____ morphogenesis
10. _____ physical constraints
11. _____ phyletic constraints
12. _____ architectural constraints

a. different cell lineages express different subsets of genes
b. the process by which cell lineages become specialized
c. signaling molecules that are the products of master genes
d. constraints based on the interactions of genes that regulate development in a lineage
e. constraints based on properties of the cell—like surface-to-volume ratio
f. embryonic cells produce signals that alter the behavior of neighboring cells
g. the process by which tissue and organs form
h. imposed by body axes
i. programmed cell death
j. the process by which certain body parts form in certain places
k. apical ectodermal ridge—induces mesoderm under it to form limb
l. genes that regulate development of specific body parts

## True–False

13. All differentiated cells retain the entire genome.

14. During morphogenesis, cells migrate to specific locations and sheets of cells expand and fold as cells change in shape.

## 39.5. REPRODUCTIVE SYSTEM OF HUMAN MALES [pp.660–661]

## 39.6. SPERM FORMATION [pp.662–663]

### Selected Words

*Scrotum* [p.660], vas deferentia [p.660], seminal vesicles [p.660], prostate gland [p.660], *PSA* [p.661], *Leydig cells* [p.662], *Sertoli cells* [p.662]

### Boldfaced Terms

[p.660] gonads _____

_____

[p.660] testosterone _____

_____

[p.660] puberty _____

_____

[p.659] LH _____

_____

[p.659] FSH _____

_____

[p.659] GnRH _____

_____

## Labeling

Label each of the indicated structures in the diagram below. [pp.660–661]

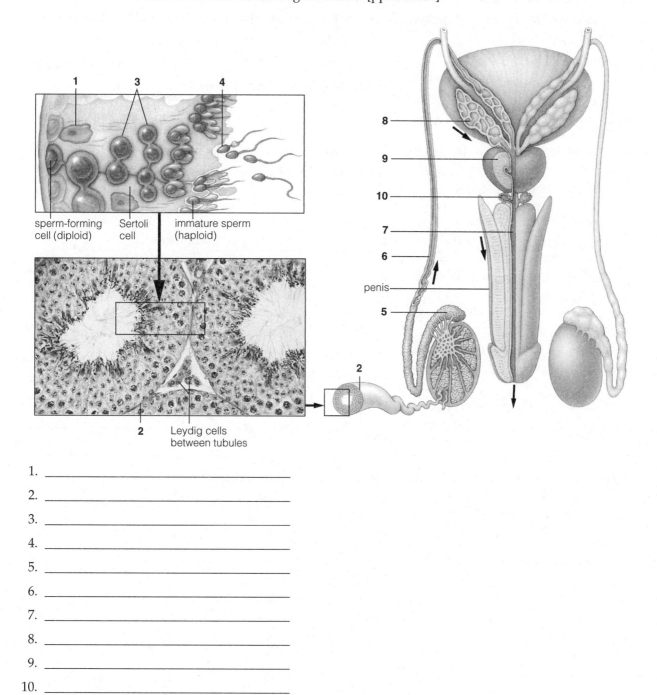

sperm-forming cell (diploid)   Sertoli cell   immature sperm (haploid)

2   Leydig cells between tubules

8
9
10
7
6
penis
5
2

1. _____

2. _____

3. _____

4. _____

5. _____

6. _____

7. _____

8. _____

9. _____

10. _____

## Dichotomous Choice

Circle one of two possible answers given between parentheses in each statement.

11. Testosterone is secreted by (Leydig/hypothalamus) cells. [p.662]

12. (Testosterone/FSH) governs the growth, form, and functions of the male reproductive tract. [p.660]

13. Sexual behavior, aggressive behavior, and secondary sexual traits are associated with (LH/testosterone). [p.660]

14. LH and FSH are secreted by the (anterior/posterior) lobe of the pituitary gland. [p.663]

15. The (testes/hypothalamus) governs sperm production by controlling secretion of testosterone, LH, and FSH. [p.660]

16. When blood levels of testosterone (increase/decrease), the hypothalamus secretes GnRH that stimulates the pituitary to step up the release of LH and FSH, which travel the bloodstream to targets in the testes. [p.663]

17. Within the testes, (LH/FSH) acts on Leydig cells; in response, they secrete testosterone, which helps stimulate sperm formation and development. [p.663]

18. (Sertoli/Leydig) cells have receptors for FSH, which are necessary for initiation of normal spermatogenesis at puberty. [p.663]

19. Feedback loops to the hypothalamus lead to (increased/decreased) testosterone secretion and sperm formation. [p.663]

20. A(n) (elevated/depressed) testosterone level in blood slows down the release of GnRH. [p.663]

21. (Leydig/Sertoli) cells release inhibin. This protein hormone acts on the hypothalamus and pituitary to cut back on the release of GnRH and FSH. [p.663]

## Labeling

Identify each numbered part of the illustration below. [p.659]

22. _____

23. _____

24. _____

25. _____

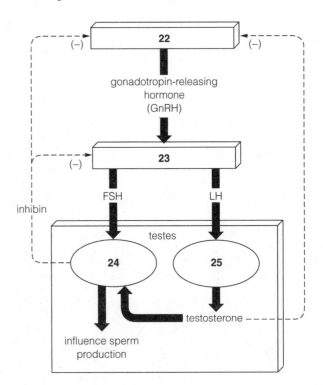

## 39.7. REPRODUCTIVE SYSTEM OF HUMAN FEMALES [pp.664–665]

## 39.8. FEMALE TROUBLES [p.665]

## 39.9. THE MENSTRUAL CYCLE [pp.666–667]

*Selected Words*

PMS [p.665], *follicular* phase [p.666], *luteal* phase [p.666]

*Boldfaced Terms*

[p.664] ovaries _____

_____

[p.664] uterus _____

_____

[p.664] endometrium _____

_____

[p.664] vagina _____

_____

[p.665] menstrual cycle _____

_____

[p.665] menopause _____

_____

[p.666] primary oocyte _____

_____

[p.666] ovulation _____

_____

[p.666] corpus luteum _____

_____

[p.666] estrogens _____

_____

[p.666] progesterone _____

_____

[p.667] secondary oocyte _____

_____

## *Labeling*

Label each of the structures in the diagram below. [p.661]

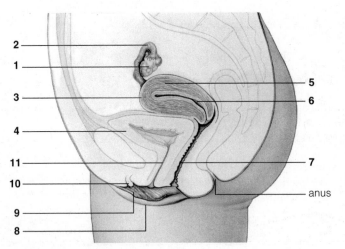

1. _____

2. _____

3. _____

4. _____

5. _____

6. _____

7. _____

8. _____

9. _____

10. _____

11. _____

Using the following diagram, label the following events that occur during a female's 28-day cycle.

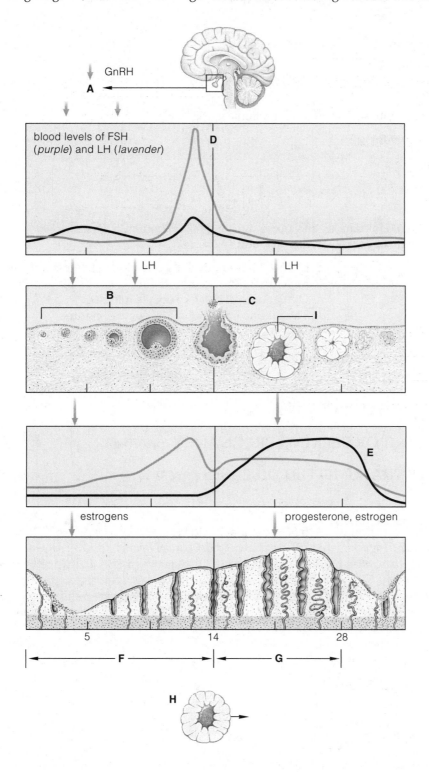

12. _____ ovulation

13. _____ mid-cycle peak of LH, which triggers ovulation

14. _____ growth of follicle

15. _____ estrogen and progesterone levels high

16. _____ follicular phase

17. _____ luteal phase

18. _____ corpus luteum persists if implantation occurs

19. _____ corpus luteum

20. _____ pituitary gland

## Ordering

Place the following events in order as they occur during the 28-day menstrual cycle.
Number 21 will be the first event, 22 the second, and so on.

21. _____
22. _____
23. _____
24. _____

    a. Corpus luteum develops causing thickening of the endometrium.
    b. Menstrual flow occurs—the follicular phase.
    c. Ovulation occurs—one oocyte released.
    d. The uterine lining rebuilds; oocytes mature.

## 39.10. FSH AND TWINS [p.668]

## 39.11. PREGNANCY HAPPENS [pp.668–669]

## 39.12. PREVENTING OR SEEKING PREGNANCY [pp.670–671]

## 39.13. SEXUALLY TRANSMITTED DISEASES [pp.672–673]

### Selected Terms

Fraternal twins [p.668], identical twins [p.668], zona pellucida [p.669], *abstinence* [p.670], *rhythm method* [p.670], *withdrawal* [p.670], *douching* [p.670], *vasectomy* [p.670], *tubal ligation* [p.670], *spermicidal foam* [p.670], *spermicidal jelly* [p.670], *diaphragm* [p.670], *cervical cap* [p.670], *condoms* [p.670], *IUD* [p.671], *birth control pill* [p.671], *birth control patch* [p.671], *progestin* injection [p.671], *emergency contraception* [p.671], *spontaneous abortion* [p.671], *induced abortion* [p.667], Mifepristone (RU-486) [p.671], *in vitro fertilization* [p.671], *Herpes* [p.671], *genital warts* [p.672], *Trichomonas vaginalis* [p.672], *Chlamydia trachomatis* [p.672], *genital herpes* [p.672], *Herpes simplex* [p.672], *Neisseria gonorrhoeae* [p.672], *Treponema pallidum* [p.673], *syphilis* [p.673], *AIDS* [p.673]

### Boldfaced Terms

[p.669] ovum _____

_____

[p.672] sexually transmitted diseases _____

_____

## Fill in the Blanks [p.669]

On average, an (1) _____ can put 150 million to 350 million sperm in the (2) _____. Fertilization occurs if they arrive a few days before or after (3) _____ or any time in between. Fewer than 30 minutes after sperm arrive in the vagina, (4) _____ move them deep into the female's reproductive tract. A few (5) _____ actually reach the upper portion of the (6) _____, where eggs are usually fertilized.

Many sperm bind to the oocyte's (7) _____. Binding triggers the release of (8) _____ from the cap over each sperm's head. Collectively, these (9) _____ enzymes make a passage through the (10) _____. Usually, only one sperm enters the secondary oocyte. Its entry triggers (11) _____ modifications that prevent other sperm from binding.

Upon sperm penetration, the secondary (12) _____ and the first (13) _____ complete meiosis II. Now, there are three polar bodies and one mature egg, or (14) _____. The egg (15) _____ fuses with the sperm (16) _____. Together, chromosomes of both nuclei restore the (17) _____ number for a brand new zygote.

## Matching

Choose the most appropriate answer for each term. [pp.670–671]

_____ 18. abstinence

_____ 19. rhythm method

_____ 20. withdrawal

_____ 21. douching

_____ 22. vasectomy

_____ 23. tubal ligation

_____ 24. spermicidal foam and jelly

_____ 25. diaphragm

_____ 26. condoms

_____ 27. birth control pill

_____ 28. progestin injections or implants

_____ 29. Mifepristone (RU-486)

_____ 30. IUDs

a. a flexible, dome-shaped device inserted into the vagina and positioned over the cervix before intercourse; 84 percent effective

b. progestin injections or implants that inhibit ovulation; 95–96 percent effective

c. a woman's oviducts are cauterized or cut and tied off; extremely effective

d. rinsing the vagina with a chemical right after intercourse; next to useless

e. thin, tight-fitting sheaths worn over the penis during intercourse; 85–93 percent reliable

f. avoiding intercourse during a woman's fertile period; 74 percent effective

g. the morning-after pill that interferes with hormones that control events between ovulation and implantation; use is still controversial in the United States

h. cutting and tying off each vas deferens of a man; extremely effective

i. an oral contraceptive made of synthetic estrogens and progestins that suppress oocyte maturation and ovulation; 94 percent effective

j. no sexual intercourse; foolproof

k. removing the penis from the vagina before ejaculation; an ineffective method

l. chemicals toxic to sperm are transferred from an applicator into the vagina just before intercourse; 82–83 percent effective

m. coils inserted into the uterus by a physician; sometimes invite pelvic inflammatory disease

## Matching

Match each of the following with *all* applicable diseases. [pp.668–669]

_____ 31. Can severely damage the brain and spinal cord in ways leading to general paralysis

_____ 32. Has no cure

_____ 33. Can cause severe cramps, fever, vomiting, and possibly sterility due to scarring and blocking of the oviducts

_____ 34. Caused by the spirochete *Treponema pallidum*

_____ 35. Infected women typically have miscarriages, stillbirths, or syphilitic newborns.

_____ 36. Caused by *Neisseria gonorrhoeae*

_____ 37. Infection requires direct contact with the viral agent; about 25 million people in the United States are infected by it.

_____ 38. Produces a flattened, painless chancre (localized ulcer) 1–8 weeks following infection

_____ 39. The virus is reactivated sporadically, producing painful sores at or near the original infection site.

_____ 40. Treponemes can cross the placenta of an infected, pregnant woman.

_____ 41. Acyclovir decreases the healing time and may also decrease the pain and viral shedding from the blisters.

_____ 42. Eight weeks following infection, a flattened, painless chancre harbors many motile bacteria

_____ 43. Contrary to common belief, it can be contracted over and over again.

_____ 44. Prompt doses of penicillin cure it, but not NGU, which requires both tetracycline and sulfonamides.

_____ 45. Generally preventable by correct condom usage

a. AIDS
b. chlamydial infection
c. genital herpes
d. gonorrhea
e. pelvic inflammatory disease
f. syphilis

## 39.14. FORMATION OF THE EARLY EMBRYO [pp.674–675]

## 39.15. EMERGENCE OF THE VERTEBRATE BODY PLAN [p.676]

## 39.16. THE FUNCTION OF THE PLACENTA [p.677]

## 39.17. EMERGENCE OF DISTINCTLY HUMAN FEATURES [pp.678–679]

## 39.18. MOTHER AS PROVIDER, PROTECTOR, POTENTIAL THREAT [pp.680–681]

## 39.19. BIRTH AND LACTATION [p.681]

### Selected Terms

*Embryonic* period [p.674], *fetal* period [p.674], trimester [p.674], embryonic disk [p.674], amniotic cavity [p.674], HCG [p.675], spina bifida [p.676], chorionic villi [p.677], lanugo [p.679], folic acid [p.680], cretinism [p.680], toxoplasmosis [p.680], fetal alcohol syndrome [p.680], thalidomide [p.680], Accutane [p.680], Paxil [p.680], oxytocin [p.681]

## Boldfaced Terms

[p.674] blastocyst _____

_____

[p.674] fetus _____

_____

[p.674] implantation _____

_____

[p.675] chorion _____

_____

[p.675] placenta _____

_____

[p.675] allantois _____

_____

[p.676] somites _____

_____

[p.677] placenta _____

_____

[p.681] labor _____

_____

[p.681] lactation _____

_____

[p.681] prolactin _____

_____

## Matching

Choose the most appropriate answer for each term. [pp.674–675]

_____ 1. embryonic period

_____ 2. fetus

_____ 3. fetal period

_____ 4. implantation

_____ 5. allantois

_____ 6. amnion

_____ 7. chorion

_____ 8. yolk sac

a. Outermost membrane; will become part of the spongy blood-engorged placenta

b. Directly encloses the embryo and cradles it in a buoyant protective fluid

c. In humans, some of this membrane becomes a site for blood cell formation, and some will give rise to germ cells, the forerunners of gametes.

d. In humans, the urinary bladder and the blood vessels form from it.

e. The blastocyst adheres to the uterine lining and sends out projections that invade the mother's tissues.

f. The time span from the third to the end of the eighth week of pregnancy

g. Term for the new, distinctly human individual at the end of the embryonic period

h. From the start of the ninth week until birth, organs enlarge and become specialized.

## Sequence

Arrange the following human developmental stages of early human embryo formation in the proper chronological sequence (day 1 to day 14). Write the letter of the first stage next to 9. The letter of the final process is written next to 16. [pp.674–675]

_____ 9.

_____ 10.

_____ 11.

_____ 12.

_____ 13.

_____ 14.

_____ 15.

_____ 16.

a. The first cleavage furrow divides the cell.

b. The chorionic cavity starts to form.

c. The morula forms.

d. Implantation begins.

e. The yolk sac, embryonic disk, and amniotic cavity have started to form.

f. The structures of the placenta start to form.

g. The cells huddle into a compact ball, with junctions allowing communication.

h. The blastocoel forms.

## Sequence

Arrange the following human developmental stages in the emerging vertebrate body plan in the proper chronological sequence (day 15 to days 24–25). Write the letter of the first stage next to 17, and so on. [p.676]

_____ 17.

_____ 18.

_____ 19.

a. Morphogenesis occurs; the neural tube and somites form; mesoderm somites give rise to the axial skeleton, skeletal muscles, and much of the dermis.

b. Pharyngeal arches form to contribute to formation of the face, neck, mouth, nasal cavities, larynx, and pharynx.

c. The primitive streak forms to mark the onset of gastrulation in vertebrate embryos.

## Labeling

Identify each numbered placental component in the illustration below. [p.673]

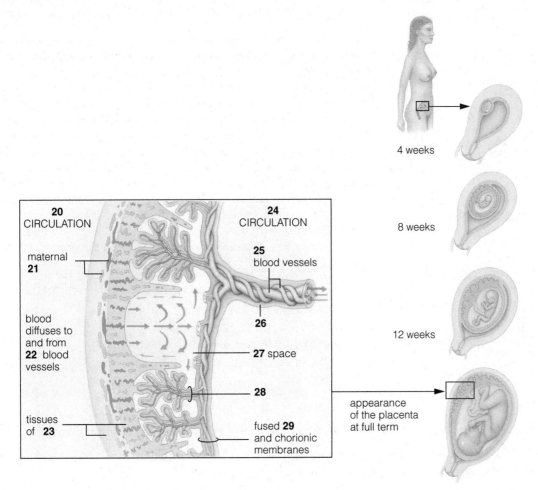

20. _____
21. _____
22. _____
23. _____
24. _____
25. _____
26. _____
27. _____
28. _____
29. _____

## Sequence

Arrange the following human developmental stages demonstrating the emergence of distinctly human features in the proper chronological sequence (four weeks to full term or completion of the fetal period). Write the letter of the first stage next to 30. The letter of the final process should be written next to 34. [pp.678–679]

_____ 30.

_____ 31.

_____ 32.

_____ 33.

_____ 34.

a. Movements begin as nerves make functional connections with developing muscles; legs kick, arms wave, fingers grasp, and the mouth puckers.

b. The length of the fetus increases from 16 centimeters to 50 centimeters, and weight increases from about 7 ounces to 7.5 pounds.

c. The human embryo has a tail and pharyngeal arches, and limbs; fingers and toes are sculpted from embryonic paddles; the circulatory system becomes more complex, and the head develops.

d. The human embryo is distinctly human as compared to other vertebrate embryos; upper and lower limbs are well formed; fingers and then toes have separated; primordial tissues of all internal and external structures have developed; the tail has become stubby.

e. Human features are visible at the boundary of the embryonic and fetal periods; the embryo floats in amniotic fluid and the chorion covers the amnion.

## Choice [p.680]

For the following questions, choose from the following threats to human development:

        a. nutrition    b. infections    c. prescription drugs    d. alcohol    e. antidepressants
        f. cigarette smoke    g. anti-acne drugs    h. all of the preceding threats

_____ 35. About 60–70 percent of newborns of these women have FAS.

_____ 36. In the case of rubella, there is a 50 percent chance some organs won't form properly.

_____ 37. In cases of exposure to the second-hand type, children were smaller, died of more post-delivery complications, and had twice as many heart abnormalities.

_____ 38. Increasing B-complex intake of the mother before conception and during early pregnancy reduces the risk that an embryo will develop severe neural tube defects.

_____ 39. Taking these early in pregnancy increases the likelihood of heart malformations.

_____ 40. Women who used the tranquilizer thalidomide during the first trimester gave birth to infants with missing or severely deformed arms and legs.

_____ 41. A pregnant woman must eat enough so that her body weight increases by 20–25 pounds, on average.

_____ 42. This causes reduced brain and head size, mental retardation, facial deformities, poor growth and poor coordination, and often heart defects.

Match each of the following terms with its correct definition. [p.681]

_____ 43. oxytocin

_____ 44. lactation

_____ 45. labor

_____ 46. prolactin

a. milk production

b. a hormone that simulates milk production

c. binds to smooth muscle of the uterus, causing contractions

d. the birth process

## 39.20. MATURATION, AGING, AND DEATH [p.682]

### Boldfaced Terms

[p.682] programmed life span hypothesis _____

_____

[p.682] telomeres _____

_____

[p.670] cumulative assaults hypothesis _____

_____

### Sequence

Arrange the following stages of human development in the proper chronological sequence. Write the letter of the first stage next to 1. The letter of the final process should be written next to 12. [p.682]

1. _____
2. _____
3. _____
4. _____
5. _____
6. _____
7. _____
8. _____
9. _____
10. _____
11. _____
12. _____

a. newborn
b. morula
c. zygote
d. adolescent
e. old age
f. adult
g. blastocyst
h. fetus
i. pubescent
j. embryo
k. infant
l. child

## Self-Quiz

___ 1. The process of cleavage most commonly produces a(n) _____. [p.654]
  a. zygote
  b. blastula
  c. gastrula
  d. third germ layer
  e. organ

___ 2. The formation of three germ (embryonic) tissue layers occurs during _____. [p.654]
  a. gastrulation
  b. cleavage
  c. pattern formation
  d. morphogenesis
  e. neural plate formation

___ 3. Muscles differentiate from _____ tissue. [p.654]
   a. ectoderm
   b. mesoderm
   c. endoderm
   d. parthenogenetic
   e. yolky

___ 4. The gray crescent is _____. [p.656]
   a. formed where the sperm penetrates the egg
   b. part of only one blastomere after the first cleavage
   c. the yolky region of the egg
   d. where the first mitotic division begins
   e. formed opposite from where the sperm enters the egg

___ 5. The type of cleavage observed in mammal eggs is _____. [p.657]
   a. incomplete
   b. animal-vegetal axis
   c. radial
   d. rotational
   e. cytoplasmic localization

___ 6. Of the following, _____ is *not* an aspect of morphogenesis. [pp.658–659]
   a. active cell migration
   b. expansion and folding of whole sheets of cells
   c. apoptosis
   d. shaping the proportion of body parts
   e. initiating production of a protein by some cells and not others

___ 7. The differentiation of a body part in response to signals from an adjacent body part is _____. [p.658]
   a. contact inhibition
   b. ooplasmic localization
   c. embryonic induction
   d. pattern formation
   e. none of the above

___ 8. Morphogens are _____. [p.658]
   a. master genes
   b. agents of apoptosis
   c. agents of cell differentiation
   d. concentrations of gap gene products
   e. degradable molecules that diffuse as long-range signals

For questions 9–12, choose from the following answers: [all from pp.660–661]

   a. Leydig cells   b. seminiferous tubules   c. vas deferens   d. epididymis   e. prostate

___  9. The _____ connects a structure on the surface of the testis with the ejaculatory duct.

___ 10. Testosterone is produced by the _____.

___ 11. Sperm formation begins in the _____.

___ 12. Sperm mature in the _____.

For questions 13–15, choose from the following answers:

   a. blastocyst   b. amnion   c. yolk sac   d. oviduct   e. cervix

___ 13. The _____ lies between the uterus and the vagina. [pp.664–665]

___ 14. The _____ is a pathway from the ovary to the uterus. [pp.664–665]

___ 15. The _____ results from the process known as cleavage. [p.674]

For questions 16–20, choose from the following answers: [all from pp.672–673]

     a. AIDS     b. chlamydial infection     c. genital herpes     d. gonorrhea     e. syphilis

\_\_\_ 16. _____ is a disease caused by a species of *Neisseria*; it is curable by prompt diagnosis and treatment.

\_\_\_ 17. _____ is a disease caused by *Treponema* that produces a localized ulcer (a chancre).

\_\_\_ 18. _____ is an incurable disease caused by the human immunodeficiency virus.

\_\_\_ 19. _____ is a disease caused by an intracellular parasite that lives in the genital and urinary tracts; also causes NGU.

\_\_\_ 20. _____ is an extremely contagious viral infection that causes sores on the mouth area and genitals; acyclovir may be effective in decreasing healing time and decreasing pain.

---

# Chapter Objectives/Review Questions

1. Understand how asexual reproduction differs from sexual reproduction. Know the advantages and problems associated with having separate sexes. [p.652].
2. Explain why evolutionary trends in many groups of organisms tend toward developing more complex, sexual strategies rather than retaining simpler, asexual strategies. [pp.652–653]
3. Describe early embryonic development and distinguish among the following: gamete formation, fertilization, cleavage, gastrulation, organ formation, and growth and tissue specialization. [pp.654–655]
4. Name each of the three embryonic tissue layers and the organs formed from each. [p.654]
5. Be able to discuss regionally localized aspects of maternal messages in the oocyte. [pp.656–657]
6. Define *cytoplasmic localization*. [p.656]
7. Tell how the origin of identical twins differs from that of fraternal twins. [p.668]
8. Define *differentiation*, and give one example of cells in a multicellular organism that have undergone differentiation. [pp.658–659]
9. Define *morphogenesis*. [p.653]
10. Explain the processes of embryonic induction and pattern formation. [p.654]
11. Explain the role of master genes, homeotic genes, and the constraints that are present in the development of an embryo. [p.655]
12. Name the primary male reproductive organs; list secondary sexual traits determined by these organs. [p.656]
13. Follow the path of a mature sperm from the seminiferous tubules to the urethral exit. List every structure encountered along the path, and state the contribution to the nurturing of the sperm. [pp.656–657]
14. Name the four hormones that directly or indirectly control male reproductive function. Diagram the negative feedback mechanisms that link the hypothalamus, anterior pituitary, and testes in controlling gonadal function. [pp.662–663]
15. Diagram the structure of a sperm, label its components, and state the function of each. [p.663]
16. Name the major structures of the female reproductive system and give the reproductive function of each. [p.664]
17. Distinguish the follicular phase of the menstrual cycle from the luteal phase, and explain how the two cycles are synchronized by hormones from the anterior pituitary, hypothalamus, and ovaries. [pp.666–667]
18. State which hormonal event brings about ovulation and which other hormonal events bring about the onset and finish of menstruation. [pp.666–667]

19. List the physiological factors that bring about erection of the penis during sexual stimulation and the factors that bring about ejaculation. [p.668]
20. List the similar events that occur in both male and female orgasm. [p.668]
21. Describe two types of sterilization. [p.670]
22. Identify the three most effective birth control methods used in the United States and the four least effective birth control methods. [pp.670–671]
23. For each STD described in the text, know the disease's causative organism and the symptoms. [pp.672–673]
24. Distinguish between the embryonic period and the fetal period. [p.674]
25. Describe the events that occur during the first month of human development; include the developed structures of the blastocyst as seen at about 15 days. [pp.674–675]
26. Describe the process of implantation. [p.674]
27. List and describe the four extraembryonic membranes and their functions. [p.675]
28. Characterize the human developmental events occurring from day 15 to day 25. [p.675]
29. List the roles of the placenta in pregnancy. [p.675]
30. Explain why the mother must be particularly careful of her diet, health habits, and lifestyle during the first trimester after fertilization (especially during the first six weeks). [p.680]
31. Generally describe the process of labor or delivery. [p.681]
32. Explain the role of the hormones oxytocin and prolactin. [p.681]
33. List the factors that bring about aging of an animal. [p.682]

## Chapter Summary

Biologically speaking, (1) _____ reproduction is much more "expensive" than (2) _____ reproduction. The rewards are great, however, in that sexual reproduction produces genetically (3) _____ offspring.

The six stages of animal life cycles are, in order; (4) _____, (5) _____, (6) _____, (7)_____, (8) _____ and (9) _____. The primary human reproductive organs are the (10) _____-producing testes in the male and the oocyte-producing (11)_____ in the female. The (12) _____ and the pituitary gland control the production and release of sex hormones in both males and females. From puberty onward, females are fertile on a (13) _____ basis. (14) _____ _____ leads to pregnancy.

Human sexual behavior can transmit (15) _____, which can cause disease. A pregnancy begins with (16) _____ and the implantation of a (17) _____ in the uterine lining. A (18) _____ connects the embryo with its mother. Aging and death are the result of (19) _____ factors and the ongoing environmental assaults on (20) _____ and tissues.

## Integrating and Applying Key Concepts

1. If embryonic induction did not occur in a human embryo, how would the eye region appear? What would happen to the forebrain and epidermis? If controlled cell death did not happen in a human embryo, how would its hands appear? Its face?
2. What rewards do you think a society should give a woman who has at most two children during her lifetime? In the absence of rewards or punishments, how can a society encourage women not to have abortions and yet ensure that the human birth rate does not continue to increase?

# 40

# POPULATION ECOLOGY

## INTRODUCTION

Population biologists and population geneticists are interested in how groups of a species interact with their environment and how the environment plays a role population size. This chapter explores some of the principles of population ecology. The chapter concludes with a look at human population growth.

## FOCAL POINTS

- Figure 40.5 [p.690] illustrates exponential growth.
- Figure 40.8 [p.692] illustrates logarithmic growth and carry capacity.
- Figure 40.16 [p.701] diagrams the four general age structure diagrams.
- Figure 40.17 [p.702] presents the demographic transition model.

## Interactive Exercises

*The Numbers Game* [pp.686–687]

## 40.1. POPULATION DEMOGRAPHICS [p.688]

## 40.2. ELUSIVE HEADS TO COUNT [p.687]

*Selected Words*

*Pre-reproductive*, *reproductive*, and *post-reproductive* ages [p.688], habitat [p.688], *crude* density [p.688], *interspecific* interactions [p.688], clumped distribution pattern [p.688], nearly uniform distribution pattern [p.688], random dispersion [p.688]

*Boldfaced Terms*

[p.688] demographics _____

_____

[p.688] population size _____

_____

[p.688] age structure _____

_____

[p.688] reproductive base _____

_____

[p.688] population density _____

_____

[p.688] population distribution _____

_____

[p.689] quadrats _____

_____

[p.689] capture-recapture method _____

_____

## Short Answer

1. Briefly state the lesson to be learned from the habitation of Easter Island. [pp.686–687]

_____

_____

_____

## Matching

Match each term with its description. [pp.688–689]

2. ___ clumped distribution
3. ___ demographics
4. ___ population size
5. ___ age structure
6. ___ pre-reproductive, reproductive, and post-reproductive ages
7. ___ reproductive base
8. ___ population density
9. ___ population distribution
10. ___ crude density
11. ___ nearly uniform distribution
12. ___ quadrats
13. ___ capture-recapture method
14. ___ random distribution

a. includes pre-reproductive and reproductive age categories
b. a measured number of individuals in some specified area
c. population sampling areas of the same size and shape, such as rectangles, squares, and hexagons
d. the number of individuals in some specified area or volume of a habitat
e. the number of individuals in each of several age categories
f. the vital statistics of the population
g. an example of age categories in a population
h. a way of sampling population density; capturing mobile individuals and marking them in some fashion and recapturing them after some period of time
i. a type of distribution that is usually the result of limited conditions or resources
j. a calculation of density without taking into consideration distribution
k. the general pattern in which individuals are dispersed in a specified area
l. individuals are more evenly spaced than expected by chance alone
m. caused by uniform conditions and resources are plentiful

## Population Problem

15. To determine the size of a fruit fly population, a researcher captures 400 flies and marks them with a fluorescent tag. Five days later the researcher repeats the capture. Out of 500 flies, 80 have tags. Based on this information, what is the estimated population size? [p.689]

_____

## 40.3. POPULATION SIZE AND EXPONENTIAL GROWTH [pp.690–691]

## 40.4. LIMITS ON POPULATION GROWTH [pp.692–693]

### Selected Words

*Capita* [p.690], *r* [p.690], J-shaped curve [p.691], *sustainable* supply of resources [p.692], S-shaped curve [p.693]

### Boldfaced Terms

[p.690] immigration _____

_____

[p.690] emigration _____

_____

[p.690] migration _____

_____

[p.690] zero population growth _____

_____

[p.690] per capita growth rate _____

_____

[p.690] exponential growth _____

_____

[p.691] doubling time _____

_____

[p.691] biotic potential _____

_____

[p.692] limiting factor _____

_____

[p.692] carrying capacity _____

_____

[p.692] logistic growth _____

_____

[p.693] density-dependent factors _____

_____

[p.693] density-independent factors _____

_____

## Growth Rate Problem

Consider the equation $G = rN$, where $G$ = the population growth rate per unit time, $r$ = the net population growth rate per individual per unit time, and $N$ = the number of individuals in the population. [p.690]

1. Assume that 5,000 rats live in the streets of a city. The rats give birth to 2,000 rats every month. Every month 500 of the 5,000 rats die. Calculate the following:

   a. birth rate = _____

   b. death rate = _____

   c. the value of $r$ = _____

## Matching

Match each of the following terms to the correct statement.

_____ 2. immigration [p.690]

_____ 3. zero population growth [p.691]

_____ 4. biotic potential [p.691]

_____ 5. emigration [p.690]

_____ 6. logistic growth [p.692]

_____ 7. limiting factor [p.692]

_____ 8. per capita growth rate [p.290]

_____ 9. doubling time [p.691]

_____ 10. exponential growth [p.290]

_____ 11. carrying capacity [p.292]

a. an essential resource that is in short supply
b. an S-shaped population growth curve
c. the time that it takes for a population to double in size
d. the maximum number of individuals that a given environment can sustain indefinitely
e. the arrival of new individuals from other populations
f. individuals leaving a population to live elsewhere
g. a J-shaped population curve
h. the difference between the per capita birth rate and per capita death rate
i. the maximum per capita rate of increase for a species
j. when the number of births equals the number of deaths

## Choice

Choose from the following. [p.693]

        a. density-dependent factors      b. density-independent factors

12. ___ a sudden freeze

13. ___ overcrowding

14. ___ bubonic plague

15. ___ application of pesticides in your backyard

16. ___ predators, parasites, and pathogens

17. ___ froughts, floods, and earthquakes

## 40.5. LIFE HISTORY PATTERNS [pp.694–695]

## 40.6. NATURAL SELECTION AND LIFE HISTORIES [pp.696–697]

### Selected Words

*Type I* curve [p.695], *type II* curve [p.695], *type III* curve [p.695]

## Boldfaced Terms

[p.694] life history pattern _____

_____

[p.694] cohort _____

_____

[p.695] survivorship curve _____

_____

## Matching

Match each term with its description. [pp.694–695]

1. ___ life history pattern
2. ___ cohort
3. ___ survivorship curve
4. ___ type I curves
5. ___ type II curves
6. ___ type III curves

a. shows, for example, the number of individuals actually reaching specified ages
b. reflect high survivorship until fairly late in life, then a large increase in deaths
c. a group of individuals, recorded from the time of birth until the last one dies; data is gathered, such as the number of offspring born to individuals in each age interval
d. signify a death rate that is highest early on; typical of species that produce many small offspring and do little, if any, parenting
e. for each species, a set of adaptations that influence survival, fertility, and age at first reproduction
f. reflect a fairly constant death rate at all ages; typical of organisms just as likely to be killed or die of disease at any age, such as lizards, small mammals, and large birds

## True–False

If the statement is true, write "T" in the blank. If the statement is false, correct it by writing the correct word(s) for the underlined word(s) in the answer blank. [pp.696–697]

_____ 7. Reznick and Endler showed that life history traits, like other characteristics, <u>cannot</u> be inherited.

_____ 8. Guppy populations deal with predators, especially pike-cichlids and <u>killifish</u>.

_____ 9. Smaller killifish prey on smaller, immature guppies but not on the larger adults; pike-cichlids live in other streams and prey on <u>larger</u> and sexually mature guppies.

_____ 10. Reznick and Endler hypothesized that predation is a selective agent acting on <u>killifish</u> life history patterns.

_____ 11. In pike-cichlid streams, guppies grow <u>slower</u> and are smaller at maturity, compared to guppies in killifish streams.

_____ 12. Guppies hunted by pike-cichlids reproduce earlier in life, reproduce <u>more</u> often, and have more young per brood.

_____ 13. Following laboratory experiments in the United States, the researchers concluded the differences between guppies preyed upon by different predators have a(n) <u>environmental</u> foundation.

_____ 14. Guppies introduced to the upstream experimental site were taken from a population that had evolved downstream from the waterfall, with the larger <u>killifishes</u>.

_____ 15. Eleven years (30 to 60 guppy generations) later, researchers revisited the stream. They found that the experimental population had not evolved.

_____ 16. Later laboratory experiments involving two generations of guppies confirmed that the differences studied in guppies has a genetic basis.

## 40.7.  HUMAN POPULATION GROWTH [pp.698–699]

## 40.8.  FERTILITY RATES AND AGE STRUCTURE [pp.700–701]

## 40.9.  POPULATION GROWTH AND ECONOMIC EFFECTS [pp.702–703]

## 40.10.  A NO-GROWTH SOCIETY [p.703]

*Selected Words*

*Preindustrial stage* [p.702], *transitional stage* [p.702], *industrial stage* [p.702], *postindustrial stage* [p.702]

*Boldfaced Terms*

[p.700] total fertility rate (TFR) _____

_____

[p.702] demographic transition model _____

_____

*Short Answer*

1.  List the three possible reasons humans began sidestepping controls. [pp.698–699]

_____

_____

_____

_____

_____

2.  Explain why continued human population growth cannot continue indefinitely. [p.699]

_____

_____

3.  What is meant by *TFR*? [p.700]

_____

_____

_____

## Matching

Match each age structure diagram with its description below. [p.701]

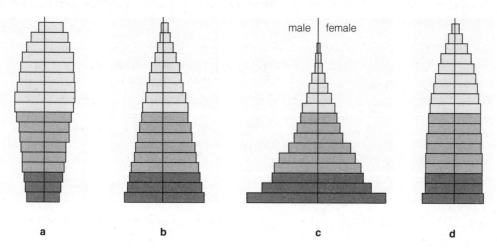

male | female

a       b       c       d

4. ___ zero growth

5. ___ rapid growth

6. ___ negative growth

7. ___ slow growth

## Choice

Refer to Figure 40.16 [p.701] in the text. Using the age structure diagrams shown above, choose the diagram that best fits the countries listed below.

8. ___ China

9. ___ United States

10. ___ India

11. ___ Mexico

12. ___ Australia

## Sequence

Arrange the following stages of the demographic transition model in correct chronological sequence. Write the letter of the first step next to 13, the letter of the second step next to 14, and so on. [p.700]

13. ___

14. ___

15. ___

16. ___

a. Industrial stage: population growth slows dramatically and industrialization is in full swing; a slowdown starts mainly because people want to move from the country to cities, and urban people want smaller families

b. Preindustrial stage: living conditions are the harshest before technological and medical advances become widespread; birth and death rates are high, so the growth rate is low

c. Postindustrial stage: population growth rates become negative; the birth rate falls below the death rate, and population size slowly decreases

d. Transitional stage: industrialization begins, food and health care improve; death rates drop, birth rates stay high in agricultural societies

*TNOW:* For an interactive exercise of this material, refer to Chapter 40: Art Labeling: Demographic Transition Model.

# Self-Quiz

_d_ 1. The number of individuals that contribute to a population's gene pool is _____. [p.688]
   a. the population density
   b. the population growth
   c. the population birth rate
   d. the population size
   e. the population distribution

_a_ 2. How are the individuals in a population _most often_ dispersed? [p.688]
   a. clumped
   b. very uniform
   c. nearly uniform
   d. random
   e. very random

_d_ 3. Assuming immigration is balancing emigration over time, _____ may be defined as an interval in which the number of births is balanced by the number of deaths. [p.690]
   a. the lack of a limiting factor
   b. exponential growth
   c. saturation
   d. zero population growth
   e. logistic growth

_b_ 4. A population that is growing exponentially in the absence of limiting factors can be illustrated accurately by a(n) _____. [p.691]
   a. S-shaped curve
   b. J-shaped curve
   c. curve that terminates in a plateau phase
   d. tolerance curve
   e. resource curve

_b_ 5. The population growth rate ($G$) is equal to the _____ net population growth rate per individual ($r$) and number of individuals ($N$). [p.690]
   a. sum of
   b. product of
   c. doubling of
   d. difference between
   e. tripling of

_a_ 6. Assuming the birth and death rate remain constant, both can be combined into a single variable, $r$, or _____. [p.690]
   a. the per capita rate
   b. the minus migration factor
   c. the number of individuals
   d. the net reproduction per individual per unit
   e. population growth

_c_ 7. Which of the following is _not_ characteristic of logistic growth? [p.692]
   a. S-shaped curve
   b. leveling off growth as carrying capacity is reached
   c. unrestricted growth
   d. slow growth of a low-density population followed by rapid growth
   e. limiting factors

_a_ 8. The maximum number of individuals of a population (or species) that a given environment can sustain indefinitely defines _____. [p.692]
   a. the carrying capacity of the environment
   b. exponential growth
   c. the doubling time of a population
   d. density-independent factors
   e. logistic growth

_a_ 9. The survivorship curve typical of industrialized human populations is type _____. [p.695]
   a. I
   b. II
   c. III
   d. both I and II
   e. both I and III

_b_ 10. "Industrialization begins, food production and health care improve and death rates slow," describes the _____ stage of the demographic transition model. [p.702]
   a. preindustrial
   b. transitional
   c. industrial
   d. postindustrial
   e. includes both preindustrial and transitional

# Chapter Objectives/Review Questions

1. Explain what lessons can be learned from the human population living on Easter Island. [pp.686–687]
2. Be able to define: *demographics, population size, age structure, population size, age structure, reproductive base, population density, population distribution, crude density, quadrats,* and *capture-recapture methods.* [pp.688–689]
3. List and describe the three patterns of dispersion illustrated by populations in a habitat. [p.688]
4. Be able to calculate population size using the mark-recapture method. [p.689]
5. Distinguish immigration from emigration; define migration and zero population growth. [p.690]
6. Understand how to calculate the per capita growth rate (r). [p.690]
7. Understand the characteristics of a population in exponential growth. [pp.691–692]
8. Understand the concept of biotic potential. [p.691]
9. List several examples of limiting factors and explain how they influence population curves. [p.692]
10. Define *carry capacity.* [p.692]
11. Tell how the pattern of logistic growth shows how carrying capacity can affect population size. [pp.692–693]
12. Explain how density-dependent factors and density-independent factors relate to population size; cite two examples of each. [p.693]
13. Understand the concept of life history patterns and survivorships curves. [pp.694–695]
14. After consideration of the research results obtained on guppies and their predators by Reznick and Endler, provide an explanation for these differences. [pp.696–697]
15. Be able to list three possible reasons why growth of the human population is out of control. [p.698]
16. List reasons why continued growth of the human population cannot be sustained indefinitely. [p.699]
17. Define *total fertility rate.* [p.700]
18. Understand the age structure diagrams for populations growing at different rates. [pp.700–701]
19. List and describe the four stages of the demographic transition model. [p.702]

# Chapter Summary

Ecologists explain population (1) _____ in terms of population (2) _____, density, (3) _____, and number of individuals in different (4) _____ categories. They have methods of estimating population size and (5) _____ in the field.

A population's size and (6) _____ base influence its rate of growth. When the population is increasing at a rate proportional to its size, it is undergoing (7) _____ growth.

Over time, exponential growth typically overshoots the (8) _____—the maximum number of individuals of a population that (9) _____ resources can sustain indefinitely. Some populations (10) _____ after a crash. Others never recover.

Resource availability, (11) _____, and predation are major factors that can (12) _____ population growth. These (13) _____ factors differ among species and shape their life (14) _____ patterns.

Human populations have (15) _____ limits to growth by way of (16) _____ expansion into new habitats, (17) _____ interventions, and (18) _____ innovations. No population can expand indefinitely when (19) _____ dwindle.

## Integrating and Applying Key Concepts

1. Following your study of this chapter and considering what you have learned about the problems facing planet Earth, prepare a list of positive changes you could make in your own living habits. These changes would help solve the serious problems we face as a human population. Share your knowledge and your list with family and friends and request that they each prepare a similar list. Above all, encourage everyone to *act* on the personal changes they propose.

2. Some researchers have proposed that we are rapidly approaching "peak oil"—a point where the discovery of new reserves can no longer meet demand. Explain the effect that peak oil may have on the demographic transition model.

# 41

# COMMUNITY STRUCTURE
# AND BIODIVERSITY

## INTRODUCTION

We have all heard stories about "radical" environmentalists who take drastic measures to save a species—the spotted owl, for example, or even a single tree. What drives these people to risk so much for seemingly so little? This chapter describes the relationships that exist in a community of species. These relationships include the shaping of communities and the many interactions that go on within a single community. The chapter then goes on to describe the effects of introduced organisms, such as kudzu, on a community and population. Historically, many introduced species have had devastating results, and these examples certainly serve as a warning for future activities. Finally, the text discusses biodiversity and introduces the field of conservation biology, which attempts to promote and sustain environmental diversity while also accommodating the survival needs of man. The focus is on the fact that all members of a community are critical to that community, and maintaining balance is paramount to species survival.

## FOCAL POINTS

- Table 41.1 [p.708] compares the different types of direct two-species interactions.
- Figure 41.6 [p.711] illustrates competitive exclusion as shown by Gause and his *Paramecium spp.*
- Figure 41.9 [p.712] shows the three models for responses of predators to prey density.
- Figure 41.21 [p.720] illustrates the effect of competition and predation in a specific environment.
- Table 41.2 [p.721] summarizes the outcomes of some species that have been introduced into the United States.
- Figure 41.24 [p.724] illustrates the correlation between species diversity and latitude.
- Figure 41.30 [p.728] maps the world's most vulnerable regions of land and sea.

## Interactive Exercises

*Fire Ants in the Pants* [pp.706–707]

### 41.1. WHICH FACTORS SHAPE COMMUNITY STRUCTURE? [p.708]

*Selected Words*

*Solenopsis invicta* [p.707], fire ants [p.707]

*Boldfaced Terms*

[p.708] habitat _____

_____

[p.708] community _____

_____

[p.708] species diversity _____

_____

[p.708] niche _____

_____

[p.708] commensalism _____

_____

[p.708] mutualism _____

_____

[p.708] interspecific competition _____

_____

[p.708] predation _____

_____

[p.708] parasitism _____

_____

[p.708] symbiosis _____

_____

[p.708] coevolution _____

_____

## Short Answer

1. Describe the biological controls being enlisted by ecologists in an attempt to control *Solenopsis*, the genus name of the dreadful fire ants. [pp.706–707]

_____

_____

_____

## Fill in the Blanks [p.708]

Every community has a characteristic structure, which arises as an outcome of five factors. First, the

(2) _____ and topography interact to dictate temperatures, rainfall, type of soil, and other conditions.

Second, the kinds and quantities of (3) _____ and other resources that become available through the

year affect which species live there. Third, (4) _____ traits of each species help it (5) _____ and

exploit specific resources on the habitat. Fourth, species interact in ways that cause shift in (6) _____ and

abundances. Fifth, the timing and history of disturbances, both natural and human-induced, affect

(7) _____ structure.

## Matching

Match each term with its description. [p.708]

8. ___ habitat

9. ___ niche

10. ___ fundamental niche

11. ___ realized niche

12. ___ indirect interactions

13. ___ commensalism

14. ___ mutualism

15. ___ interspecific competition

16. ___ predation

17. ___ parasitism

18. ___ symbiosis

a. An interaction that directly helps one species but does not affect the other much, if at all
b. Both species can be harmed.
c. Constraining factors do come into play; is dynamic and may shift over time, in small or large ways.
d. Defined as a close association between two or more species during part or all of the life cycle
e. Possess physical and chemical features, such as temperature, and an array of species; an organism's "address"
f. Kill prey
g. The one that might prevail in the absence of competition and any other factors that could limit how individuals get and use resources.
h. The sum of an organism's activities and interactions as it goes about acquiring and using the resources required to survive and reproduce; the "profession" a species has in the community.
i. Weaken hosts
j. Some species interactions in a community where some species help others in roundabout ways; an example is Canadian lynx eating hares fattened by plants.
k. Both species benefit; a two-way exploitation.

## Complete the Table [p.708]

Complete the following table, which summarizes the effect of different types of species interactions on both species involved.

| Type of Interaction | Effect on Species 1 | Effect on Species 2 |
|---|---|---|
| 19. | helpful | none |
| mutualism | 20. | helpful |
| interspecific competition | harmful | 21. |
| Predation | helpful | 22. |
| 23. | helpful | harmful |

## 41.2. MUTUALISM [p.709]

## 41.3. COMPETITIVE INTERACTIONS [pp.710–711]

### Selected Words

*Heteractis magnifica* [p.709], *Amphiprion perideraion* [p.709], *Paramecium* [pp.710–711]

## Boldfaced Terms

[p.710] interference competition _____

_____

[p.710] exploitive competition _____

_____

[p.710] competitive exclusion _____

_____

[p.711] resource partitioning _____

_____

[p.711] character displacement _____

_____

## Complete the Table

1. Complete the following table by describing how each organism listed is intimately dependent on the other for survival and reproduction in a mutualistic interaction. [p.709]

| Organism | Dependency |
|---|---|
| a. Yucca moth and yucca plant | |
| b. Fungal hyphae and plant roots in a mycorrhiza | |
| c. Anemone fish and sea anemones | |
| d. Endosymbiosis; phagocytic cells and aerobic bacterial cells | |

## Choice

Choose from the following: [pp.710–711]

   a. interference competition     b. exploitative competition     c. competitive exclusion
   d. resource partitioning        e. interspecific competition

2. ___ Bristly foxtail grasses, Indian mallow plants, and smartweed plants coexist in the same habitat.

3. ___ Two species of *Paramecium* are grown in the same culture but cannot coexist indefinitely.

4. ___ One species of chipmunk chases another species of chipmunk out of its habitat.

5. ___ Nine species of chipmunks live in different habitats on the slopes of the Sierra Nevada.

6. ___ Deer have no acorns because blue jays ate most of them.

7. ___ Many sages exude aromatic compounds from their leaves that taint the soil around the plant and prevent potential competitors from taking root.

8. ___ Two species of *Paramecium* do not overlap much in requirements; the two continued to coexist.

## 41.4.  PREDATOR–PREY INTERACTIONS [pp.712–713]

## 41.5.  AN EVOLUTIONARY ARMS RACE [pp.714–715]

### Selected Words

Type I, II, and III models [p.712]

### Boldfaced Terms

[p.712] predators _____

_____

[p.712] prey _____

_____

[p.714] camouflage _____

_____

[p.714] warning coloration _____

_____

[p.714] mimicry _____

_____

### Matching

Match each term with its description. [pp.712–713]

1. ___ coevolution
2. ___ type I model
3. ___ type II model
4. ___ type III model
5. ___ other factors besides individual predator response to prey density
6. ___ lynx and snowshoe hare cycle

a. Predator response is lowest when prey density is low; it is highest at intermediate prey densities, and then levels off.
b. A wolf that has just killed a caribou will not hunt another until it has eaten and digested the first one.
c. The joint evolution of two or more species that exert selection pressure on one another through their close ecological interaction over the generations.
d. Predator and prey reproductive rates affect the interaction as do hiding places for prey, the presence of other prey or predator species, and carrying capacities.
e. Spiders as passive predators; the more flies there are, the more get caught in webs.
f. Oscillations in population densities are explained only by multilevel interactions.

### Choice

Choose from the following: (*Note:* The same letter may be used more than once, but use only one letter per blank.) [pp.714–715]

a. camouflaging     b. warning coloration     c. mimicry     d. cornered animals and plants under attack
e. little or no attempt at concealment     f. adaptive responses of predators

7. ___ Grasshopper mice plunging the noxious chemical-spraying tail end of their beetle prey into the ground to feast on the head end

8. ___ Aggressively stinging yellow jackets are the likely model for nonstinging edible wasps, beetles, and flies.

9. ___ Polar bears stalking seals over ice, striped tigers crouched in tall-stalked, golden grasses, and scorpionfish well hidden on the seafloor

10. ___ Dangerous or repugnant species, such as skunks who spray repellants and poisonous frogs of the genus *Dendrobates*

11. ___ Resemblance of *Lithops*, a desert plant, to a small rock

12. ___ Yellow-banded wasps or an orange-patterned monarch butterfly; predators learn to avoid

13. ___ Least bitten with coloration similar to that of surrounding withered reeds

14. ___ Peach, apricot, and rose seeds loaded with cyanide; lethal ricin in castor bean plants.

15. ___ Animals may startle the predator by hissing, puffing up, showing teeth, or flashing big-eye-shaped spots; opossum and hognose snakes pretend to be dead; many animals secrete of squirt irritating chemical repellents or toxins.

## 41.6. PARASITE–HOST INTERACTIONS [pp.716–717]

## 41.7. COWBIRD CHUTZPAH [p.717]

### Selected Words

*Myxobolus cerebralis* [p.716], *biological controls* [p.717], *Molothrus ater* [p.717]

### Boldfaced Terms

[p.716] parasites _____

_____

[p.716] parasitoids _____

_____

[p.716] social parasites _____

_____

### True–False [pp.716–717]

If the statement is true, write "T" in the blank. If the statement is false, correct it by writing the correct word(s) for the underlined word(s) in the answer blank.

_____ 1. All viruses, some bacteria, protists, fungi, tapeworms, flukes, and some roundworms are notorious invertebrate <u>parasitoids</u>.

_____ 2. Dodder and mistletoe are types of <u>predatory</u> plants.

_____ 3. Social parasites are animals that take advantage of the <u>social</u> behavior of a host species in order to carry out their life cycle.

_____ 4. Brown cowbirds alter the <u>digestive</u> behavior of another species to complete their life cycle.

_____ 5. Because parasites and parasitoids can help control the population growth of other species, many are raised commercially and then selectively released as <u>biological controls</u>.

## Short Answer

6. List the five attributes displayed by effective biological controls. [p.717]

_____

_____

_____

_____

_____

7. List 5 ways parasites can adversely affect their host.

_____

_____

_____

_____

_____

# 41.8. ECOLOGICAL SUCCESSION [pp.718–719]

## Selected Words

*Nature in balance* [p.718], climax community [p.718]

## Boldfaced Terms

[p.718] pioneer species _____

_____

[p.716] primary succession _____

_____

[p.716] secondary succession _____

_____

[p.716] ecological succession _____

_____

[p.719] intermediate disturbance hypothesis _____

_____

## Choice

For questions 1–11, choose from the following: [pp.718–719]

    a. primary succession    b. secondary succession    c. climax model
    d. intermediate disturbance hypothesis

____ 1. A process that begins when pioneer species colonize a barren habitat, such as a new volcanic island and land exposed when a glacier retreats

_____ 2. When trees fall in tropical forests, small gaps open in the dense tree canopy, and more light reaches patches on the forest floor; then conditions favor growth of previously suppressed small trees and germination of pioneers or shade-intolerant species.

_____ 3. Included are lichens and mosses that are small, have short life cycles, and can survive intense sunlight, extreme temperature changes, and nutrient-poor soil.

_____ 4. A disturbed area within a community recovers; if improved soil is still there, secondary succession can be quite rapid.

_____ 5. This pattern is common in abandoned fields, burned forests, and areas cleared by volcanic eruptions.

_____ 6. Many types are mutualists with nitrogen-fixing bacteria, so they can grow in nitrogen-poor habitats.

_____ 7. Many environmental factors vary in their effects across a region.

_____ 8. Once established, the pioneers improve conditions for other species and often set the stage for their own replacement.

_____ 9. In time, organic wastes and remains accumulate, adding volume and nutrients to soil that help more species take hold.

_____ 10. The type that would occur when land is exposed by the retreat of a glacier.

_____ 11. Many small-scale changes recur in patches of habitats; they contribute to the community's internal dynamics.

12. The modern view of succession cites three factors that affect the species composition of a community. List these three factors.

_____

_____

_____

## 41.9. SPECIES INTERACTIONS AND COMMUNITY INSTABILITY [pp.720–721]

## 41.10. EXOTIC INVADERS [pp.722–723]

### Selected Words

_Jump_ dispersal [p.721], kudzu [p.722], _Caulerpa taxifolia_ [p.722], European rabbits [p.723], _myxomatosis_ [p.723]

### Boldfaced Terms

[p.720] keystone species _____

_____

[p.721] geographic dispersal _____

_____

[p.721] exotic species_____

_____

Match each term with its description.

1. ___ community stability [p.718]
2. ___ keystone species [pp.718–719]
3. ___ geographic dispersal [pp.719–721]
4. ___ jump dispersal [p.719]
5. ___ exotic species [p.719]

a. A rapid geographic dispersal mechanism, as when an insect might travel in a ship's cargo hold from an island to the mainland; often takes individuals across regions where they could not survive on their own.
b. An outcome of forces that have come into uneasy balance
c. When residents of established communities move out from their home range and successfully take up residence elsewhere; kudzu, *Caulerpa*, and European rabbits are examples
d. A species that has a disproportionately large effect on a community relative to its abundance; the periwinkle is an example.
e. Residents of established communities disperse from their home range and become established elsewhere; permanently insinuates itself into a new community.

## Choice

Choose from the following. [p.722]

a. kudzu       b. *Caulerpa taxifolia*       c. European rabbits

6. ___ Researchers from Germany's Stuttgart Aquarium developed a hybrid, sterile strain.

7. ___ In 1876, this vine was introduced to the United States; grows faster in the southeast where it now blankets streambanks, trees, telephone poles, houses, and almost everything else.

8. ___ It grows asexually by runners; it has also invaded the coastal waters of the United States.

9. ___ An ideal animal with no natural predators; imported to Australia in 1859

10. ___ Asians use an extracted starch that is used in drinks, herbal medicines, and candy; it may also help save trees as it is an alternative source of paper.

11. ___ The government introduced a myxoma virus that causes myxomatosis; resistant organisms survived to continue a rise in populations.

## 41.11. BIOGEOGRAPHIC PATTERNS IN COMMUNITY STRUCTURE [pp.724–725]

### Selected Word

Surtsey [p.724]

### Boldfaced Terms

[p.724] biogeography _____

_____

[p.725] distance effect _____

_____

[p.725] area effect _____

_____

*Fill in the Blanks* [p.724]

The most striking pattern of biodiversity corresponds to distance from the (1) _____. For most groups of plants and animals, on land and in the seas, the number of coexisting species is greatest in the (2) _____, and it systematically declines toward the poles. Tropical latitudes intercept more incoming intense (3) _____ and receive more rainfall, and their growing season is longer. As one outcome, (4) _____ availability tends to be greater and more reliable in the tropics than elsewhere.

Tropical communities have been (5) _____ for a longer time than temperate ones, some of which did not start forming until the end of the last Ice Age.

Species diversity might be self-reinforcing. Diversity of tree species is much (6) _____ than in comparable forests at higher latitudes. When more plant species compete and coexist, more species of (7) _____ similarly compete and coexist, partly because no single (8) _____ can overcome all of the chemical defenses of all the different plants. More (9) _____ and (10) _____ species tend to evolve in response to a diversity of prey and hosts. The same applies to diversity on tropical (11) _____ [p.722].

*Species Diversity Problem*

12. Study of the distance effect, the area effect, and species diversity patterns as they relate to the equator, and answer the following question. There are two islands (b and c) of the same size and topography that are equidistant from the African coast (a), as shown in the diagram. Which will have the higher species diversity values? [p.724]

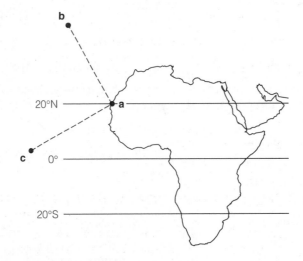

## 41.12. THREATS TO BIODIVERSITY [pp.726–727]

## 41.13. SUSTAINING BIODIVERSITY AND HUMAN POPULATIONS [pp.728–729]

*Selected Words*

Ecotourism [p.729], strip logging [p.729]

*Boldfaced Terms*

[p.726] endangered species _____

_____

[p.726] habitat loss _____

_____

[p.726] indicator species _____

_____

[p.726] conservation biology _____

_____

[p.728] hot spots _____

_____

[p.728] ecoregion _____

_____

[p.729] riparian zone _____

_____

## Matching

Match each term with its description. [pp.726–729]

1. ___ sixth major extinction event
2. ___ endangered species
3. ___ habitat loss
4. ___ indicator species
5. ___ conservation biology
6. ___ hot spots
7. ___ ecoregion
8. ___ strip logging
9. ___ riparian zone

a. Its major goal is to conserve and utilize, in sustainable ways, as much biodiversity as possible.
b. This is a broad land or ocean region defined by climate, geography, and producer species; 25 are high-priority targets for conservation efforts.
c. This is used in portions of forests that are sloped and have a number of streams; removal can be sustained even while the maximum biodiversity is being maintained.
d. This is any endemic species that is highly vulnerable to extinction.
e. These are species that help warn us of changes in habitats and impending loss of diversity; birds are a good example.
f. This is a narrow corridor of vegetation along a stream or river; defense against flood damage and shade to help conserve water are major benefits.
g. These are habitats of a large number of species that are found nowhere else and in greatest danger of extinction; widely separated forests of Mexico are an example.
h. This is the physical reduction in suitable places to live, as well as the loss of suitable habitat through pollution; puts major pressure on more than 90 percent of the endemic species that now face extinction.
i. This is underway; there is evidence that rapid population growth of humans is driving many species over the edge with large-scale habitat losses, species introductions, and overharvesting.

# Self-Quiz

___ 1. _____ have physical and chemical features, such as temperature, and an array of species. [p.708]
   a. habitats
   b. niches
   c. interactions
   d. ecosystems
   e. climates

___ 2. A lopsided interaction that directly benefits one species but does not affect the other much, if at all, is _____. [p.708]
   a. commensalism
   b. competitive exclusion
   c. predation
   d. mutualism
   e. parasitism

___ 3. The relationship between the yucca plant and the yucca moth that pollinates it is best described as _____. [p.709]
   a. camouflage
   b. commensalism
   c. competitive exclusion
   d. obligatory mutualism
   e. mimicry

___ 4. _____ is represented by foxtail grass, mallow plants, and smartweed, because their root systems subdivide different areas of the soil in a field. [p.711]
   a. succession
   b. resource partitioning
   c. a climax community
   d. a disturbance
   e. competitive exclusion

___ 5. G. Gause grew two species of *Paramecium* separately and then together. They required identical resources and could not coexist indefinitely. This is an example of _____. [p.710]
   a. warning coloration
   b. mimicry
   c. camouflage
   d. competitive exclusion
   e. obligatory mutualism

___ 6. Edible insect species often resemble toxic or unpalatable species that are not at all closely related. This is an example of _____. [p.714]
   a. mimicry
   b. camouflage
   c. a prey defense
   d. warning coloration
   e. both c and d

___ 7. _____ is among the pervasive influences that parasites have on populations. [p.716]
   a. draining host nutrients
   b. predator vulnerability
   c. sterility
   d. death
   e. all of the above

___ 8. During the process of primary succession, _____. [p.718]
   a. pioneer populations colonize a barren habitat
   b. pioneers include lichens and mosses
   c. pioneers improve conditions for other species
   d. pioneers set the stage for their own replacement
   e. all of the above

___ 9. The _____ is an example of a keystone species. [p.720]
   a. kudzu vine
   b. sea star
   c. mussel
   d. European rabbit
   e. *Caulerpa*

___ 10. The most striking patterns of biodiversity corresponds to _____. [p.724]
   a. distance effect
   b. area effect
   c. immigration rate for new species
   d. distance from the equator
   e. resource availability

# Chapter Objectives/Review Questions

1. Discuss possible controls for the two species of invading Argentine fire ants. [p.706]
2. List the five factors that shape the structure of a biological community. [p.708]

3. A _____ niche is the one that might prevail in the absence of competition and other factors that can constrain how individuals get and use resources. [p.708]
4. Be able to define the following terms and cite examples: *habitat, community, niche, fundamental niche, realized niche, indirect interactions, commensalism, mutualism, interspecific competition, predation, parasitism,* and *symbiosis*. [p.708]
5. Lichens and mycorrhizae are cases of _____ mutualism. [p.709]
6. Describe and cite examples of interference competition, exploitative competition, competitive exclusion, and resource partitioning. [pp.710–711]
7. Describe the interactions of predator-prey, warning coloration, mimicry, camouflage, when an animal is cornered, and predator responses to prey; cite examples. [pp.712–715]
8. Define and list one example each of types I, II, and III models for predator—prey reactions. [p.712]
9. Be able to list the adaptations of prey and the adaptive responses of predators. [p.712]
10. _____ are insects that develop inside another species of insect, which they devour from the inside out as they mature. [p.716]
11. List pervasive influences that parasites have on populations, the kinds of parasites and tell how parasites may be potential biological control agents. [pp.716–717]
12. Describe the patterns of ecological succession; include the following terms: *pioneer species, climax community, primary* and *secondary succession*. [pp.718–719]
13. List and explain examples of forces leading to community instability. [pp.720–721]
14. Explain why kudzu, *Caulerpa*, and European rabbits are exotic invaders that lead to community instability. [pp.722–723]
15. Describe mainland, marine, and island patterns of diversity. [pp.724–725]
16. Be able to define the following terms and cite examples as they relate to threats to biodiversity and sustaining biodiversity: *hot spots, ecoregion, sixth major extinction, endangered species, habitat loss, indicator species, conservation biology, strip logging,* and *riparian zone*. [pp.726–728]

## Chapter Summary

A community consists of all (1) _____ in a habitat. Each species has a (2) _____, which is the sum of its

activities and relationships. A habitat's history, its biological and physical characteristics, and interactions

among species in the habitat affect (3) _____ structure. Interspecific interactions include (4) _____,

mutualism, competition, (5) _____, and parasitism. They influence the population (6) _____ of

participating species, which in turn influences the community's structure. A community changes over time.

Factors that affect this change include: physical characteristics of the habitat, (7) _____ _____,

(8) _____, and chance events.

(9) _____ identify regional patterns in species distribution. The (10) _____

regions hold the greatest number of species. The characteristics of islands can be used to predict how many

(11) _____ an island will hold. (12) _____ biologists are working to preserve species richness by

resource management that is not in conflict with the human need for survival.

## Integrating and Applying Key Concepts

1. Is there a *fundamental niche* that is occupied by humans? If you think so, describe the minimal abiotic and biotic conditions required by populations of humans in order to live and reproduce. (Note that *thrive* and *be happy* are not criteria.) If you do not think so, state why.

# 42

# ECOSYSTEMS

## INTRODUCTION

Ecosystems represent an interaction between the biotic (living) and abiotic (nonliving) factors in a given area. This chapter explores how energy moves through an ecosystem. This is followed by an introduction to the cycling of biologically important elements in ecosystems.

## FOCAL POINTS

- Figure 42.2 [p.734] illustrates the difference between nutrient and energy cycling in an ecosystem.
- Figure 42.11 [p.741] provides an overview of biogeochemical cycles.
- Figure 42.12 [p.742] illustrates the water cycle.
- Figure 42.15 [pp.744–745] illustrates the carbon cycle.

## Interactive Exercises

*Bye-Bye, Blue Bayou* [pp.732–733]

## 42.1. THE NATURE OF ECOSYSTEMS [pp.734–735]

## 42.2. BIOLOGICAL MAGNIFICATION IN FOOD WEBS [pp.738–739]

### Selected Words

"Self-feeders" [p.734], *herbivores* [p.734], *carnivores* [p.734], *parasites* [p.734], *omnivores* [p.734], *troph* [p.734], *herbicides* [p.738], *insecticides* [p.738], *fungicides* [p.738]

### Boldfaced Terms

[p.734] ecosystem _____

_____

[p.734] primary producers _____

_____

[p.734] consumers _____

_____

[p.734] detritivores _____

_____

[p.734] decomposers _____

_____

[p.734] trophic levels _____

_____

[p.737] food chain _____

_____

[p.737] food webs _____

_____

[p.735] grazing food web _____

_____

[p.735] detrital food web _____

_____

[p.739] biological magnification _____

_____

## Matching

Match each of the following terms with the correct description. [pp.734–735]

1. _____ herbivores
2. _____ carnivores
3. _____ omnivores
4. _____ parasites
5. _____ consumers
6. _____ ecosystem
7. _____ primary producers
8. _____ detritivores
9. _____ trophic levels
10. _____ food chain
11. _____ food web
12. _____ decomposers

a. autotrophic organisms that obtain energy from the sun
b. consumers that eat plants
c. consumers that live inside a host and eat its tissues
d. consumers that eat the flesh of animals
e. consume small organic matter called detritus
f. hierarchy of feeding relationships
g. a food chain that has cross-connected levels
h. consumers that eat both plants and animals
i. feed on organic wastes and break them down into inorganic building blocks
j. sequence of steps that transfers energy from producer to organisms in higher trophic levels
k. the general term for heterotrophs that eat other organisms
l. an array of organisms and the physical environment

*Choice*

The figure below represents a theoretical food web. The numbers in the circles represent different species. Organisms 1, 2, and 3 are producers. The arrows indicate the direction of energy flow. For each statement, list the correct numbers. [pp.736–737]

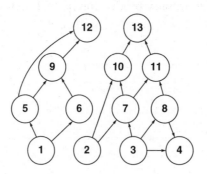

13. Which organisms are the decomposers? _____

14. Which organisms represent only the first trophic level? _____

15. Which organisms represent only the second trophic level? _____

16. Which organisms represent only the third trophic level? _____

17. Which organism(s) represent(s) only the fourth trophic level? _____

18. Which organisms represent more than one trophic level? _____

19. Which organism(s) could be classified as an herbivore? _____

20. Which organism(s) could be classified as an omnivore? _____

*TNOW:* For an animation of this material, refer to Chapter 42: Co-op Activity 18.1: Food Chains and Chapter 42: Co-op Activity 18.2: Food Webs.

*Short Answer*

21. Distinguish between a grazing and detrital food web. [p.737]

_____

_____

_____

*TNOW:* For an animation of this question, refer to Chapter 42: Art Labeling: Models of Energy Transfer in an Ecosystem.

*Short Answer*

21. Define biological magnification. [p.738]

_____

_____

_____

*TNOW:* For an animation of this question, refer to Chapter 42: Co-op Activity 18.5: Bioaccumulation.

## 42.3. STUDYING ENERGY FLOW THROUGH ECOSYSTEMS [pp.740–741]

## 42.4. BIOGEOCHEMICAL CYCLES [p.741]

### Selected Words

*Gross* primary production [p.740], *net* primary production [p.740], *hydrologic* cycle [p.741], *atmospheric* cycle [p.741], *sedimentary* cycle [p.741]

### Boldfaced Terms

[p.740] primary productivity _____

_____

[p.740] net ecosystem production _____

_____

[p.740] biomass pyramid _____

_____

[p.741] energy pyramid _____

_____

[p.741] biogeochemical cycle _____

_____

### Matching

Match each of the following statements to the correct term. [pp.740–741]

1. _____ rate at which producers capture and store energy in a given interval

2. _____ dry weight of all organisms at each trophic level of an ecosystem

3. _____ the movement of elements from reservoirs to ecosystems and back to reservoirs

4. _____ biogeochemical cycle that involves the movement of carbon and hydrogen

5. _____ biogeochemical cycle that involves elements that do not have a gaseous form

6. _____ gross primary production minus energy that the producers, detritivores, and decomposers require

7. _____ biogeochemical cycle that involves the gaseous form of a nutrient

8. _____ illustrates how usable energy is reduced through an ecosystem

a. atmospheric cycles
b. sedimentary cycles
c. hydrologic cycles
d. energy pyramid
e. biogeochemical cycles
f. primary productivity
g. net ecosystem production
h. biomass pyramid

## Labeling

Provide the missing term for each number in the diagram below. [p.741]

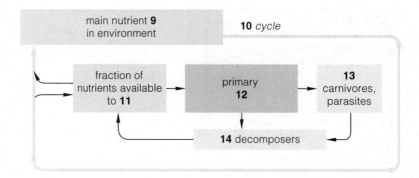

9. _____     12. _____

10. _____     13. _____

11. _____     14. _____

## 42.5. THE WATER CYCLE [pp.742–743]

### Selected Words

*Evaporation* [p.742], *transpiration* [p.742], *condensation* [p.742], *precipitation* [p.742]

### Boldfaced Terms

[p.742] hydrologic cycle _____

_____

[p.742] watershed _____

_____

[p.742] aquifers _____

_____

[p.742] groundwater _____

_____

[p.742] runoff _____

_____

[p.742] salinization _____

_____

[p.743] desalinization _____

_____

## Concept Map

Provide the missing terms for the numbers in the below concept map of the water cycle. [pp.742–743]

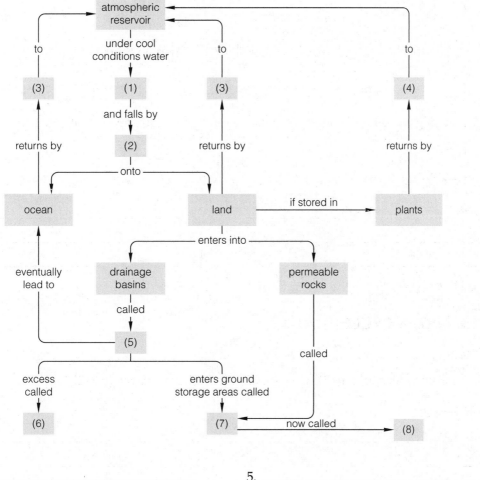

1. _____
2. _____
3. _____
4. _____

5. _____
6. _____
7. _____
8. _____

## 42.6.  CARBON CYCLE [pp.744–745]

## 42.7.  GREENHOUSE GASES [pp.746–747]

### Selected Words

"Fix" [p.745], carbon dioxide [p.746], methane [p.746], chlorofluorocarbons [p.746], "greenhouse gases" [p.746]

### Boldfaced Terms

[p.744] carbon cycle _____

_____

[p.744] carbon-oxygen cycle _____

_____

[p.746] greenhouse effect _____

_____

[p.746] global warming _____

_____

## Concept Map

Provide the missing terms for the numbers in the following concept map of the carbon cycle. [pp.744–745]

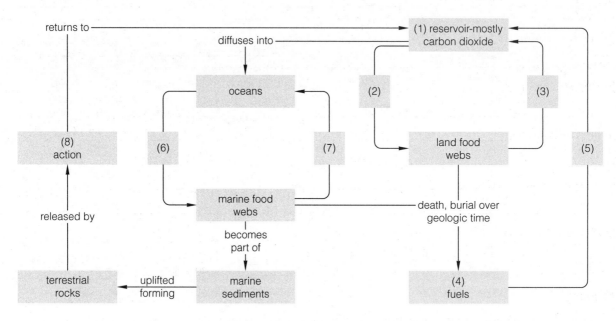

1. _____    5. _____
2. _____    6. _____
3. _____    7. _____
4. _____    8. _____

## Short Answer

9. Describe the relationship between greenhouse gases and global warming. [pp.746–747]

_____

_____

_____

_____

_____

## 42.8. NITROGEN CYCLE [pp.748–749]

## 42.9. PHOSPHOROUS CYCLE [pp.750–751]

### Selected Words

*Rhizobium* [p.748]

### Boldfaced Terms

[p.748] nitrogen cycle _____

_____

[p.748] nitrogen fixation _____

_____

[p.749] ammonification_____

_____

[p.749] nitrification _____

_____

[p.749] denitrification _____

_____

[p.750] phosphorous cycle _____

_____

[p.751] eutrophication

_____

### Choice

For each of the following terms or statements, choose the appropriate cycle from the list below.

> a. nitrogen cycle [pp.748–749]     b. phosphorous cycle [pp.750–751]
> c. both the nitrogen and phosphorous cycles [pp.748–751]

1. _____ Earth's crust is the largest reservoir.

2. _____ Eutrophication may change this nutrient's concentration in an ecosystem.

3. _____ The atmosphere is the largest reservoir.

4. _____ *Rhizobium* bacteria are involved in fixing this for use by organisms.

5. _____ Deforestation and conversion to grassland cause losses.

6. _____ Use of fertilizers disrupts normal cycles.

7. _____ Ammonium is one form.

8. _____ Moved by hydrologic cycle through ecosystems.

## Self-Quiz

1. Which of the following breaks down organic material into inorganic building blocks? [p.734]
   a. herbivores
   b. detritivores
   c. decomposers
   d. omnivores
   e. none of the above

2. In an ecosystem, energy _____ the ecosystem while nutrients _____ the ecosystem. [p.734]
   a. flows through; flow through
   b. flows through; cycle within
   c. cycle within; flow through
   d. cycle within; cycle within

3. The members of feeding relationships are structured in a hierarchy, the steps of which are called _____. [p.734]
   a. predator-prey relationships
   b. trophic levels
   c. feeding groups
   d. functional groups

4. Which of the following would be considered to be an example of biological magnification? [p.739]
   a. carbon
   b. nitrogen
   c. DDT
   d. glucose

5. In a _____ food web, energy flows from producers to herbivores, then to carnivores, and then decomposers. [p.737]
   a. primary
   b. secondary
   c. detrital
   d. grazing
   e. atmospheric

6. Which of the following does not typically have a major reservoir in the atmosphere? [p.750]
   a. nitrogen
   b. carbon
   c. water
   d. phosphorous

7. The dry weight of all of the organisms in an ecosystem is called the _____. [p.740]
   a. biomass
   b. biototal
   c. gross productivity
   d. net energy

8. Global warming is directly associated with which of the following cycles? [pp.746–747]
   a. phosphorous
   b. water
   c. carbon
   d. nitrogen

9. Which of the following makes up the majority of the atmosphere? [p.748]
   a. carbon dioxide
   b. water
   c. nitrogen
   d. oxygen

10. Which of the following captures solar energy for an ecosystem? [p.734]
    a. detritivores
    b. decomposers
    c. primary producers
    d. herbivores
    e. parasites

## Chapter Objectives/Review Questions

1. Be able to define the term ecosystem. [p.734]
2. Understand the role of producers, consumers, detritivores, and decomposers in an ecosystem. [p.734]
3. Understand the concept of a trophic level. [p.734]
4. Distinguish between a food chain and a food web. [p.735]
5. Distinguish between a detrital and grazing food web. [p.737]
6. Understand the relationship between a food web and biological magnification. [p.739]
7. Understand the relationship between productivity and biomass. [p.740]
8. Understand the principles of the biogeochemical cycles. [p.741]

9. Be able to recognize terms associated with the water cycle. [pp.742–743]
10. Understand the major components of the carbon cycle. [pp.744–745]
11. Be able to define a greenhouse gas. [p.746]
12. Recognize the relationship between greenhouse gases and global warming. [pp.746–747]
13. Understand the terminology and major steps of the nitrogen cycle. [pp.748–749]
14. Understand the major steps of the phosphorous cycle. [pp.750–751]
15. Define *eutrophication*. [p.751]

## Chapter Summary

An (1) _____ consists of a community and its physical environment. A one-way flow of (2) _____ and a cycling of raw materials among its interacting participants maintain it. It is an open system, with inputs and outputs of energy and nutrients.

Food (3) _____ are linear sequences of feeding relationships, from producers through consumers, decomposers, and (4) _____. The chains cross-connect as food (5) _____. Most of the energy that enters a food web returns to the environment, mainly as metabolic heat. Most the nutrients are cycled; some reenter the environment. (6) _____ is the increasing concentration of a substance in tissues of organisms as it moves up food chains.

Primary (7) _____ is the rate at which an ecosystem's producers capture and store energy in their tissues during a given interval. The number of producers and the balance between (8) _____ and aerobic respiration influence the amount stored.

The availability of water, (9) _____, (10) _____, phosphorous, and other substances influences primary productivity. Ions or molecules of these substances move slowly in global cycles, from environmental (11) _____, into food webs, then back to reservoirs. (12) _____ activities can disrupt these cycles.

## Integrating and Applying Key Concepts

In 1971, *Diet for a Small Planet* was published. Frances Moore Lappé, the author, felt that people in the United States wasted protein and ate too much meat. She said, "We have created a national consumption pattern in which the majority, who can pay, overconsume the most inefficient livestock products [cattle] well beyond their biological needs (even to the point of jeopardizing their health), while the minority, who cannot pay, are inadequately fed, even to the point of malnutrition." Cases of marasmus (a nutritional disease caused by prolonged lack of food calories) and kwashiorkor (caused by severe, long-term protein deficiency) have been found in Nashville, Tennessee, and on an Indian reservation in Arizona, respectively. Lappé's partial solution to the problem was to encourage people to get as much of their protein as possible directly from plants and to supplement that with less meat from the more efficient converters of grain to protein (chickens, turkeys, and hogs) and with seafood and dairy products. Most of us realize that feeding the hungry people of the world is not just a matter of distributing the abundance that exists—it is also a matter of political, economic, and cultural factors. Yet, it is still valuable to consider applying Lappé's idea to our everyday living. Devise two full days of breakfasts, lunches, and dinners that would enable you to exploit the lowest acceptable trophic levels to sustain yourself healthfully.

# 43

# THE BIOSPHERE

## INTRODUCTION

This chapter takes a look at the biosphere, or Earth. The chapter begins with a discussion of climate and then introduces how climate relates to the major ecosystems, both terrestrial and aquatic, on the planet.

## FOCAL POINTS

- Figure 43.12 [pp.762–763] presents a worldwide view of the distribution of major terrestrial ecosystems.
- Figure 43.35 [p.778] diagrams the major oceanic zones.

## Interactive Exercises

*Surfers, Seals, and the Sea* [pp.754–755]

## 43.1. GLOBAL AIR CIRCULATION PATTERNS [pp.756–757]

## 43.2. CIRCULATING AIRBORNE POLLUTANTS [pp.758–759]

## 43.3. THE OCEAN, LANDFORMS, AND CLIMATES [pp.760–761]

*Selected Words*

Biosphere [p.755], *hydrosphere* [p.755], *lithosphere* [p.755], *atmosphere* [p.755], "visual pollution" [p.757], "ozone hole" [p.758], chlorofluorocarbons [p.758], *industrial* smog [p.758], *phytochemical* smog [p.759], *leeward* [p.761], monsoon [p.761]

## Boldfaced Terms

[p.756] climate _____

_____

[p.758] pollutants _____

_____

[p.758] ozone layer _____

_____

[p.758] smog _____

_____

[p.759] acid rain _____

_____

[p.761] rain shadow _____

_____

## Matching

Match each of the following terms to its correct definition.

_____ 1. biosphere [p.755]

_____ 2. climate [p.756]

_____ 3. atmosphere [p.757]

_____ 4. solar-hydrogen energy [p.757]

_____ 5. ozone layer [p.758]

_____ 6. hydrosphere [p.755]

_____ 7. phytochemical smog [p.759]

_____ 8. industrial smog [p.758]

_____ 9. acid rain [p.759]

_____ 10. rain shadow [p.761]

_____ 11. monsoons [p.761]

a. gases and airborne particles that envelop the Earth

b. average humidity, cloud cover, temperature, and wind speed, over time

c. cold, wet winters and thermal inversions cause this form of pollution

d. the influence of mountains on precipitation patterns

e. a thin layer of the atmosphere that contains high levels of $O_3$

f. Nitrous oxides and sulfur dioxides in the atmosphere produce this form of pollution.

g. the ocean, ice caps, and other bodies of water, liquid or frozen

h. an alternation of dry and wet seasons, producing periods of heavy rainfall

i. this occurs in cities that are geographically located in basins, such as Los Angeles

j. the sum of all places on Earth where life is found

k. the use of energy from the sun to split water and generate hydrogen for use as a fuel

## Labeling

Label each of the indicated portions of the diagram below. [pp.756–757]

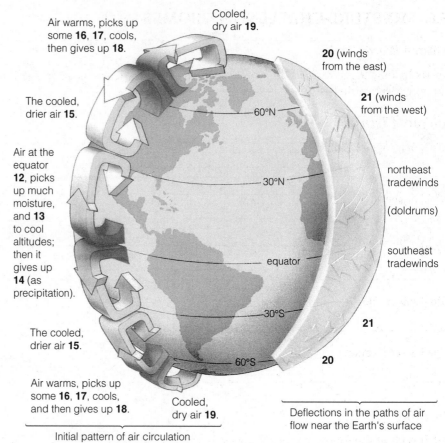

Air warms, picks up some **16**, **17**, cools, then gives up **18**.

Cooled, dry air **19**.

**20** (winds from the east)

**21** (winds from the west)

The cooled, drier air **15**.

60°N

northeast tradewinds

Air at the equator **12**, picks up much moisture, and **13** to cool altitudes; then it gives up **14** (as precipitation).

30°N

(doldrums)

equator

southeast tradewinds

The cooled, drier air **15**.

30°S

**21**

Air warms, picks up some **16**, **17**, cools, and then gives up **18**.

60°S

**20**

Cooled, dry air **19**.

Initial pattern of air circulation

Deflections in the paths of air flow near the Earth's surface

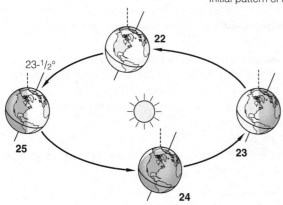

**22**

23-1/2°

**25**

**23**

**24**

12. _____     19. _____

13. _____     20. _____

14. _____     21. _____

15. _____     22. _____

16. _____     23. _____

17. _____     24. _____

18. _____     25. _____

## 43.4. BIOGEOGRAPHIC REALMS AND BIOMES [pp.762–763]

## 43.5. AVAILABILITY OF SUNLIGHT, SOILS, AND MOISTURE [pp.764–765]

## 43.6. MOISTURE-CHALLENGED BIOMES [pp.766–767]

### Selected Words

*Deserts* [p.766], *dry shrublands* [p.766], *dry woodlands* [p.766], *grasslands* [p.766], savannas [p.766], shortgrass and tallgrass prairie [p.767]

### Boldfaced Terms

[p.762] biogeographic realms _____

_____

[p.762] biomes _____

_____

[p.764] soil _____

_____

[p.764] soil profile _____

_____

[p.766] desertification _____

_____


### Matching

Match each of the following terms with the correct definition.

_____ 1. soil [p.764]

_____ 2. soil profile [p.764]

_____ 3. biogeographic realms [p.762]

_____ 4. biomes [p.762]

a. mixtures of mineral particles and decomposing organic matter
b. a layered structure that develops over time
c. a division of the Earth into six large regions where you can expect to find certain plants and animals
d. a subdivision of a realm, represents evolutionary processes

*TNOW:* For an interactive exercise on biogeographic realms, refer to Chapter 43: Art Labeling: Biogeographic Realms.

*Choice*

For each statement, choose the correct biome. Some answers may be used more than once. [pp.766–767]

>     a. savannas      b. shortgrass and tallgrass prairies      c. dry shrublands
>     d. dry woodlands      e. grasslands

_____ 5. precipitation is between 25–60 cm per year

_____ 6. precipitation is between 40–100 cm per year

_____ 7. tall trees, but not a continuous canopy

_____ 8. warm summers and cold winters, not enough precipitation to support forests

_____ 9. broad belts of grasslands with few shrubs and trees

_____ 10. vulnerable to lightening-sparked, wind-driven firestorms

_____ 11. a form of grassland in North America, now largely destroyed

## 43.7.  MORE RAIN, BROADLEAF FORESTS [p.768]

## 43.8.  YOU AND THE TROPICAL FORESTS [p.769]

## 43.9.  CONIFEROUS FORESTS [p.770]

## 43.10.  BRIEF SUMMERS AND LONG, ICY WINTERS [p.772]

*Selected Words*

*Evergreen broadleaf* forests [p.768], *semi-evergreen* forests [p.768], *tropical deciduous* forests [p.768], "deciduous" [p.768], *temperate* deciduous forests [p.768], *coniferous* forests [p.770], *taigas* [p.770], "swamp forests" [p.770], *pine barrens* [p.770], *artic* tundra [p.771], *alpine* tundra [p.771]

*Boldfaced Terms*

[p.771] permafrost _____

---

*Matching*

Match each of the following to its correct definition.

_____ 1. evergreen broadleaf forests [p.768]

_____ 2. semi-evergreen forests [p.768]

_____ 3. coniferous forests [p.770]

_____ 4. taigas [p.770]

_____ 5. artic tundra [p.771]

_____ 6. alpine tundra [p.771]

_____ 7. permafrost [p.771]

_____ 8. pine barrens [p.770]

_____ 9. tropical deciduous forests [p.768]

_____ 10. temperate deciduous forests [p.768]

a. trees dormant in winter; rainfall between 50 and 150 cm per year

b. a mixed forest of pines and shrubs in a sandy, acidic soil

c. rainfall between 130 and 200 cm per year, average temperature of 25°C

d. longer dry seasons than evergreen broadleaf forests; slow decomposition

e. a frozen layer with anaerobic growing conditions

f. dry and cold, but occurs in the high mountains

g. dry and cold for most of year, rainfall less than 25 cm per year

h. also known as swamp forests

i. dominated by evergreen trees with needle-shaped cones

j. trees that shed leaves at the start of the dry season once a year

## 43.11. FRESHWATER ECOSYSTEMS [pp.772–773]

## 43.12. "FRESH" WATER? [p.774]

## 43.13. LIFE AT LAND'S END [pp.774–775]

## 43.14. THE ONCE AND FUTURE REEFS [pp.776–777]

## 43.15. THE OPEN OCEAN [pp.778–779]

### Selected Words

Spring overturn [p.773], *oligotrophic* [p.773], *eutrophic* [p.773], "dead zone" [p.773], *mangrove wetland* [p.774], "mangrove" [p.774], *upper* littoral zone [p.775], *mid*littoral zone [p.775], *lower* littoral zone [p.775], *phyto*plankton [p.778], *zoo*plankton [p.778], *sandy* shores [p.771], *muddy* shores [p.771], *coral bleaching* [p.772]

### Boldfaced Terms

[p.772] lake _____

_____

[p.772] thermocline _____

_____

[p.772] fall overturn _____

_____

[p.773] eutrophication _____

_____

[p.773] streams _____

_____

[p.773] river _____

_____

[p.774] estuary _____

_____

[p.776] coral reefs _____

_____

[p.778] marine snow _____

_____

[p.778] seamounts _____

_____

[p.778] hydrothermal vents _____

_____

[p.778] upwelling _____

_____

[p.778] downwelling _____

_____

[pp.778–779] El Niño _____

_____

## Matching

Match each of the following terms to the correct definition.

_____ 1. fall overturn [p.772]

_____ 2. eutrophication [p.773]

_____ 3. upwelling [p.776]

_____ 4. downwelling [p.776]

_____ 5. estuaries [p.774]

_____ 6. spring overturn [p.772]

_____ 7. marine snow [p.776]

a. Water flowing into a coastline is forced downward and away from it.

b. A natural or artificial process that enriches a body of water with nutrients

c. The action of winds move dissolved oxygen into the depths of a lake.

d. The Earth's rotational force deflects slow-moving water from the coast; it is replaced vertically by cold, deep water.

e. The downward movement of small bits of organic matter; the basis of deep-sea food webs

f. Partially enclosed areas where sea water mixes with nutrient-rich fresh water

g. As lake water cools in the fall, it becomes more dense and sinks.

## Choice

For the following questions, choose from the following answers. Some answers may be used more than once.

    a. stream ecosystems [p.773]    b. lake ecosystems [pp.772–773]
    c. wetlands and intertidal zones [pp.772–773]    d. coral reefs [pp.775–776]
    e. coastlines [p.775]

_____ 8. May be either oligotrophic or eutrophic

_____ 9. In response to environmental stress, the symbiotic relationships may break.

_____ 10. Contain riffles, pools, and runs

_____ 11. Consist of estuaries with complex food webs

_____ 12. These consist of three zones, the upper, middle, and lower littoral zones

_____ 13. Nutrients may be cycled by spring and fall overturn.

_____ 14. Represents the accumulated remains of marine organisms

_____ 15. Geography, altitude, and precipitation influence the characteristics of this system.

## Labeling

Label the indicated areas of the ocean in the diagram below. [p.778]

16. _____

17. _____

18. _____

19. _____

20. _____

## 43.16.  CLIMATE, COPEPODS, AND CHOLERA [pp.780–781]

### Selected Terms

"Southern oscillation" [p.780], cholera [p.780]

### Boldfaced Terms

[p.780] ENSO _____

_____

[p.780] La Niña _____

_____

### Short Answer

1. What is the difference between an El Niño and La Niña event? [p.780] _____

_____

_____

_____

_____

2. What is the cause of cholera? [pp.780–781] _____

_____

_____

_____

## Self-Quiz

1. Tropical rain forests belong to the _____ biome. [pp.768–769]
   a. evergreen broadleaf
   b. deciduous broadleaf
   c. coniferous
   d. taigas

2. Permafrost would be found in which of the following? [p.770]
   a. taiga
   b. artic tundra
   c. alpine tundra
   d. all of the above

3. Which one of these has the lowest amount of annual rainfall? [pp.766–767]
   a. grassland
   b. dry woodland
   c. dry shrubland
   d. prairies
   e. all of the above are equal

4. A rain shadow is associated with _____. [p.763]
   a. eutrophication
   b. acid rain
   c. mountains
   d. El Niño
   e. La Niña

5. This aquatic ecosystem starts as a seep. [p.773]
   a. lakes
   b. wetlands
   c. oceans
   d. streams
   e. all of the above

6. CFCs are directly related to the severity of _____. [p.758]
   a. global warming
   b. ozone thinning
   c. eutrophication
   d. thermal inversions
   e. acid rain

7. Sulfur dioxides and nitrogen oxides contribute to the problem of _____. [p.759]
   a. global warming
   b. ozone thinning
   c. eutrophication
   d. thermal inversions
   e. acid rain

8. The average of all weather conditions for a region is called the _____. [p.756]
   a. temperature zone
   b. rain shadow
   c. climate
   d. biome
   e. biogeographic realm

9. Desertification affects which of the following? [p.766]
   a. grasslands
   b. intertidal zones
   c. wetlands
   d. broadleaf deciduous forests

10. Which of the following soil horizons is typically located closest to the surface? [pp.764–756]
    a. A
    b. B
    c. C
    d. O

## Chapter Objectives/Review Questions

1. Explain how climate is related to the heating and movement of the atmosphere. [p.756]
2. Explain the potential benefits of solar-hydrogen energy. [p.757]
3. Describe the causes of ozone thinning, temperature inversions, and acid rain. [pp.758–759]
4. Understand the concept of an ocean current. [p.760]

5. Describe the relationship between mountains and rainfall patterns. [p.761]
6. Explain the relationship between a biogeographic realm and a biome. [p.762]
7. Recognize what is meant by a soil profile. [pp.764–765]
8. List the key characteristics of deserts, dry shrublands, dry woodlands, grasslands, and prairies. [pp.766–767]
9. Distinguish among evergreen broadleaf forests, deciduous broadleaf forests, and coniferous forests. [pp.768–770]
10. Understand the concept of permafrost and its relationship to the types of tundra. [p.771]
11. Define a *lake* and *stream* ecosystem. [pp.772–773]
12. Explain why eutrophication is dangerous for a lake ecosystem. [p.773]
13. Give the key characteristics of a wetland, intertidal zone, estuary, coastline, and coral reef. [pp.774–775]
14. Explain the importance of marine snow, upwelling, and downwelling with regards to the movement of ocean nutrients. [pp.778–779]
15. Explain the process of ENSOs and La Niña events. [pp.780–781]

# Chapter Summary

Air circulation patterns start with (1) _____ differences in (2) _____ inputs from the sun, Earth's rotation and orbit, and the (3) _____ of land and seas. These factors give rise to great weather systems and regional (4) _____.

Interactions among (5) _____ currents, air (6) _____ patterns, and landforms produce regional climates, which affect the distribution and dominant features of (7) _____.

(8) _____ realms are vast regions characterized by species that evolved nowhere else. They are subdivided into (9) _____, or regions characterized mainly by the dominant vegetation. Sunlight intensity, moisture, (10) _____ type, species interactions, and evolutionary history vary within and between biomes.

Water provinces cover more than (11) _____ percent of the Earth's surface. All freshwater and marine ecosystems have gradients in (12) _____ availability, temperature, and dissolved (13) _____ that vary daily and seasonally. The variations influence primary (14) _____.

Understanding interactions among the atmosphere, (15) _____ and land can lead to discoveries about specific events—in one case, recurring cholera (16) _____ — that impact human life.

# Integrating and Applying Key Concepts

1. One species, *Homo sapiens*, uses about 40 percent of all of Earth's productivity, and its representatives have invaded every biome, either by living there or by dumping waste products there. Many of Earth's residents are being denied the minimal resources they need to survive, while human populations continue to increase exponentially. Can you suggest a better way of keeping Earth's biomes healthy while providing at least the minimal needs of all Earth's residents (not just humans)? If so, outline the requirements of such a system and devise a way in which it could be established.
2. Evidence is mounting that we are experiencing a period of global climate change that is frequently called "global warming." Choose the type of ecosystem that you live in and suggest what would occur if temperature were to increase? What type of ecosystem would your be turning into?

# 44

# BEHAVIORAL ECOLOGY

## INTRODUCTION

The last chapter of this text examines the concept of behavior, exploring the differences between instinctive and learned behavior and giving examples of many different types of behaviors in the animal kingdom.

## FOCAL POINTS

- Section 44.2 [p.787] compares instinctive and learned behaviors.

## Interactive Exercises

*My Pheromones Made Me Do It* [pp.784–785]

### 44.1. BEHAVIOR'S HERITABLE BASIS [p.786]

### 44.2. INSTINCT AND LEARNING [p.787]

*Selected Words*

"Killer bees" [p.784], *intermediate* response [p.786]

*Boldfaced Terms*

[p.786] stimulus _____

_____

[p.787] instinctive behavior _____

_____

[p.787] fixed action pattern _____

_____

[p.787] learned behavior _____

_____

[p.787] imprinting _____

_____

## Matching

Choose the most appropriate answer for each term. [pp.786–787]

_____ 1. intermediate response

_____ 2. instinctive behavior

_____ 3. stimulus

_____ 4. fixed action pattern

_____ 5. learned behavior

_____ 6. imprinting

a. Animals process and integrate information gained from experiences, then use it to vary or change responses to stimuli.
b. Information from the environment that a receptor has detected
c. Term applied to genetically based behavioral reactions of hybrid offspring
d. Time-dependent form of learning; triggered by exposure to sign stimuli and usually occurring during sensitive periods of young animals
e. A behavior performed without having been learned by actual environmental experience
f. A program of coordinated muscle activity that runs to completion independently of feedback from the environment

## Dichotomous Choice

Circle one of two possible answers given between parentheses in each statement. [pp.786–787]

7. For garter snake populations living along the California coast, the food of choice is (the banana slug/tadpoles and small fishes).

8. In Stevan Arnold's experiments, newborn garter snakes that were offspring of coastal parents usually (ate/ignored) a chunk of slug as the first meal.

9. Newborn garter snake offspring of (coastal/inland) parents ignored cotton swabs drenched in essence of slug and only rarely ate the slug meat.

10. The differences in the behavioral eating responses of coastal and inland snakes (were/were not) learned.

11. Hybrid garter snakes with coastal and inland parents exhibited a feeding response that indicated a(n) (environmental/genetic) basis for this behavior.

## Complete the Table

12. Complete the following table to consider examples of instinctive and learned behavior.

| Category | Examples |
| --- | --- |
| [p.787] a. Instinctive behavior | |
| [p.787] b. Learned behavior | |

## 44.3. THE ADAPTIVE VALUE OF BEHAVIOR [p.788]

## 44.4. COMMUNICATION SIGNALS [pp.788–789]

## 44.5. MATES, OFFSPRING, AND REPRODUCTIVE SUCCESS [pp.790–791]

### Selected Words

Pheromones [p.788], _chemical_ cues [p.788], _acoustical_ signals [p.788], _visual_ signal [p.789], tactile display [p.789], "nuptial gift" [p.790]

### Boldfaced Terms

[p.788] communication signals _____

_____

[p.790] lek _____

_____

## Choice

Choose the appropriate type of communication cue for each statement. [pp.788–789]

a. visual    b. chemical    c. tactile    d. acoustical

_____ 1. baboon threat display

_____ 2. touch

_____ 3. pheromones

_____ 4. bird songs and prairie dog barks

_____ 5. courtship displays in birds

## Complete the Table

Complete the following table to supply the common names of the animals that fit the text examples of sexual selection. [pp.790–791]

| Animals | Descriptions of Sexual Selection |
| --- | --- |
| 6. | Females select the males that offer them superior material goods; females permit mating only after they have eaten the "nuptial gift" for about five minutes. |
| 7. | Males congregate in a lek or communal display ground; each male stakes out a few square meters as his territory; females are attracted to the lek to observe male displays and usually select and mate with only one male. |
| 8. | Females of a species cluster in defendable groups at a time when they are sexually receptive; males compete for access to the clusters; combative males are favored. |
| 9. | Extended parental care improves the likelihood that the current generation of offspring will survive; this behavior comes at a reproductive cost to the adults. |

## 44.6. LIVING IN GROUPS [pp.792–793]

## 44.7. WHY SACRIFICE YOURSELF? [pp.794–795]

## 44.8. HUMAN BEHAVIOR [p.795]

### Selected Words

"Fishing sticks" [p.792], _eusocial_ [p.794], "self-sacrifice" [p.795]

### Boldfaced Terms

[p.792] selfish-herd_____

_____

[p.794] altruistic behavior _____

_____

[p.794] theory of inclusive fitness _____

_____

## Matching

Match each of the following terms to its correct definition. [pp.792–794]

_____ 1. theory of inclusive fitness

_____ 2. selfish herd

_____ 3. eusocial

_____ 4. altruistic behavior

a. enhances another's reproductive success at the expense of the individual

b. individuals hiding behind one another

c. groups that stay together for generations; typically have a division of labor

d. genes associated with altruistic behavior are selected for because they promote the success of close relatives

---

# Self-Quiz

___ 1. The observable, coordinated responses that animals make to stimuli are what we call _____. [p.785]
   a. imprinting
   b. instinct
   c. behavior
   d. learning

___ 2. In _____, a particular behavior is performed without having been learned by actual experience in the environment. [p.787]
   a. natural selection
   b. altruistic behavior
   c. sexual selection
   d. instinctive behavior

___ 3. Newly hatched goslings follow any large moving objects to which they are exposed shortly after hatching; this is an example of _____. [p.787]
   a. homing behavior
   b. imprinting
   c. piloting
   d. migration

___ 4. A young toad flips its sticky-tipped tongue and captures a bumblebee that stings its tongue; in the future, the toad leaves bumblebees alone. This is _____. [p.787]
   a. instinctive behavior
   b. a fixed reaction pattern
   c. altruistic
   d. learned behavior

___ 5. The claiming of the more protected central locations of the bluegill colony by the largest, most powerful males suggests _____. [p.792]
   a. cooperative predator avoidance
   b. the selfish herd
   c. a huge parent cost
   d. self-sacrificing behavior

___ 6. A chemical odor in the urine of male mice triggers and enhances estrus in female mice. The source of stimulus for this response is a _____. [p.788]
   a. tactical signal
   b. pheromone
   c. acoustical signal
   d. visual signal

___ 7. Female insects often attract mates by releasing sex pheromones. This is an example of a(n) _____ signal. [p.788]
   a. chemical
   b. visual
   c. acoustical
   d. tactile

## Chapter Objectives/Review Questions

1. Define a *stimuli*. [p.786]
2. Distinguish between instinctive and learned behaviors. [p.787]
3. Explain the process of imprinting. [p.783]
4. Define *adaptive behavior*. [p.788]
5. List the forms of communication signals and give an example of each. [pp.788–789]
6. Explain the purpose of a pheromone. [p.784]
7. Explain the similarities and differences between sexual selection and natural selection. [p.790]
8. Why is parental care important? [p.791]
9. Define *selfish herd*; cite an example. [p.792]
10. Explain why a dominance hierarchy may be beneficial. [p.793]
11. What are the basic premises of inclusive fitness? [p.794]
12. Explain the benefits of altruistic behavior. [p.794]

## Chapter Summary

An individual's (1) _____ starts with interactions among gene products, such as hormones and (2) _____.
Most forms of behavior have innate components, but they can be modified by (3) _____ factors.

When behavioral traits have a (4) _____ basis, they may evolve by way of (5) _____.

(6) _____ behavior depends on evolved modes of (7) _____. Communication signals hold clear meaning for both the sender and the receiver of signals.

Life in social groups has reproductive (8) _____ and costs. Not all environments favor the (9) _____ of such groups. Self-sacrificing behavior has evolved among a few kinds of animals that live in large family groups.

(10) _____ behavior was shaped by the same evolutionary forces that shape the behavior of other (11) _____.
Only humans consistently make (12) _____ choices about their behavior.

## Integrating and Applying Key Concepts

Think about communication signals that humans use and list them. Do you believe a dominance hierarchy exists in human society? Think of examples.

# ANSWERS TO STUDENT INTERACTIVE WORKBOOK

## Chapter 1  Invitation to Biology

*Lost Worlds and Other Wonders* [pp.2–3]
**1.1. Life's Levels of Organization** [pp.4–5]
1. g; 2. d; 3. a; 4. c; 5. i; 6. f; 7. b; 8. e; 9. k; 10. l; 11. h;
12. molecules; 13. organs; 14. multicelled organism;
15. population; 16. community; 17. ecosystem
**1.2. Overview of Life's Unity** [pp.6–7]
**1.3. If So Much Unity, Why So Many Species?** [pp.8–9]
**1.4. An Evolutionary View of Diversity.** [p.10]
1. f; 2. h; 3. i; 4. c; 5. j; 6. e; 7. d; 8. b; 9. a; 10. g; 11. a; 12. b;
13. a; 14. b; 15. c; 16. a; 17. f; 18. b; 19. c; 20. e; 21. d; 22. b;
23. a; 24. b; 25. a; 26. b; 27. b; 28. a
**1.5. Critical Thinking and Science** [p.11]
**1.6. How Science Works** [pp.12–13]
1. h; 2. a; 3. f; 4. b; 5. g; 6. d; 7. e; 8. c; 9. observation;
10. hypothesis; 11. prediction; 12. experiments; 13.
results; 14. scientific community
**1.7. The Power of Experimental Tests** [pp.14–15]
**1.8. Sampling Error in Experiments** [p.16]
1. b; 2. e; 3. d; 4. a; 5. c; 6. b; 7. c; 8. d; 9. e
**Self-Quiz**
1. b; 2. c; 3. a; 4. c; 5. c; 6. d; 7. a; 8. b; 9. c; 10. a
**Chapter Summary**
1. organization; 2. cells; 3. unity; 4. cells; 5. environment;
6. DNA; 7. reproduction; 8. form; 9. natural selection;
10. diversity; 11. Observations

## Chapter 2  Life's Chemical Basis

*What Are You Worth?* [pp.20–21]
**2.1. Start With Atoms** [p.22]
**2.2. Putting Radioisotopes to Work** [p.23]
1. b; 2. j; 3. i; 4. a; 5. k; 6. g; 7. e; 8. m; 9. h; 10. f; 11. l; 12. c;
13. d
**2.3. Why Electrons Matter** [pp.24–25]
**2.4. What Happens When Atoms Interact?** [pp.26–27]
1. c; 2. e; 3. b; 4. d; 5. a; 6. sodium; 7. chlorine; 8. carbon;
9. oxygen; 10. helium; 11. d; 12. c; 13. a; 14. c; 15. a; 16. b;
17. a; 18. b
**2.5. Water's Life-Giving Properties** [pp.28–29]
**2.6. Acids and Bases** [pp.30–31]
1. hydrogen; 2. hydrophilic; 3. hydrophobic; 4. sphere of
hydration; 5. temperature; 6. freezing; 7. evaporation;
8. c; 9. e; 10. g; 11. j; 12. a; 13. h; 14. I; 15. f; 16. b; 17. d;
18. 8, base; 19. 2, acid; 20. 10, base; 21. 5, acid; 22. 1, acid;
23. 5, acid

**Self Quiz**
1. b; 2. d; 3. b; 4. d; 5. b; 6. d; 7. a; 8. d; 9. c; 10. c
**Chapter Summary**
1. matter; 2. neutrons; 3. protons; 4. isotopes; 5. electrons;
6. sharing; 7. ionic; 8. hydrogen; 9. water; 10. cohesion;
11. ions

## Chapter 3  Molecules of Life

*Science or Supernatural?* [pp.34–35]
**3.1. Molecules of Life—From Structure to Function**
[pp.36–37]
1. methyl; 2. hydroxyl; 3. carbonyl (ketone); 4. amino;
5. phosphate; 6. carboxyl; 7. carbonyl (aldehyde);
8. rearrangement; 9. two molecules covalently bonded
into a larger one; 10. one or more electrons taken from
one molecule are donated to another molecule;
11. cleavage; 12. functional group transfer;
13. a. hydrolysis, b. condensation
**3.2. Carbohydrates—The Most Abundant Ones**
[pp.38–39]
1. lactose (oligosaccharide); 2. chitin (polysaccharide);
3. glucose (monosaccharide); 4. cellulose
(polysaccharide); 5. starch (polysaccharide); 6. sucrose
(monosaccharide); 7. deoxyribose (monosaccharide);
8. ribose (monosaccharide); 9. glycogen (polysaccharide)
**3.3. Greasy, Oily—Must Be Lipids** [pp.40–41]
1. a. unsaturated (circle one double bond), b. saturated,
c. unsaturated (circle three double bonds); 2. a. phospholipids, b. sterol, c. fatty acid, d. triglyceride; 3. b; 4. d; 5. a;
6. c; 7. a; 8. c; 9. a; 10. c; 11. b; 12. c; 13. a; 14. c
**3.4. Proteins—Diversity in Structure and Function**
[pp.42–43]
**3.5. Why Is Protein Structure So Important?** [pp.44–45]
1. a. secondary, b. tertiary, c. quaternary, d. tertiary,
e. primary, f. secondary; 2. h; 3. f; 4. e; 5. d; 6. b; 7. l; 8. m;
9. i; 10. c; 11. j; 12. a; 13. k; 14. g
**3.6. Nucleotides, DNA, and the RNAs** [pp.46–47]
1. ATP; 2. coenzymes; 3. nucleic acids; 4. DNA; 5. RNA
**Self-Quiz**
1. b; 2. b; 3. a; 4. d; 5. c; 6. c; 7. d; 8. a; 9. a; 10. b; 11. b;
12. c; 13. a; 14. a; 15. d
**Chapter Summary**
1. proteins; 2. fatty acids; 3. carbon; 4. functional;
5. carbohydrates; 6. energy; 7. complex; 8. reservoirs;
9. membranes; 10. proteins; 11. enzymes; 12. nucleic;
13. DNA; 14. proteins

# Chapter 4  Cell Structure and Function

*Animalcules and Cells Fill'd with Juices* [pp.50–51]

**4.1. What Is a Cell?** [pp.52–53]

**4.2. How Do We See Cells?** [pp.54–55]

1. h; 2. a; 3. i; 4. d; 5. e; 6. f; 7. c; 8. g; 9. b; 10. Surface-to-volume ratio constrains cell size because past a certain point of growth the surface area is not sufficient to allow the movement of nutrients and wastes. As a cell grows, the surface area increases with the square of its diameter, whereas the volume increases with the cube of its diameter; 11. Every organism consists of one or more cells. The cell is the smallest unit that has all the properties of life. Each new cell arises from a preexisting cell; 12. a; 13. c; 14. b; 15. c; 16. d; 17. b

**4.3. Membrane Structure and Function** [pp.56–57]

**4.4. Introducing Prokaryotic Cells** [pp.58–59]

**4.5. Microbial Mobs** [p.59]

1. f; 2. h; 3. g; 4. b; 5. e; 6. d; 7. c; 8. a; 9. flagellum; 10. pilus; 11. capsule; 12. cell wall; 13. plasma membrane; 14. nucleoid; 15. cytoplasm

**4.6. Introducing Eukaryotic Cells** [p.60]

**4.7. The Nucleus** [p.61]

**4.8. The Endomembrane System** [pp.62–63]

1. b; 2. d; 3. f; 4. g; 5. h; 6. e; 7. c; 8. a; 9. a. nucleolus, b. nuclear envelope, c. chromatin, d. nucleoplasm, e. chromosome; 10. ribosomes; 11. rough; 12. smooth; 13. Golgi body; 14. lipid; 15. vesicles; 16. plasma membrane

**4.9. Mitochondria and Chloroplasts** [p.64]

**4.10. Visual Summary of Eukaryotic Cell Components** [p.65]

1. b; 2. a; 3. a; 4. a; 5. b; 6. a; 7. b; 8. c; 9. a; 10. b; 11. o; 12. i, w; 13. e, s; 14. h, v; 15. n; 16. j, x; 17. c, q; 18. g, u; 19. l, z; 20. f, t; 21. y; 22. a, aa; 23. k; 24. m; 25. b, p; 26. d, r

**4.11. Cell Surface Specializations** [pp.66–67]

**4.12. The Dynamic Cytoskeleton** [pp.68–69]

1. b; 2. d; 3. c; 4. f; 5. a; 6. g; 7. e; 8. h; 9. b; 10. a; 11. c; 12. a; 13. b; 14. a; 15. b

**Self-Quiz**

1. c; 2. d; 3. e; 4. b; 5. a; 6. d; 7. a; 8. b; 9. d; 10. b

**Chapter Summary**

1. plasma membrane; 2. cytoplasm; 3. cells; 4. bilayer; 5. proteins; 6. water-soluble; 7. bacteria; 8. smallest; 9. eukaryotic; 10. nucleus; 11. protein; 12. chromosomes

# Chapter 5  Ground Rules of Metabolism

*Alcohol, Enzymes, and Your Liver* [pp.72–73]

**5.1. Energy and the World of Life** [pp.74–75]

**5.2. ATP in Metabolism** [p.76]

1. b; 2. a; 3. a; 4. b; 5. a; 6. b; 7. b; 8. g; 9. b; 10. f; 11. e; 12. d; 13. c; 14. a; 15. releases; 16. ADP; 17. phosphorylation; 18. energy

**5.3. Enzymes in Metabolism** [pp.76–77]

**5.4. Enzymes Don't Work Alone** [pp.78–79]

1. b; 2. c; 3. a; 4. d; 5. electron transfer; 6. condensation; 7. cleavage; 8. one molecule is converted to another;

9. movement of functional groups between molecules; 10. c; 11. f; 12. a; 13. e; 14. b; 15. d; 16. e

**5.5. Metabolism-Organized, Enzyme-Mediated Reactions** [pp.80–81]

1. b; 2. g; 3. e; 4. f; 5. a; 6. d; 7. c; 8. a; 9. b; 10. b; 11. a; 12. a; 13. b

**5.6. Diffusion, Membranes, and Metabolism** [pp.82–83]

**5.7. Working With and Against Gradients** [pp.84–85]

1. e; 2. d; 3. b; 4. a; 5. c; 6. size of the molecule, steepness of the concentration gradient, temperature; 7. passive transport does not expend energy and occurs down the concentration gradient; 8. a; 9. b; 10. a; 11. a; 12. b; 13. c; 14. a; 15. b; 16. d; 17. e; 18. d; 19. a; 20. b; 21. c

**5.8. Which Way Will Water Move?** [pp.86–87]

**5.9. Membrane Traffic To and From the Cell Surface** [pp.88–89]

**5.10. Night Lights** [p.89]

1. c; 2. I; 3. e; 4. a; 5. g; 6. d; 7. f; 8. h; 9. b; 10. hypotonic; 11. isotonic; 12. hypertonic; 13. T; 14. T; 15. F, hypotonic; 16. F, isotonic; 17. F, hypertonic; 18. T

**Self-Quiz**

1. a; 2. d; 3. d; 4. c; 5. c; 6. b; 7. d; 8. a; 9. d; 10. d

**Chapter Summary**

1. energy; 2. organization; 3. ATP; 4. enzymes; 5. environmental; 6. metabolic pathways; 7. enzymes; 8. concentration; 9. proteins; 10. solute; 11. metabolism; 12. nature

# Chapter 6  Where It Starts—Photosynthesis

*Sunlight and Survival* [pp.92–93]

**6.1. Sunlight as an Energy Source** [pp.94–95]

**6.2. Exploring the Rainbow** [p.96]

1. f; 2. e; 3. a; 4. c; 5. d; 6. g; 7. b; 8. the peaks represent those wavelengths that the pigment absorbs the most

**6.3. Overview of Photosynthesis** [p.97]

**6.4. Light-Dependent Reactions** [pp.98–99]

1. sunlight; 2. $O_2$; 3. $H_2O$; 4. $CO_2$; 5. sugars; 6. ATP; 7. $NADP^+$; 8. T; 9. F, light dependent; 10. F, thylakoid membrane; 11. F, stroma; 12. T; 13. light energy; 14. electron transfer chain; 15. $H_2O$; 16. oxygen; 17. ATP; 18. stroma; 19. thylakoid compartment; 20. NADPH

**6.5. Energy Flow in Photosynthesis** [p.100]

**6.6. Light-Independent Reactions: The Sugar Factory** [p.101]

1. c; 2. b; 3. b; 4. c; 5. a; 6. b; 7. a; 8. $CO_2$ (e); 9. PGA (c); 10. ATP (g); 11. NADPH (a); 12. PGAL (d): 13. glucose (f): 14. PGAL (b)

**6.7. Adaptations: Different Carbon-Fixing Pathways** [p.102]

**6.8. A Burning Concern** [p.103]

1. c; 2. a; 3. b; 4. c; 5. b; 6. b; 7. a; 8. a; 9. c; 10. b; 11. c; 12. a; 13. b; 14. a; 15. b; 16. The primary cause of global warming is the burning of fossil fuels which releases carbon dioxide into the atmosphere. Carbon dioxide acts as a greenhouse gas that traps heat in the atmosphere.

**Self-Quiz**

1. c; 2. d; 3. c; 4. d; 5. d; 6. e; 7. e; 8. a; 9. d; 10. e

## Chapter Summary

1. chlorophylls; 2. glucose; 3. chloroplasts; 4. light;
5. chemical; 6. carbohydrates; 7. photosynthesis; 8. ATP;
9. NADPH; 10. synthesis; 11. carbon dioxide; 12. photosynthesis; 13. electrons; 14. autotrophs; 15. metabolism;
16. global warming

---

# Chapter 7  How Cells Release Chemical Energy

*When Mitochondria Spin Their Wheels* [pp.106–107]

**7.1. Overview of Carbohydrate Breakdown Pathways**
[pp.108–109]

1. c; 2. b; 3. a; 4. b; 5. a; 6. a; 7. b; 8. a; 9. c; 10. oxygen;
11. glycolysis; 12. Krebs cycle; 13. $CO_2$; 14. ATP; 15.
NADH; 16. $FADH_2$; 17. 32

**7.2. Glycolysis—Glucose Breakdown Starts** [pp.110–111]

**7.3. Second Stage of Aerobic Respiration** [pp.112–113]

1. ATP; 2. PGAL; 3. NADH; 4. ATP; 5. phosphorylation;
6. ATP; 7. 2; 8. 2; 9. pyruvate; 10. NADH; 11. $CO_2$;
12. $NAD^+$; 13. ATP; 14. $FADH_2$; 15. oxaloacetate; 16. a. 6,
b. 8, c. 2, d. 2

**7.4. Aerobic Respiration's Big Energy Payoff**
[pp.114–115]

**7.5. Anaerobic Energy-Releasing Pathways** [pp.116–117]

**7.6. The Twitchers** [p.117]

1. NADH; 2. pyruvate; 3. cytoplasm; 4. mitochondrial;
5. $CO_2$; 6. $FADH_2$; 7. ATP; 8. Krebs cycle; 9. oxygen;
10. a. 6, b. 10, c. 2, d. 36; 11. it serves as an electron acceptor;
12. c; 13. a; 14. c; 15. c; 16. b; 17. c; 18. a; 19. c; 20. b; 21. a

**7.7. Alternative Energy Sources in the Body**
[pp.118–119]

**7.8. Reflections on Life's Unity** [p.120]

1. a; 2. b; 3. c; 4. b; 5. a; 6. a; 7. c; 8. a; 9. the outputs of
photosynthesis are used in aerobic respiration, the outputs of aerobic respiration are used in photosynthesis;
10. photosynthesizers convert inorganic materials into
organic materials for use in the energy-releasing
pathways

**Self-Quiz**

1. d; 2. c; 3. d; 4. a; 5. a; 6. e; 7. d; 8. c; 9. e; 10. d

**Chapter Summary**

1. degradative; 2. chemical; 3. aerobic; 4. mitochondria;
5. glycolysis; 6. $NAD^+$; 7. electrons; 8. two; 9. Krebs cycle;
10. carbon dioxide; 11. phosphorylation; 12. oxygen; 13.
glycolysis; 14. electron; 15. small; 16. glucose; 17. lipids;
18. cellular; 19. one-way

---

# Chapter 8  How Cells Reproduce

*Henrietta's Immortal Cells* [pp.124–125]

**8.1. Overview of Cell Division Mechanisms**
[pp.126–127]

**8.2. Introducing the Cell Cycle** [pp.128–129]

1. HeLa cells are tumor cells taken from and named for a
cancer patient, Henrietta Lacks, in 1951. Henrietta Lacks
died, at age 31, two months following her diagnosis of
cancer. HeLa cells have continued to divide in culture
and are used for cancer research in laboratories all over

the world. The legacy of Henrietta Lacks continues to
benefit humans everywhere. 2. h; 3. j; 4. g; 5. b; 6. k;
7. d; 8. a; 9. i; 10. c; 11. e; 12. f; 13. G1 interval;
14. S interval; 15. G2 interval; 16. prophase;
17. metaphase; 18. anaphase; 19. telophase;
20. cytoplasmic division; 21. interphase; 22. mitosis;
23. daughter cells; 24. 15; 25. 22; 26. 14; 27. 13; 28. 21;
29. 20; 30. 21

**8.3. A Closer Look at Mitosis** [pp.130–131]

**8.4. Division of the Cytoplasm** [pp.132–133]

1. interphase-daughter cells (F); 2. anaphase (A); 3. late
prophase (G); 4. metaphase (D); 5. cells at interphase (E);
6. early prophase (C); 7. transition to metaphase (B);
8. telophase (H); 9. a; 10. b; 11. a; 12. a; 13. b; 14. b; 15. b;
16. a; 17. a

**8.5. When Control Is Lost** [pp.134–135]

1. First, cancer cells grow and divide abnormally. Second,
cancer cells have profoundly altered cytoplasm and
plasma membranes. Third, cancer cells have a weakened
capacity for adhesion. Fourth, cancer cells usually have a
lethal effect.

**Self-Quiz**

1. a; 2. d; 3. c; 4. e; 5. c; 6. d; 7. d; 8. a; 9. d; 10. c; 11. c

**Chapter Summary**

1. species; 2. chromosomes; 3. shape; 4. hereditary;
5. daughter; 6. cytoplasm; 7. cycle; 8. mitosis; 9. mass;
10. DNA; 11. nucleus; 12. anaphase; 13. microtubular;
14. nuclei; 15. nuclear; 16. cytoplasm; 17. animal;
18. plant; 19. timing; 20. rate; 21. tumor; 22. cancer

**Chapter Objectives/Review Questions**

2. nucleus; 5. S, or synthesis; 6. DNA; 7. chromosome;
8. diploid

---

# Chapter 9  Meiosis and Sexual Reproduction

*Why Sex?* [pp.138–139]

**9.1. Introducing Alleles** [p.140]

**9.2. What Meiosis Does** [pp.140–141]

**9.3. Visual Tour of Meiosis** [pp.142–143]

1. b; 2. a; 3. a; 4. b; 5. a; 6. b; 7. a; 8. b; 9. a; 10. b;
11. Meiosis; 12. gamete; 13. Diploid; 14. Diploid;
15. Meiosis; 16. sister chromatids; 17. sister chromatids;
18. one; 19. four; 20. two; 21. interphase preceding
meiosis I; 22. chromosome; 23. haploid; 24. meiosis II;
25. twenty-three; 26. anaphase II (H); 27. metaphase II (F);
28. metaphase I (A); 29. prophase II (B); 30. telophase II (C);
31. telophase I (G); 32. prophase I (E); 33. anaphase I (D);
34. E ($2n = 2$); 35. D ($2n = 2$); 36. B ($2n = 2$); 37 A ($n = 1$);
38. C ($n = 1$)

**9.4. How Meiosis Introduces Variations in Traits**
[pp.144–145]

**9.5. From Gametes to Offspring** [pp.146–147]

**9.6. Mitosis and Meiosis—an Ancestral Connection?**
[pp.148–149]

1. f; 2. g; 3. b; 4. d; 5. h; 6. c; 7. e; 8. a; 9. b; 10. a; 11. a;
12. a; 13. b; 14. b; 15. a; 16. b; 17. b; 18. a; 19. 2 (2n);
0. 5 (n); 21. 4 (n); 22. 3 (n); 23. a; 24. b; 25. e; 26. d; 27. c;

28. Fertilization also adds to variation among offspring. During prophase I, an average of two or three crossovers take place in each human chromosome. Random positioning of pairs of paternal and maternal chromosomes at metaphase I results in one of millions of possible chromosome combinations in each gamete. Of all male and female gametes produced, which two get together is a matter of chance. The sheer numbers of combinations that can exist is staggering.

**Self-Quiz**

1. d; 2. a; 3. c; 4. e; 5. b; 6. a; 7. d; 8. d; 9. b; 10. c; 11. b

**Chapter Summary**

1. asexual; 2. genetic; 3. sexual; 4. alleles; 5. gene; 6. trait; 7. diploid; 8. meiosis; 9. sexual; 10. haploid; 11. cytoplasmic; 12. chromosomes; 13. alleles; 14. shuffled; 15. chance; 16. variation; 17. gametes; 18. spore; 19. molecular; 20. mitosis; 21. DNA

**Chapter Objectives/Review Questions**

5. diploid; 6. haploid; 8. I, II; 12. Fertilization

---

# Chapter 10  Observing Patterns in Inherited Traits

*In Pursuit of a Better Rose* [pp.152–153]

**10.1. Mendel, Pea Plants, and Inheritance Patterns** [pp.154–155]

**10.2. Mendel's Theory of Segregation** [pp.156–157]

**10.3. Mendel's Theory of Independent Assortment** [pp.158–159]

1. b; 2. a; 3. c; 4. 5; 5. genotype: $1/2\ Tt$; $1/2\ tt$, phenotype: 1/2 tall; 1/2 short; 6.a. 1 tall : 1 short; 1 heterozygous tall : 1 homozygous short; b. all tall; 1 homozygous tall : 1 heterozygous tall; c. all short; all homozygous short; d. 3 tall : 1 short; 1 homozygous tall : 2 heterozygous tall : 1 homozygous short; e. 1 tall : 1 short; 1 homozygous short : 1 heterozygous tall; f. all tall; all heterozygous tall; g. all tall; all homozygous tall; h. all tall; 1 heterozygous tall : 1 homozygous tall; 7. a. 9/16; pigmented eyes, right-handed; b. 3/16 pigmented eyes, left-handed; c. 3/16 blue-eyed, right-handed; d. 1/16 blue-eyed, left-handed; 8.a. $F_1$: black trotter; $F_2$: nine black trotters, three black pacers, three chestnut trotters, one chestnut pacer; b. black pacer; c. *BbTt*; d. *bbtt*, chestnut pacers and *BBTT*, black trotters

**10.4. Beyond Simple Dominance** [pp.160–161]

**10.5. Linkage Groups** [pp.162–163]

**10.6. Genes and the Environment** [pp.164–165]

**10.7. Complex Variations in Traits** [pp.166–167]

1. Genotypes: 1/4 *AA*; 1/4 *AB*; 1/4 *AO*; 1/4 *BO*, phenotypes: 1/2 A; 1/4 AB; 1/4 B; 2. Genotypes: 1/4 *AB*; 1/4 *BO*; 1/4 *AO*; 1/4 *OO*, phenotypes: 1/4 AB; 1/4 B; 1/4 A; 1/4 O; 3. Genotypes: 1/4 *AA*; 1/2 *AB*; 1/4 *BB*, phenotypes: 1/4 A; 1/2 AB; 1/4 B; 4.a. phenotype: all pink; genotype: all *RR*; b. phenotypes: 1/4 red; 1/2 pink; 1/4 white; genotypes: 1/4 RR; 1/2 RR′; 1/4 R′R′; 5. The genotype of the male parent is *RrPp* and the genotype of the female parent is *rrpp*. The offspring are 1/4 walnut comb, *RrPp*; 1/4 rose comb, *Rrpp*; 1/4 pea comb, *rrPp*; 1/4 single comb, rrpp; the process is called epistasis

6. 3/8 black; 1/2 yellow; 1/8 brown; 7.a. incomplete dominance; b. codominance; c. multiple alleles; d. gene interactions; e. pleiotropy; 8. b; 9. b; 10. a; 11. b; 12. a; 13. a; 14. b; 15. An individual's phenotype is an outcome of complex interactions among its genes, enzymes and other gene products, and environmental factors (the phenotype is a product of genetic and environmental factors).

**Self-Quiz**

1. d; 2. b; 3. a; 4. c; 5. d; 6. b; 7. e; 8. a; 9. a; 10. c; 11. d

**Chapter Objectives/Review Questions**

3. self; 4. genes; 8. homozygous, heterozygous; 10. genotype, phenotype; 11. monohybrid; 15. segregation; 17. independent assortment

**Chapter Review**

1. experimental; 2. inheritance; 3. heritable; 4. units; 5. genes; 6. segregation; 7. homologous; 8. gametes; 9. independent; 10. distributed; 11. recessive; 12. codominant; 13. traits; 14. environment

**Integrating and Applying Key Concepts**

DdPp × Ddpp.

---

# Chapter 11  Chromosomes and Human Inheritance

*Strange Genes, Tortured Minds* [pp.168–169]

**11.1. Human Chromosomes** [pp.170–171]

**11.2. Examples of Autosomal Inheritance patterns** [pp.172–173]

1. Schizophrenia is characterized by delusions, hallucinations, disorganized speech and behavior, and social dysfunction. In bipolar disorder, people show extreme swings in mood, thoughts, energy, and behavior. Change in any step of a crucial biochemical pathway could impair the brain's wiring. Therefore, we can expect that alterations in genes contribute to the abnormal neurochemistry in NBDs; 2. The human X and Y chromosomes differ physically. The Y is a lot shorter, almost a remnant of the other in appearance. The two also differ in which genes they carry. They still synapse in a small region; this allows them to interact as homologues during meiosis; 3. In autosomal dominant inheritance, if one parent does not have a mutated allele, but the other parent is heterozygous, every child has a 50 percent chance of getting the allele. The trait usually appears every generation. In autosomal recessive inheritance, if both parents are heterozygous, each child will have 50 percent chance of being heterozygous, and 25 percent will be homozygous. The heterozygous individuals often show no characteristics; 4. Assuming the father is heterozygous with Huntington disorder and the mother normal, the chances are 1/2 that the son will develop the disease; 5. The woman's mother is heterozygous normal, *Gg*, the woman is also heterozygous normal, *Gg*. The galactosemic man, *gg*, has two heterozygous normal parents, *Gg*. The two normal children are heterozygous normal, *Gg*; the galactosemic child is *gg*; 6. sons; 7. mothers; 8. daughters; 9. *SRY* is one of 330 genes in a human Y chromosome and is the master gene for male

sex determination. Its expression in XY embryos triggers the formation of testes, the primary male reproductive organs; 10. c; 11. e; 12. d

**11.3. Young, Yet Old [p.173]**

**11.4. Examples of X-Linked Inheritance Patterns [pp.174–175]**

1. If only male offspring are considered, the probability is 1/2 that the couple will have a color-blind son; 2. The probability is that 1/2 of the sons will have hemophilia; the probability is 0 that a daughter will express hemophilia; the probability is that 1/2 of the daughters will be carriers; 3. If the woman marries a normal male, the chance that her son would be color-blind is 1/2. If she marries a color-blind male, the chance that her son would be color blind is also 1/2; 4. The frequency of progeria is one in 8 million; it is caused by a spontaneous mutation of an autosomal dominant gene. The mutation grossly disrupts interactions among genes that bring about growth and development. Observable symptoms begin to materialize before age two. The skin thins, skeletal muscles weaken, and tissues in limb bones that should lengthen and grow stronger soften. Hair loss is pronounced with inevitable premature baldness. Most progeriacs expect to die in their early teens as a result of strokes or heart attacks brought on by hardening of the arteries.

**11.5. Heritable Changes in Chromosome Structure [pp.176–177]**

**11.6. Heritable Changes in the Chromosome Number [pp.178–179]**

1. duplication; 2. inversion; 3. deletion; 4. translocation; 5. deletion; 6. Structural alterations in DNA may have been the origin of species differences among closely related organisms, such as apes and humans. Eighteen of the twenty-three pairs of human chromosomes are almost identical with chromosomes of chimpanzees and gorillas. The other five differ at inverted and translocated regions; 7.a. With aneuploidy, individuals have one extra or one less chromosome; a major cause of human reproductive failure; b. With polyploidy, individuals have three or more of each type of chromosome; common in flowering plants and some animals but lethal in humans; c. Nondisjunction is due to a failure of one or more pairs of chromosomes to separate in mitosis or meiosis; some or all forthcoming cells will have too many or too few chromosomes; 8. All gametes will be abnormal; 9. One-half of the gametes will be abnormal; 10. Half of all species of flowering plants, some insects, fishes, and other animals are polyploid; 11. If a normal gamete fuses by chance with an $n + 1$ gamete (one extra chromosome), the new individual will be trisomic ($2n + 1$), with three of one type of chromosome and two of every other type. Should an $n - 1$ gamete fuse with a normal $n$ gamete, the new individual will be monosomic ($2n - 1$). Mitotic divisions perpetuate the mistake as an early embryo forms; 12. d; 13. b; 14. c; 15. e; 16. a; 17. f; 18. d; 19. Crossing over would be expected to occur twice as often between genes A and B as it would between genes C and D; 20. The son has genotype $Ab/Ab$ because a crossover

must have occurred during meiosis in his mother, resulting in the recombination $Ab$.

**11.7. Human Genetic Analysis [pp.180–181]**

**11.8. Prospects in Human Genetics [pp.182–183]**

1. a; 2. e; 3. f; 4. b; 5. c; 6. g; 7. d; 8. A genetic abnormality is nothing more than a rare, uncommon version of a trait that society may judge as abnormal or merely interesting; a genetic disorder is an inherited condition that sooner or later causes mild to severe medical problems; a syndrome is a recognized set of symptoms that characterize a given disorder; a genetic disease is illness caused by a person's genes increasing susceptibility to infection or weakening the response to it; 9a. Autosomal recessive; b. Autosomal dominant; c. X-linked recessive; d. Autosomal dominant; e. Changes in chromosome number; f. Changes in chromosome structure; g. Changes in chromosome number; h. X-linked recessive; i. Changes in chromosome number; j. X-linked recessive; k. Autosomal dominant; l. Changes in chromosome number; m. Autosomal recessive; 10a. Prenatal diagnosis; b. Abortion; c. Genetic counseling; d. Preimplantation diagnosis; e. Phenotypic treatments

**Self-Quiz**

1. d; 2. d; 3. b; 4. c; 5. d; 6. a; 7. c; 8. b; 9. b; 10. c

**Chapter Objectives/Review Questions**

1. brain; 5. SRY; 6. karyotype; 7. linkage; 9. pedigree; 10. abnormality, disorder; 11. syndrome; 15. deletion, inversion, duplication, translocation; 17. aneuploidy; 18. polyploidy; 19. nondisjunction; 20. Down, XO, Klinefelter, XYY; 23. in vitro

**Chapter Summary**

1. autosomes; 2. genes; 3. sex; 4. gene; 5. karyotyping; 6. chromosomes; 7. Mendelian; 8. dominance; 9. traits; 10. males; 11. females; 12. structure; 13. duplicated; 14. translocated; 15. number; 16. genetic; 17. diagnostic; 18. ethical

**Integrating and Applying Key Concepts**

1. Karyotype, this probably occurred as an outcome of nondisjunction during meiosis.

2. Yes, the husband has a case because he could not have supplied either of his daughter's recessive genes. His only X chromosome bears the gene for normal iris, the dominant gene.

---

# Chapter 12  DNA Structure and Function

*Here Kitty, Kitty, Kitty, Kitty, Kitty* [pp.186–187]

**12.1. The Hunt for Fame, Fortune, and DNA [pp.188–189]**

1. Surviving clones typically have health problems. Like Dolly, many become overweight as they age. Others are unusually large from birth or have some enlarged organs. Cloned mice have lung and liver problems, and almost all die prematurely. Cloned pigs have heart problems, they limp, and one never did develop an anus or a tail; 2. DNA from a dead pathogenic strain of bacteria was taken up by a live nonpathogenic strain of the same bacteria. The live bacteria expressed the "new" DNA and became pathogenic, causing disease in animals; 3.a. In

the 1800s, Miescher discovered what came to be known as deoxyribonucleic acid, or DNA; b. Griffith; c. Avery; d. Pauling deduced the structure of the protein collagen; why not use his methods to open the secrets of DNA?; e. Hershey and Chase; f. In 1953, Watson and Crick built a model of DNA that fit all the pertinent biochemical rules and insights they had gleaned from other sources; they discovered the structure of DNA.

**12.2. The Discovery of DNA's Structure** [pp.190–191]
1. A five-carbon sugar called deoxyribose, a phosphate group, and one of the four nitrogen-containing bases (A, T, G, or C); 2. guanine (pu); 3. cytosine (py); 4. adenine (pu); 5. thymine (py); 6. deoxyribose; 7. phosphate; 8. purine; 9. pyrimidine; 10. purine; 11. pyrimidine; 12. nucleotide; 13. Erwin Chargaff; 14. x-ray diffraction; 15. T; 16. Watson and Crick; 17. T

**12.3. Fame and Glory** [p.192]

**12.4. Replication and Repair** [pp.192–193]
1. d; 2. c; 3. b; 4. replication; 5. Enzymes; 6. strand; 7. nucleotides; 8. T; 9. C; 10. double helix; 11. conserved; 12. semiconservative; 13. Replication; 14. rewinding; 15. Enzyme; 16. Hydrogen; 17. polymerases; 18. phosphate; 19. energy; 20. ligases; 21. double helix; 22. ligases; 23. polymerases; 24. survival; 25. mutation

26. 
| T- | A | T | -A |
|----|---|---|----|
| G- | C | G | -C |
| A- | T | A | -T |
| C- | G | C | -G |
| C- | G | C | -G |
| C- | G | C | -G |

**Self-Quiz**
1. d; 2. a; 3. b; 4. a; 5. a; 6. c; 7. d; 8. e; 9. d; 10. d

**Chapter Objectives/Review Questions**
4. Watson, Crick; 9. T-A-A-G-C-G; 12. polymerases; 13. ligases, double helix; 14. ligases, polymerases

**Chapter Summary**
1. DNA; 2. nucleotides; 3. hydrogen; 4. double; 5. cytosine; 6. order; 7. species; 8. structure; 9. copy; 10. errors; 11. mutations; 12. function; 13. cloning

**Integrating and Applying Key Concepts**
2. The amount of DNA in the daughter cells would be half of what should be in each cell, since DNA was not replicated prior to division.

---

# Chapter 13 From DNA to Proteins

*Ricin and Your Ribosomes* [pp.196–197]

*13.1. Transcription* [pp.198–199]
1. d; 2. d; 3. b; 4. c; 5. b; 6.a. mRNA, the only kind of RNA that carries protein-building instructions; b. rRNA, a component of ribosomes, where polypeptide chains are built; c. tRNA, delivers amino acids one at a time to a ribosome; 7. RNA is single-stranded whereas most DNA is double-stranded. RNA contains four types of ribonucleotides. Each consists of a five-carbon sugar, ribose (not DNA's deoxyribose), a phosphate group, and a base. Three bases—adenine, cytosine, and guanine—are the same in both RNA and DNA. In RNA, however, the fourth base is uracil, not thymine; 8. DNA; 9. RNA polymerase; 10. polymerase; 11. RNA; 12. one; 13. promoter;

14. transcription; 15. RNA polymerase; 16. polymerase; 17. ribonucleotide; 18. RNA; 19. base; 20. transcript; 21. d; 22. f; 23. g; 24. b; 25. c; 26. a; 27. e

**13.2. The Genetic Code** [pp.200–201]

**13.3. tRNA and rRNA** [pp.201–202]
1. i; 2. b; 3. j; 4. k; 5. a; 6. g; 7. e; 8. f; 9. h; 10. c; 11. mRNA transcript; AUG UUC UAU UGU AAU AAA GGA UGG CAG UAG; 12. tRNA anticodons: UAC AAG AUA ACA UUA UUU CCU ACC GUC AUC; 13. amino acids: (start) met phe tyr cys asn lys gly try gln stop; 14. AUG, TAC, Met; 15. UGA, UAA, or UAG

**13.4. The Three Stages of Translation** [pp.202–203]
1.a. termination; b. elongation; c. initiation

**13.5. Mutated Genes and Their Protein Products** [pp.204–205]
1. b; 2. c; 3. d; 4. b; 5. c

**Self-Quiz**
1. b; 2. b; 3. d; 4. c; 5. d; 6. e; 7. a; 8. b; 9. e; 10. d; 11. d

**Chapter Objectives/Review Questions**
1. Ricin; 2. Ribosomal, messenger, transfer; 9. codon; 10. genetic code; 12. hook, anticodon

**Chapter Summary**
1. enzymes; 2. polypeptide; 3. nucleotide; 4. genes; 5. transcription; 6. translation; 7. DNA; 8. template; 9. RNA; 10. messenger; 11. three; 12. code; 13. conserved; 14. translation; 15. messenger; 16. transfer; 17. ribosomal; 18. peptide; 19. mutations; 20. variations

---

# Chapter 14 Controls Over Genes

*Between You and Eternity* [pp.208–209]

**14.1. Gene Expression in Eukaryotic Cells** [pp.210–211]

**14.2. A Few Outcomes of Gene Controls** [pp.212–213]

**14.3. There's a Fly in my Research** [pp.214–215]

**14.4. Prokaryotic Gene Control** [pp.212–213]
1. f; 2. a; 3. c; 4. d; 5. e; 6. b; 7. genes; 8. structure; 9. specialized; 10. function; 11. differentiation; 12. genes; 13. a; 14. b; 15. d; 16. c; 17. d; 18. a; 19. b; 20. b; 21. a; 22. c; 23. e

**Self-Quiz**
1. c; 2. d; 3. a; 4. b; 5. c; 6. d; 7. d; 8. a; 9. c; 10. b

**Chapter Summary**
1. control; 2. expressed; 3. conditions; 4. outside; 5. transcription; 6. product; 7. master; 8. selective; 9. differential; 10. function; 11. localized; 12. multicellular; 13. inactivated; 14. complex; 15. activated; 16. suppress; 17. nutrient; 18. environment; 19. transcription

---

# Chapter 15 Studying and Manipulating Genomes

*Golden Rice or Frankenfood?* [pp.218–219]

**15.1. A Molecular Toolkit** [pp.222–223]

**15.2. From Haystacks to Needles** [pp.224–225]
1. 2; 2. no; 3. 1; 4. c; 5. f; 6. a; 7. d; 8. b; 9. b; 10. f; 11. e; 12. a; 13. d; 14. c

**15.3. DNA Sequencing** [p.226]

**15.4. Analyzing DNA Fingerprints** [p.227]
1. separates DNA fragments based upon molecular size; 2. identifies a unique array of DNA sequences inherited in a Mendelian pattern; 3. used to replicate genes, or part

of genes, into multiple copies; 4. uses probes to distinguish one DNA sequence from others in a library; 5. provides a rapid method of obtaining the DNA sequence of a gene; 6. d; 7. f; 8. b; 9. c; 10. a; 11. e; 12. DNA fingerprinting is used to compare genetic sequences. In the case of a crime seen, the DNA found at the scene is compared to the DNA of a suspect. If the DNA fingerprints match, then that suspect can be convicted if the other evidence supports the conviction.

**15.5. The Rise of Genomics** [pp.228–229]
**15.6. Genetic Engineering** [p.230]
**15.7. Designer Plants** [pp.230–231]
**15.8. Biotech Barnyards** [pp.232–233]
**15.9. Safety Issues** [p.233]
**15.10. Modified Humans?** [p.234]
1. F, transgenics; 2. F, genetic engineering; 3. T; 4. T; 5. b; 6. a; 7. d; 8. a; 9. c; 10. d

**Self-Quiz**
1. c; 2. b; 3. d; 4. a; 5. a; 6. d; 7. a; 8. c; 9. e; 10. a
**Chapter Review**
1. recombinant; 2. species; 3. vectors; 4. PCR; 5. fragments; 6. sequencing; 7. fingerprint; 8. genomes; 9. species; 10. genes; 11. ethical

---

# Chapter 16  Evidence of Evolution

*Measuring Time* [pp.238–239]
**16.1. Early Beliefs, Confounding Discoveries** [pp.240–241]
**16.2. A Flurry of New Theories** [pp.242–243]
**16.3. Darwin, Wallace, and Natural Selection** [pp.244–245]
1. f; 2. c; 3. e; 4. a; 5. b; 6. d; 7. c; 8. b; 9. a; 10. d; 11. c; 12. b; 13. b; 14. d; 15. b; 16. b; 17. b; 18. d; 19. b; 20. c; 21. a
**16.4. Fossils-Evidence of Ancient Life** [pp.246–247]
**16.5. Dating Pieces of the Puzzle** [pp.248–249]
1. g; 2. c; 3. a; 4. b; 5. f; 6. e; 7. d; 8. 0.5 grams, 16,110 years; 9. prokaryotes; 10. eukaryotes; 11. early fishes; 12. mammals, dinosaurs; 13. flowering plants; 14. mammals; 15. d; 16. e; 17. e; 18. d; 19. d; 20. d; 21. a; 22. c; 23. e
**16.6. Drifting Continents, Changing Seas** [pp.250–251]
**16.7. Divergences from a Shared Ancestor** [pp.252–253]
**16.8. Changes in Patterns of Development** [pp.254–255]
**16.9. Clues in DNA, RNA, and Proteins** [pp.256–257]
1. d; 2. b; 3. a; 4. c; 5. d; 6. d; 7. b; 8. a; 9. d; 10. c; 11. a; 12. a
**16.10. Organizing Information About Species** [pp.258–259]
**16.11. How to Construct a Cladogram** [pp.260–261]
**16.12. Preview of Life's Evolutionary History** [p.261]
1. a; 2. f; 3. e; 4. b; 5. d; 6. h; 7. c; 8. g; 9. hair; 10. shark; 11. gizzard; 12. lamprey; 13. notochord
**Self-Quiz**
1. d; 2. d; 3. a; 4. b; 5. d; 6. c; 7. c; 8. c; 9. a; 10. d
**Chapter Summary**
1. global; 2. fossils; 3. evolution; 4. Charles Darwin; 5. natural selection; 6. heritable; 7. physical; 8. geologic; 9. species; 10. lineages; 11. ancestor; 12. molecular; 13. tree; 14. domains; 15. Eukarya

---

# Chapter 17  Processes of Evolution

*Rise of the Super Rats* [pp.264–265]
**17.1. Individuals Don't Evolve, Populations Do** [pp.266–267]
1. a; 2. h; 3. m; 4. i; 5. l; 6. d; 7. g; 8. b; 9. e; 10. j; 11. c; 12. k; 13. f; 14. n
**17.2. When Is a Population Not Evolving?** [pp.268–269]
**17.3. Natural Selection Revisited** [p.269]
1. 0.16; 2. 0.4; 3. 0.6; 4. 0.36; 5. 0.48; 6. 0.4; 7. no mutation, large population size, isolated population, random mating, equal reproductive success; 8. directional; 9. disruptive; 10. stabilizing
**17.4. Directional Selection** [pp.270–271]
**17.5. Selection Against or in Favor of Extreme Phenotypes** [pp.272–273]
1. a; 2. c; 3. c; 4. b; 5. a; 6. a; 7. c; 8. b; 9. c
**17.6. Maintaining Variation** [pp.274–275]
**17.7. Genetic Drift—The Chance Changes** [pp.276–277]
**17.8. Gene Flow** [p.277]
1. a; 2. f; 3. g; 4. h; 5. c; 6. d; 7. b; 8. e; 9. F, loss; 10. F, balanced polymorphism; 11.T; 12. T; 13. F, inbreeding; 14. F, decreases
**17.9. Reproductive Isolation** [pp.278–279]
**17.10. Allopatric Speciation** [pp.280–281]
**17.11. Other Speciation Models** [pp.282–283]
1. prezygotic; 2. temporal isolation; 3. mechanical isolation; 4. behavioral isolation; 5. ecological isolation; 6. gamete incompatibility; 7. postzygotic; 8. hybrid inviability; 9. hybrid sterility; 10. c; 11. b; 12. a; 13. b; 14. a; 15. c; 16. b
**17.12. Macroevolution** [pp.284–285]
**17.13. For the Birds** [p.286]
**17.14. Adaptation to What?** [pp.286–287]
1. a; 2. I; 3. d; 4. h; 5. e; 6. g; 7. f; 8. f; 9. c; 9. b
**Self-Quiz**
1. a; 2. d; 3. b; 4. b; 5. c; 6. d; 7. c; 8. a; 9. b; 10. b
**Chapter Summary**
1. populations; 2. alleles; 3. phenotypes; 4. frequency; 5. mutation; 6. genetic; 7. flow; 8. interbreed; 9. natural; 10. fertile; 11. isolated; 12. flow; 13. independently; 14. divergence; 15. reproductive; 16. macroevolution; 17. adaptive; 18. extinctions; 19. adaptation; 20. behavior; 21. survive; 22. environment

---

# Chapter 18  Life's Origin and Early Evolution

*Looking for Life in All the Wrong Places* [pp.290–291]
**18.1. In the Beginning** [pp.292–293]
**18.2. How Did Cells Emerge?** [pp.294–295]
1. e; 2. a; 3. d; 4. b; 5. c; 6. d; 7. e; 8. a; 9. b; 10. e; 11. a; 12. c; 13. b; 14. c; 15. e; 16. abiotic; 17. lipid; 18. DNA; 19. RNA; 20. selectively permeable; 21. protocells
**18.3. The First Cells** [pp.296–297]
**18.4. Where Did Organelles Come From?** [pp.298–299]
**18.5. Time Line for Life's Origin and Evolution** [pp.300–301]
1. a; 2. d; 3. e; 4. b; 5. c; 6. life could no longer arise spontaneously, aerobic respiration evolved, ozone layer

formed; 7. a; 8. c; 9. b; 10. a; 11. c; 12. a; 13. d; 14. I; 15. g; 16. b; 17. c; 18. a; 19. f; 20. j; 21. e; 22. h

**Self-Quiz**

1. a; 2. d; 3. a; 4. a; 5. d; 6. b; 7. d; 8. d; 9. b; 10. d

**Chapter Summary**

1. 4 billion; 2. seas; 3. organic; 4. meteorites; 5. experiments; 6. chemical; 7. membrane; 8. anaerobic; 9. bacteria; 10. eukaryotic; 11. photosynthetic; 12. atmosphere; 13. selection; 14. membrane-bound; 15. plasma membrane; 16. chloroplasts; 17. time line; 18. organisms

# Chapter 19 Prokaryotes and Viruses

*West Nile Virus Takes Off* [pp.304–305]

**19.1. Characteristics of Prokaryotic Cells** [pp.306–306]

1. DNA (b); 2. pilus (c); 3. flagellum (f); 4. capsule (a); 5. cell wall (d); 6. plasma membrane (g); 7. cytoplasm (e); 8. b; 9. a; 10. c; 11. d; 12. b; 13. prokaryotic fission; 14. DNA; 15. membrane; 16. cell wall; 17. genetically; 18–20. rods, cocci, spirals

**19.2. The Bacteria** [pp.308–309]

**19.3. The Archaeans** [pp.310–311]

1. a; 2. e; 3. b; 4. c; 5. d; 6. a; 7. e; 8. cyanobacteria; 9. proteobacteria; 10. cyanobacteria; 11. gram-positive; 12. Archae; 13. proteobacteria; 14. proteobacteria; 15. chlamydias; 16. proteobacteria; 17. proteobacteria; 18. spirochete; 19. gram positive; 20. Archae; 21. gram positive; 22. b, c; 23. b, c; 24. c; 25. a; 26. b

**19.4. The Viruses** [pp.312–313]

**19.5. Viroids and Prions** [pp.314–315]

1. infectious; 2. RNA; 3. protein; 4. host; 5. bacterial; 6. tail fibers; 7. head; 8. receptors; 9. f; 10. j; 11. a; 12. l; 13. b; 14. i; 15. c; 16. h; 17. g; 18. d; 19. e; 20. a; 21. d; 22. b; 23. c; 24. b; 25. a

**19.5. Evolution and Infectious Diseases** [p.315]

1. infection; 2. disease; 3. contagious; 4. sporadic; 5. few; 6. endemic; 7. epidemic; 8. pandemic; 9. SARS; 10. AIDS; 11. accidental; 12. penicillin; 13. resistance; 14. penicillin; 15. resistance

**Self-Quiz**

1. a; 2. c; 3. c; 4. d; 5. c; 6. d; 7. e; 8. d; 9. c; 10. d

**Chapter Summary**

1. Bacteria; 2. Archaea; 3. nucleus; 4. organelles; 5. metabolic; 6. fission; 7. plasmid; 8. prokaryotic; 9. extreme; 10. replicate; 11. protein; 12. prions; 13. RNA; 14. proteins; 15. pathogens; 16. natural selection

# Chapter 20 Protists—the Simplest Eukaryotes

*Tiny Critters, Big Impacts* [pp.318–319]

**20.1. An Evolutionary Road Map** [p.320]

**20.2. Evolutionarily Ancient Flagellates** [p.321]

1. eukaryotic; 2. nucleus; 3. ER; 4. chloroplasts; 5. kingdom; 6. prokaryotes; 7. smaller; 8. f; 9. h; 10. a; 11. e; 12. g; 13. d; 14. b; 15. c; 16. autotroph

**20.3. Shelled Amoebas** [p.322]

**20.4. Alveolates** [p.322]

**20.5. Malaria and the Night Feeding Mosquitoes** [p.324]

1. g; 2. b; 3. d; 4. a; 5. e; 6. f; 7. c; 8. f; 9. c; 10. e; 11. b; 12. a; 13. f; 14. f; 15. c; 16. e; 17. d; 18. e; 19. e; 20. b

**20.6. Single-Celled Stramenophiles** [p.325]

**20.7. Brown Algae** [p.326]

**20.8. Green Algae** [pp.326–327]

**20.9. Red Algae** [p.328]

1. a, e; 2. c; 3. b; 4. b; 5. e; 6. e; 7. b; 8. b; 9. c; 10. b; 11. green algae; 12. red algae; 13. green algae; 14. green algae; 15. oomycota

**20.10. Amoeboid Cells at the Crossroads** [p.329]

1. ciliate; 2. apicomplexa; 3. green algae; 4. flagellate; 5. amoeba; 6. flagellate; 7. green algae; 8. flagellate; 9. amoeba; 10. flagellate; 11. NP; 12. malaria; 13. NP; 14. dysentery; 15. dysentery; 16. African sleeping sickness; 17. NP; 18. vaginosis; 19. NP; 20. NP

**Self-Quiz**

1. c; 2. d; 3. d; 4. e; 5. d; 6. a; 7. d; 8. a; 9. c; 10. e

**Chapter Summary**

1. eukaryotic; 2. multicelld; 3. fungi; 4. animals; 5. protozoans; 6. radiolarians; 7. dinoflagellates; 8. photoautotrophs; 9. sacs; 10. strameophiles; 11. photoautotrophs; 12. oomycotes; 13. photosynthetic; 14. plants; 15. amoebozoans; 16. fungi

# Chapter 21 Plant Evolution

*Beginnings and Endings* [pp.344–335]

**21.1. Evolutionary Trends among Plants** [pp.336–337]

1. g; 2. i; 3. h; 4. c; 5. a; 6. b; 7. d; 8. e; 9. j; 10. f

**21.2. The Bryophytes—No Vascular Tissues** [pp.336–337]

1. T; 2. lack; 3. rhizoids; 4. T; 5. sporophyte; 6. T; 7. acid; 8. T; 9. Bryophytes; 10. diploid, sporophyte; 11. sporophyte, diploid, gametophyte, haploid; 12. meiosis, haploid spores; 13. gametophyte, haploid; 14. gametophyte; 15. gametophyte; 16. haploid sperms; 17. haploid eggs

**21.3. Seedless Vascular Plants** [pp.338–339]

1. b; 2. a; 3. d; 4. c; 5. c; 6. d; 7. b; 8. c; 9. c; 10. b; 11. b; 12. d; 13. a; 14. b; 15. b; 16. d; 17. gametophyte, haploid; 18. sporophyte, diploid; 19. haploid, mitosis; 20. gametophyte, haploid; 21. diploid, gametophyte; 22. spore production, diploid, sporophyte

**21.4. Ancient Carbon Treasures** [p.340]

**21.5. The Rise of Seed-Bearing Plants** [p.341]

1. h; 2. e; 3. j; 4. f; 5. b; 6. i; 7. a; 8. g; 9. c; 10. d; 11. i

**21.6. Gymnosperms—Plants with "Naked" Seeds** [pp.342–343]

1. b; 2. d; 3. b; 4. c; 5. a; 6. d; 7. e; 8. b; 9. e; 10. c; 11. d; 12. e; 13. sporophyte, diploid; 14. seed; 15. sporophyte, diploid; 16. pollen sac, cone; 17. ovule, diploid; 18. female gametophyte, haploid; 19. eggs, haploid, mitosis; 20. male gametophyte, haploid; 21. megaspore, meiosis, haploid; 22. microspores, haploid, pollen grain; 23. pollen grain, female cone

**21.7. Angiosperms—the Flowering Plants** [pp.344–345]

**21.8. Focus on a Flowering Life Cycle** [p.346]

1. f; 2. j; 3. e; 4. i; 5. c; 6. a; 7. h; 8. b; 9. g; 10. d; 11. sporophyte, diploid; 12. female gametophyte, haploid; 13. egg, haploid, mitosis; 14. ovule, diploid, seed; 15. male gametophyte, haploid, mitosis; 16. released pollen grain, haploid; 17. embryo sporophyte plant, diploid, mitosis, diploid zygote; 18. seed, ovule

1. c; 2. c; 3. a; 4. b; 5. e; 6. e; 7. c; 8. c; 9. d; 10. e
**Chapter Objectives/Review Questions**
4. gametophyte; 5. ferns; 6. sporophyte; 7. coal; 10.
sporophyte; 11. flower; 12. *Archaefructus*; 14. sporophyte
**Chapter Summary**
1. 475; 2. divergences; 3. radiations; 4. functional;
5. nonvascular; 6. gamete; 7. water; 8. vascular; 9. seeds;
10. internal; 11. sperm; 12. drier; 13. pollen; 14. seeds;
15. flowers; 16. diversity; 17. angiosperms

# Chapter 22 Fungi

*Food, Forest, Fungi* [pp.350–351]
**22.1. Characteristics of Fungi** [p.352]
**22.2. Zygomycetes—the Zygote Fungi** [p.353]
**22.3. Ascomycetes—the Sac Fungi** [p.354]
**22.4. Basidiomycetes—the Club Fungi** [p.355]
 **Completion**
1. zygomycete; 2. club; 3. sac; 4. sac; 5. sac; 6. sac; 7. bread
mold; 8. mushroom pizza; 9. antibiotics; 10. roundworm
control; 11. soy sauce; 12. yeast infections
**22.5. The Fungal Symbionts** [pp.356–357]
**22.6. An Unloved Few** [p.357]
**Labeling**
1. b; 2. d; 3. e; 4. f; 5. g; 6. c; 7. a
**Choice**
8. c; 9. a; 10. c; 11. b; 12. a; 13. c; 14. d
**Self-Quiz**
1. d; 2. a; 3. d; 4. c; 5. b
**Chapter Summary**
1. heterotrophs; 2. animals; 3. organic; 4. filaments;
5. zygote; 6. club; 7. sac; 8. intracellular; 9. spore;
10. sexual; 11. spore; 12. sexual; 13. cell; 14. species;
15. lichens; 16. parasites; 17. humans

# Chapter 23 Animal Evolution— the Invertebrates

*Old Genes, New Drugs* [pp.360–361]
**23.1. Animal Traits and Trends** [pp.362–363]
**22.2. Animal Origin and Early Radiations** [p.364]
1. T; 2. F, heterotrophs; 3. T; 4. T; 5. e; 6. f; 7. m; 8. h; 9. I;
10. o; 11. a; 12. n; 13. b; 14. j; 15. g; 16. c; 17. d; 18. l;
19. colonial protists; 20. flagellum; 21. division;
22. feeding; 23. reproduction; 24. gene; 25. animal;
26. Sea; 27. Cambrian; 28. c; 29. e; 30. a; 31. b; 32. f;
33. d; 34. g
**23.3. Sponges—Success in Simplicity** [p.365]
**23.4. Cnidarians—Simple Tissues, No Organs**
 [pp.366–367]
**23.5. Flatworms—Simple Organ Systems** [pp.368–369]
1. b; 2. b; 3. c; 4. a; 5. c; 6. a; 7. c; 8. b; 9. b; 10. c; 11. d; 12. a;
13. f; 14. c; 15. e; 16. d; 17. b
**23.6. Annelids—Segments Galore** [pp.370–371]
**23.7. The Pliable Mollusks** [pp.372–373]
**23.8. Roundworms** [p.374]
1. a; 2. c; 3. a; 4. a; 5. c; 6. b; 7. a; 8. b; 9. c; 10. b; 11. d;
12. e; 13. c; 14. f; 15. a; 16. b; 17. g; 18. radula; 19. mantle;
20. foot; 21. stomach; 22. digestive gland; 23. heart;

24. excretory gland; 25. gill; 26. trichinosis; 27. pinworm;
28. stomach pain, vomiting; 29. elephantitis
**23.9. Why Such Spectacular Arthropod Diversity?**
 [p.375]
**23.10. Spiders and Their Relatives** [p.376]
**23.11. A Look at the Crustaceans** [p.377]
**23.12. A Look at Insect Diversity** [pp.378–379]
**23.13. Unwelcome Arthropods** [p.380]
**23.14. The Spiny Skinned Echinoderms** [p.381]
1. F, open; 2. T; 3. F, exoskeleton; 4. T; 5. F, coelum; 6. a;
7. b; 8. c; 9. d; 10. c; 11. d; 12. b; 13. a; 14. b, c; 15. b; 16. c;
17. segmented; 18. thorax; 19. abdomen; 20. antennae;
21. thorax; 22. two; 23. invertebrates; 24. complete; 25.
food; 26. water; 27. malpighian tubules; 28. ion; 29. T;
30. F, radial; 31. T; 32. F, decentralized; 33. arthropod;
34. arthropod; 35. arthropod; 36. arthropod; 37. arthropod; 38. recluse spider—tissue destruction, can be
fatal; 39. deer tick—Lyme disease; 40. black widow
spider—neurotoxin; 41. Rocky Mountain spotted fever;
42. malaria, West Nile virus, encephalitis; 43. hardened
exoskeleton; 44. jointed appendages; 45. highly modified
segments; 46. respiratory structures; 47. sensory specializations; 48. specialized stages of development
**Self Quiz**
1. a; 2. e; 3. a; 4. b; 5. c; 6. d; 7.a; 8. c; 9. b; 10. a; 12. d; 13. b;
14. d; 15. b
**Chapter Objectives**
2. invertebrates; 21. placozoan; 22. hydras, venus flower
basket; 23. cnidarians; 24. fluke, tapeworms; 25. annelids;
26. snails, slugs, squid; 27. crustaceans, spiders, insects;
28. roundworms; 29. echinoderms
**Chapter Summary**
1. heterotrophs; 2. develop; 3. life; 4. tissues; 5. simple;
6. complex; 7. integration; 8. symmetry; 9. body;
10. segments; 11. deuterostomes; 12. symmetry;
13. tissues; 14. radially; 15. bilateral; 16. systems;
17. embryos; 18. arthropods; 19. crustaceans; 20. insects;
21. chordates; 22. invertebrates; 23. radial

# Chapter 24 Animal Evolution— the Vertebrates

*Interpreting and Misinterpreting the Facts*
 [pp.384–385]
**24.1. The Chordate Heritage** [pp.386–387]
**24.2. Evolutionary Trends Among the Vertebrates**
 [pp.388–389]
1. b; 2. c; 3. f; 4. i; 5. e; 6. d; 7. h; 8. g; 9. a; 10. a, d, e, h;
11. a; 12. c; 13. d; 14. a; 15. e; 16. a; 17. e; 18. craniates;
19. vertebrates; 20. jawed vertebrates; 21. tetrapods;
22. amniotes
**24.3. Jawed Fishes and the Rise of Tetrapods**
 [pp.390–391]
**24.4. Amphibians—the First Tetrapods on Land**
 [pp.392–393]
**24.5. Vanishing Acts** [p.393]
1. e; 2. b; 3. d; 4. c; 5. a; 6. g; 7. h; 8. d; 9. f; 10. b; 11. e;
12. a; 13. c; 14. i
**24.6. The Rise of Amniotes** [pp.394–395]

**24.7. So Long, Dinosaurs** [p.395]
**24.8. Portfolio Of Modern "Reptiles"** [pp.396–397]
**24.9. Birds—the Feathered Ones** [pp.398–399]
1. f; 2. c; 3. d; 4. e; 5. a; 6. b; 7. crocodilians; 8. lizards; 9. snakes; 10. 2; 11. shell, no teeth; 12. f; 13. g; 14. b; 15. c; 16. d; 17. e; 18. a
**24.10. The Rise of Mammals** [pp.400–401]
**24.11. From Early Primates to Hominids** [pp.402–403]
**24.12. Emergence of Early Humans** [pp.404–405]
**24.13. Emergence of Modern Humans** [pp.406–407]
1. T; 2. T; 3. monotremes; 4. marsupials; 5. T; 6. convergent; 7. tree-dwellers; 8. daytime; 9. bipedalism; 10. hands; 11. hand; 12. teeth; 13. brain; 14. culture; 15. behavioral; 16. learning; 17. traits; 18. a; 19. b; 20. e; 21. d; 22. c
**Self-Quiz**
1. c; 2. d; 3. b; 4. a; 5. c; 6. b; 7. d; 8. b; 9. b; 10. d
**Chapter Summary**
1. chordates; 2. nerve; 3. gill slits; 4. tail; 5. vertebrates; 6. backbone; 7. muscles; 8. sensory; 9. brain; 10. gills; 11. gas; 12. limbs; 13. seas; 14. fishes; 15. diversity; 16. tetrapods; 17. amniotes; 18. mammals; 19. land; 20. human; 21. climate; 22. resources; 23. Africa

## Chapter 25 Plants and Animals—Common Challenges

*A Cautionary Tale* [pp.410–411]
**25.1. Levels of Structural Organization** [pp.412–413]
**25.2. Recurring Challenges to Survival** [pp.414–415]
1. b; 2. f; 3. g; 4. d; 5. e; 6. a; 7. c; 8. fluid; 9. wastes; 10. nutrients; 11. extracellular; 12. volume; 13. ions; 14. metabolism; 15. stable; 16. T; 17. F, internal; 18. T; 19. F, habitat
**25.3. Homeostasis in Animals** [pp.416–417]
**25.4. Does Homeostasis Occur in Plants?** [pp.418–419]
**25.5. How Cells Receive and Respond to Signals** [pp.420–421]
1. d; 2. e; 3. b; 4. a; 5. c; 6. N; 7. P; 8. P; 9. N; 10. walls; 11. toxic; 12. resins; 13. wood; 14. toxins; 15. compartments; 16. poisoned; 17. compartmentalization
**Self-Quiz**
1. c; 2. c; 3. a; 4. b; 5. c; 6. c; 7. a; 8. a
**Chapter Summary**
1. anatomy; 2. tissues; 3. physiology; 4. environment; 5. growth; 6. gases; 7. transport; 8. internal; 9. cell; 10. resources; 11. fluid; 12. organs; 13. cells; 14. homeostasis; 15. feedback; 16. chemical; 17. signals

## Chapter 26 Plant Tissues

*Droughts Versus Civilization* [pp.424–425]
**26.1. Components of the Plant Body** [pp.426–427]
1. reduces photosynthesis, fewer flowers produced, lower pollination, seeds and fruits may fall off before ripening; 2. b (c); 3. a (e); 4. e (d); 5. c (a); 6. d (b); 7. primary growth occurs at apical meristems, secondary growth occurs at lateral meristems; 8. a. one, three (or multiples of three), parallel, one pore or furrow, scattered throughout stem ground tissue, b. two, four, or five (or

multiples of four or five), netlike, three pores or furrows, ring in the ground tissue
**26.2. Components of Plant Tissues** [pp.428–429]
1. parenchyma; 2. xylem, 3. phloem; 4. sclerenchyma; 5. companion cell; 6. sieve plate; 7. d; 8. b; 9. a; 10. c; 11. a; 12. d; 13. c; 14. b; 15. b
**26.3. Primary Structure of Shoots** [pp.430–431]
**26.4. A Closer Look at Leaves** [pp.432–433]
1. leaf; 2. apical meristem; 3. lateral bud; 4. cortex; 5. vascular; 6. pith; 7. leaf; 8. apical meristem; 9. meristems; 10. cortex; 11. phloem; 12. xylem; 13. pith; 14. monocot; 15. eudicot; 16. petiole; 17. blade; 18. bud; 19. node; 20. stem; 21. blade; 22. sheath; 23. node; 24. blade; 25. vein; 26. vein; 27. xylem; 28. phloem; 29. oxygen; 30. stomata; 31. cuticle; 32. epidermis; 33. palisade; 34. spongy; 35. epidermis; 36. cuticle; 37. epidermis
**26.5. Primary Structure of Roots** [pp.434–435]
1. endodermis; 2. pericycle; 3. xylem; 4. phloem; 5. cortex; 6. epidermis; 7. root hair; 8. matured; 9. elongate; 10. dividing; 11. root cap; 12. taproot; 13. fibrous; 14. lateral root
**26.6. Accumulated Secondary Growth—The Woody Plants** [pp.436–437]
1. periderm; 2. phloem; 3. heartwood; 4. sapwood; 5. bark; 6. vascular cambium; 7. early; 9. late; 10. growth rings
**Self-Quiz**
1. d; 2. c; 3. a; 4. b; 5. b; 6. e; 7. c; 8. d; 9. b; 10. e
**Chapter Summary**
1. vascular; 2. shoot; 3. root; 4. water; 5. photosynthesis; 6. environment; 7. meristems; 8. primary; 9. dermal; 10. organization; 11. sunlight; 12. water; 13. gas; 14. eudicot; 15. mineral; 16. anchor; 17. secondary; 18. wood

## Chapter 27 Plant Nutrition and Transport

*Leafy Clean-Up Crews* [pp.440–441]
**27.1. Plant Nutrients and Availability in Soil** [pp.442–443]
1. c; 2. j; 3. i; 4. b; 5. d; 6. e; 7. f; 8. g; 9. a; 10. h
**27.2. How Do Roots Absorb Water and Mineral Ions?** [pp.444–445]
1. mutualism; 2. gaseous nitrogen; 3. root nodules; 4. sugars; 5. minerals; 6. exodermis; 7. cortex; 8. endodermal cells; 9. vascular cylinder; 10. xylem; 11. phloem; 12. endodermis; 13. Casparian strip; 14. b; 15. c; 16. d; 17. a
**27.3. How Does Water Move Through Plants?** [pp.446–447]
**27.4. How Do Stems and Leaves Conserve Water?** [pp.448–44]
1. stomata; 2. xylem; 3. tension; 4. hydrogen; 5. roots; 6. e; 7. a; 8. b; 9. d; 10. c
**27.5. The Bad Ozone** [p.449]
**27.6. How Do Organic Compounds Move Through Plants?** [pp.450–451]
1. b; 2. a; 3. f; 4. e; 5. d; 6. c; 7. source; 8. active; 9. solutes; 10. water; 11. turgor; 12. sink

**Self-Quiz**
1. b; 2. a; 3. d; 4. a; 5. c; 6. c; 7. b; 8. d; 9. b; 10. d
**Chapter Summary**
1. water; 2. mineral; 3. root; 4. soil; 5. erosion; 6. solutes;
7. evaporation; 8. stomata; 9. gas; 10. photosynthesis;
11. phloem; 12. organic

# Chapter 28  Plant Reproduction and Development

*Imperiled Sexual Partners* [pp.454–455]
**28.1. Reproductive Structures of Flowering Plants**
[pp.456–457]
1. sepal; 2. petal; 3. stamen; 4. filament; 5. anther;
6. carpel; 7. stigma; 8. style; 9. ovary; 10. meiosis;
11. microspores; 12. megaspores; 13. gametophyte;
14. fertilization; 15. zygote; 16. sporophyte
**28.2. A New Generation Begins** [pp.458–459]
**28.3. From Zygotes to Seeds and Fruits** [pp.460–461]
**28.4. Seed Dispersal—the Function of Fruits** [p.462]
1. sporophyte; 2. ovule; 3. ovary; 4. meiosis;
5. megaspores; 6. mitosis; 7. eight; 8. embryo; 9. ovule;
10. gametophyte; 11. anther; 12. filament; 13. pollen sac;
14. meiosis; 15. microspores; 16. pollen grain; 17. anther;
18. pollination; 19. gametophyte; 20. pollen; 21. sperm;
22. endosperm; 23. egg; 24. ovule; 25. egg;
26. endosperm; 27. pollen; 28. endosperm; 29. embryo;
30. coat; 31. g; 32. b; 33. c; 34. d; 35. f; 36. e; 37. a
**28.5. Asexual Reproduction of Flowering Plants**
[pp.462–463]
1. f; 2. d; 3. c; 4. e; 5. b; 6. a; 7. g
**28.6. Overview of Plant Development** [pp.464–465]
1. root; 2. branch; 3. coleoptile; 4. foliage;
5. adventitious; 6. branch; 7. primary; 8. prop; 9. coat;
10. root; 11. cotyledons; 12. hypocotyls; 13. leaf;
14. foliage; 15. primary; 16. branch; 17. primary;
18. nodule
**28.7. Plant Hormones and Other Signaling Molecules**
[pp.466–467]
**28.8. Adjusting the Direction and Rates of Growth**
[pp.468–469]
1. a; 2. c; 3. c; 4. d; 5. a; 6. e; 7. d; 8. e; 9.e; 10. b;
11. e; 12. b; 13. d; 14. light; 15. gravitropism;
16. thigmotropism; 17. flavoproteins; 18. statoliths;
19. auxins
**28.9. Sensing Recurring Environmental Changes**
[pp.470–471]
**28.10. Entering and Breaking Dormancy** [p.472]
1. long-day; 2. short-day; 3. day-neutral; 4. longer;
5. shorter; 6. red; 7. far-red; 8. inactivated; 9. b; 10. a;
11. c; 12. d
**Self-Quiz**
1. d; 2. d; 3. b; 4. b; 5. d; 6. e; 7. c; 8. d; 9. d; 10. a
**Chapter Summary**
1. sexual; 2. pollinators; 3. flowers; 4. gametophytes;
5. fertilization; 6. sporophyte; 7. ovary; 8. dispersal;
9. animals; 10. asexually; 11. hormones;
12. germination; 13. root; 14. synthesis; 15. environmental; 16. night

# Chapter 29  Animal Tissues and Organ Systems

*Open or Close the Stem Cell Factories* [pp.476–477]
**29.1 Epithelial Tissue** [pp.478–479]
**29.2 Connective Tissue** [pp.480–481]
1. a; 2. b; 3. m; 4. j; 5. i; 6. g; 7. h; 8. c; 9. l; 10. f; 11. e; 12. d;
13. k; 14. b; 15. a; 16. b; 17. a; 18. a
**29.3 Muscle Tissues** [pp.482–483]
**29.4. Nervous Tissue** [p.483]
1. T; 2. F, cardiac; 3. F, muscle; 4. F. involuntary; 5. T;
6. F, smooth
**29.5. Overview of Major Organ Systems** [pp.484–485]
**29.6. Vertebrate Skin—Example of an Organ System**
[pp.486–487]
1. c; 2. b; 3. a; 4. superior; 5. distal; 6. proximal; 7. posterior; 8. transverse; 9. inferior; 10. anterior; 11. frontal;
12. circulatory (D); 13. respiratory (E); 14. urinary (A);
15. skeletal (C); 16. endocrine (J): 17. reproductive (G);
18. digestive (I); 19. muscular (H); 20. nervous (B);
21 integumentary (F); 22. lymphatic (K); 23. hair;
24. epidermis; 25. dermis; 26. hypodermis; 27. hair
follicle; 28. smooth muscle
**Self-Quiz**
1. c; 2. d; 3. c; 4. b; 5. b; 6. d; 7. a; 8. a; 9. a; 10. e
**Chapter Summary**
1. connective; 2. nervous; 3. epithelia; 4. secretory;
5. insulate; 6. cartilage; 7. adipose; 8. muscle; 9. cardiac;
10. communication; 11. neurons; 12. compartmentalize;
13. ectoderm; 14. embryo; 15. skin; 16. sensory; 17. water;
18. wastes; 19. stimuli

# Chapter 30  Neural Control

*In Pursuit of Ecstasy* [pp.490–491]
**30.1. Evolution of Nervous Systems** [pp.492–493]
**30.2. Neurons—the Great Communicators** [pp.494–495]
**30.3. A Look at Action Potentials** [pp.496–497]
1. carry information toward cell body; 2. one per cell in
most neurons; 3. branch farther from cell body; 4. no insulating sheath; 5. e; 6. f; 7. g; 8. d; 9. a; 10. b; 11. c; 12. T;
13. T; 14. dendrite; 15. input zone; 16. trigger zone; 17.
axon; 18. conducting zone; 19. output zone; 20. a; 21. f;
22. e; 23. d; 24. c; 25. g; 26. b; 27. a; 28. T; 29. F, intensifies;
30. F, sodium-potassium; 31. T; 32. T
**30.4. How Neurons Send Messages to Other Cells**
[pp.498–499]
**30.5. A Smorgasbord of Signals** [pp.500–501]
1. c; 2. b; 3. a; 4. d; 5. f; 6. e; 7. a; 8. b; 9. c; 10. d; 11. c; 12. e,
13. a; 14. e; 15. a; 16. b; 17. c; 18. d; 19. ACh; 20. receptors
attacked; 21. Alzheimer's; 22. Parkinson's
**30.6. Drugs and Disrupted Signaling** [p.501]
**30.7. Organization of Neurons in Nervous Systems**
[pp.502–503]
**30.8. What Are the Major Expressways?** [pp.504–505]
1. tolerance; 2. habituation; 3. inability to stop drug use,
even if desire to do so persists; 4. concealment;
5. dangerous actions taken to obtain drug; 6. deterioration of professional and personal relationships;

7. anger or defensiveness is someone suggest a problem;
8. drug use preferred over customary actions; 9. a; 10. c;
11. e; 12. g; 13. f; 14. d; 15. b; 16. T; 17. F, white; 18. F, central; 19. T; 20. T; 21. g; 22. a; 23. d; 24. e; 25. b; 26. c; 27. f;
28. optic; 29. vagus; 30. pelvic; 31. cervical; 32. thoracic;
33. lumbar; 34. sacral; 35. e; 36. b; 37. a; 38. d; 39. c; 40. d;
41. d; 42. d; 43. a; 44. b; 45. c; 46. a; 47. c; 48. a; 49. a;
50. a, b, c

**30.9. The Vertebrate Brain** [pp.506–507]
**30.10. The Human Cerebrum** [pp.508–509]
**30.11. Neuroglia** [p.510]
1. a; 2. d; 3. g; 4. f; 5. e; 6. c; 7. b; 8. spinal cord;
9. cerebrospinal; 10. blood capillaries; 11. tight;
12. proteins; 13. glucose; 14. ions; 15. urea; 16. carbon
dioxide; 17. neural; 18. corpus callosum; 19. hypothalamus; 20. thalamus; 21. pineal gland; 22. midbrain;
23. cerebellum; 24. pons; 25. medulla oblongata; 26. M;
27. S; 28. A; 29. M; 30. A; 31. S; 32. T; 33. F, olfactory; 34. F,
long-term; 35. F, hippocampus; 36. F, separately
**Self-Quiz**
1. d; 2. d; 3. b; 4. a; 5. d; 6. e; 7. c; 8. b; 9. b; 10. b
**Chapter Summary**
1. neurons; 2. radially; 3. bilaterally; 4. plasma membrane; 5. self; 6. electric charge; 7. chemical; 8. inhibit;
9. psychoactive; 10. reflex arc; 11. brain; 12. spinal cord;
13. peripheral; 14. cerebral cortex; 15. neuroglia

# Chapter 31  Sensory Perception

*A Whale of a Dilemma* [pp.512–513]
**31.1. Overview of Sensory Pathways** [p.514]
**31.2. Somatic Sensations** [p.515]
1. d; 2. c; 3. a; 4. f; 5. e; 6. b; 7. F, somatic; 8. F, touch
(pressure); 9. T; 10. F, frequency; 11. T
**31.3. Sampling the Chemical World** [p.516]
**31.4. Keeping the Body Balanced** [p.517]
**31.5. Collecting, Amplifying, and Sorting Out Sounds**
[pp.518–519]
1. c; 2. a; 3. b, d; 4. a; 5. c; 6. a, c; 7. d; 8. b; 9. F,
pheromones; 10. F, olfactory; 11 T; 12. T; 13. T; 14. F, static;
15. T; 16. eardrum; 17. round window; 18. cochlea;
19. auditory nerve; 20. oval window; 21. stirrup;
22. anvil; 23. hammer; 24. b; 25. f; 26. d; 27. g; 28. e;
29. c; 30.a
**31.6. Do You See What I See?** [pp.520–521]
**31.7. From the Retina to the Visual Cortex** [p.522]
**31.8. Visual Disorders** [p.523]
1. choroid (f); 2. iris (g); 3. lens (h); 4. pupil (a); 5. cornea
(e); 6. aqueous humor (b); 7. vitreous humor (i); 8. optic
nerve (c); 9. retina (d); 10. F, left; 11. F, rod; 12. F, three;
13. F, epithelium; 14. F, rod; 15. e; 16. a; 17. f; 18. c; 19. b;
20. d; 21. g
**Self-Quiz**
1. d; 2. c; 3. a; 4. a; 5. b; 6. c; 7. c; 8. b; 9. d; 10. e
**Chapter Summary**
1. nervous; 2. nerve; 3. process; 4. stimulus; 5. number;
6. frequency; 7. somatic; 8. mechanoreceptors; 9.
chemoreceptors; 10. balance; 11. gravity; 12. hearing;
13. action potentials; 14. light-sensitive; 15. photoreceptors; 16. stimuli; 17. retina; 18. visual cortex.

# Chapter 32  Endocrine Control

*Hormones in the Balance* [pp.526–527]
**32.1. Introducing the Vertebrate Endocrine System**
[pp.528–529]
1. d; 2. b; 3. c; 4. a; 5. hypothalamus (d); 6. pituitary gland
(b); 7. adrenal gland (a); 8. ovaries (f); 9. testes (i);
10. pancreatic islets (c); 11. thymus (h); 12. parathyroid
gland (g); 13. thyroid gland (j); 14. pineal gland (e)
**32.2. Nature of Hormone Action.** [pp.530–531]
1. steroid, amines, peptides, proteins; 2. must have
receptors, interaction of hormones, concentration of the
hormone, cell's metabolic and nutritional state,
environmental cues; 3. a; 4. b; 5. b; 6. a; 7. b; 8. steroid;
9. lipid; 10. nucleus; 11. transcription; 12. response;
13. peptide; 14. enzyme; 15. cyclic AMP
**32.3. The Hypothalamus and Pituitary Gland**
[pp.532–533]
**32.4. Thyroid and Parathyroid Glands** [p.534]
1. e; 2. d; 3. b; 4. a; 5. b; 6. a; 7. c; 8. d; 9. d; 10. c
**32.5. The Adrenal Glands** [p.535]
**32.6. Pancreatic Hormones** [p.536]
**32.7. Blood Sugar Disorders** [p.537]
1. b; 2. a; 3. a; 4. a; 5. a; 6. b; 7. b; 8. T; 9. F, fight-flight;
10. F, insulin; 11. F, beta; 12. T; 13. F, pancreatic islets;
14. T; 15. T; 16. F, Type I
**32.8. Other Endocrine Glands** [p.538]
**32.9. A Comparative Look at a Few Invertebrates** [p.539]
1. c; 2. d; 3. f; 4. a; 5. e; 6. g; 7. b
**Self-Quiz**
1. d; 2. b; 3. a; 4. b; 5. d; 6. d; 7. e; 8. e; 9. b; 10. a
**Chapter Summary**
1. hormones; 2. development; 3. endocrine; 4. receptors;
5. transduction; 6. hypothalamus; 7. glands; 8. negative;
9. environment; 10. molting; 11. invertebrates;
12. mammalian; 13. receptors

# Chapter 33  Structural Support
#  and Movement

*Pumping up Muscles* [pp.544–545]
**33.1. Animal Skeletons** [pp.546–547]
**33.2. Zooming in on Bones and Joints** [pp.548–549]
1. F, androstenedione; 2. F, has not; 3. T; 4. T; 5. a; 6. b;
7. a; 8. c; 9. c; 10. b; 11. a; 12. e; 13. c; 14. d; 15. c; 16. e;
17. b; 18. b; 19. c; 20. g; 21. d; 22. f; 23. a; 24. e; 25. h;
26. compact bone; 27. spongy bone; 28. spongy bone;
29. compact bone; 30. connective tissue; 31. blood vessel;
32. hormones; 33. negative; 34. thyroid; 35. calcitonin;
36. osteoclasts; 37. parathyroid; 38. parathyroid hormone
(PTH); 39. osteoclast; 40.a; 41. b; 42. e; 43. c; 44. d
**33.3. Skeletal-Muscular Systems** [pp.550–551]
**33.4. How Does Skeletal Muscle Contract?** [pp.552–553]
1. F, skeletal; 2. T; 3. F, tendon; 4. T; 5. triceps brachii (e);
6. pectoralis major (h); 7. external oblique (j); 8. rectus
abdominis (k); 9. sartorius (b); 10. quadriceps femoris (i);
11. biceps brachii (c); 12. deltoid (d); 13. trapezius (f);
14. gluteus maximus (a); 15. biceps femoris (g); 16. a;
17. c; 18. d; 19. e; 20. b; 21. 5; 22. 3; 23. 1; 24. 2; 25. 4

**33.5. From Signal to Responses** [pp.554–555]
**33.6. Oh** *Clostridium*! [p.556]
1. h; 2. b; 3. c; 4. g; 5. e; 6. f; 7. d; 8. a; 9. tetanus;
10. botulism
**Self-Quiz**
1. d; 2. c; 3. c; 4. c; 5. d; 6. b; 7. b; 8. b; 9. a; 10. d
**Chapter Summary**
1. contractile; 2. skeleton; 3. hydrostatic; 4. exoskeleton;
5. endoskeleton; 6. collagen; 7. minerals; 8. blood;
9. joints; 10. skeletal; 11. reverse; 12. tendons;
13. myofibrils; 14. sarcomeres; 15. actin; 16. ATP;
17. contraction; 18. motor units; 19. neuron; 20. tension;
21. contracts; 22. exercise

# Chapter 34  Circulation

*And Then My Heart Stood Still* [pp.558–559]
**34.1. The Nature of Blood Circulation** [pp.560–561]
1. a; 2. c; 3. e; 4. d; 5. b; 6. c; 7. a; 8. b; 9. c; 10. a; 11. b;
12. b
**34.2. Characteristics of Blood** [pp.562–563]
**34.3. Blood Disorders** [p.564]
**34.4. Blood Typing** [pp.564–565]
1. a; 2. c; 3. b; 4. d; 5. c; 6. b; 7. a; 8. a; 9. c; 10. a; 11. water;
12. plasma proteins; 13. 1–2 percent; 14. red blood cells;
15. neutrophils; 16. lymphocytes; 17. phagocytosis;
18. defense against parasitic worms; 19. basophils;
20. platelets; 21. b; 22. a; 23. b; 24. a; 25. a; 25. a; 26. a;
27. self; 28. ABO; 29. B; 30. A; 31. AB; 32. receive;
33. foreign; 34. O; 35. Rh; 36. marker; 37. Rh+; 38. Rh-;
39. antibodies; 40. Rh+; 41. Charles
**34.5. Human Cardiovascular System** [pp.566–567]
**34.6. The Heart Is a Lonely Pumper** [pp.568–569]
1. jugular veins (h); 2. superior vena cava ( j);
3. pulmonary veins (d); 4. hepatic portal vein (a); 5. renal
vein (o); 6. inferior vena cava (q); 7. iliac veins (f); 8.
femoral vein (b); 9. femoral artery (c); 10. iliac arteries (i);
11. abdominal aorta (p); 12. renal artery (g); 13. brachial
artery (n); 14. coronary arteries (e); 15. pulmonary arteries (k); 16. ascending aorta (m); 17. carotid arteries (l);
18. superior vena cava; 19. semilunar valve;
20. pulmonary veins; 21. right atrium; 22. AV valve;
23. right ventricle; 24. inferior vena cava; 25. septum;
26. left ventricle; 27. AV valve; 28. left atrium;
29. pulmonary veins; 30. semilunar valve; 31. ventricles;
32. SA; 33. T; 34. T; 35. pericardium
**34.7. Pressure, Transport, and Flow Distribution**
   [pp.570–571]
**34.8. Diffusion at Capillaries, Then Back to the Heart**
   [pp.572–573]
**34.9. Cardiovascular Disorders** [pp.574–575]
**34.10. Interactions with the Lymphatic System**
   [pp.576–577]
1. h; 2. j; 3. k; 4. f; 5. I; 6. g; 7. a; 8. d; 9. e; 10. b; 11. c; 12. b
13. d 14. c 15. a 16. g; 17. d; 18. e; 19. f; 20. a; 21. b; 22. c;
23. lymph vascular; 24. water; 25. interstitial; 26. lymph;
27. lymph; 28. plasma; 29. blood; 30. fats; 31. small;
32. pathogens; 33. nodes

**Self-Quiz**
1. e; 2. a; 3. d; 4. b; 5. c; 6. a; 7. c; 8. a; 9. d; 10. d
**Chapter Summary**
1. circulatory; 2. tissues; 3. connective; 4. platelets;
5. plasma; 6. gas; 7. defend; 8. four; 9. ventricles; 10. lung;
11. heart; 12. rhythmically; 13. ventricles; 14. atria;
15. arterioles; 16. capillaries; 17. vessels; 18. rhythms;
19. lifestyles; 20. lymph; 21. infectious

# Chapter 35  Immunity

*Viruses, Vulnerability, and Vaccines* [pp.580–581]
**35.1. Integrated Responses to Threats** [pp.582–583]
**35.2. Surface Barriers** [p.583]
1. b; 2. c; 3. f; 4. e; 5. a; 6. d; 7. lysozyme, enzyme; 8. low
pH, *Lactobacillus* secretions; 9. dead keratin-packed cells,
harmless bacteria and yeast; 10. harmless microbes,
saliva lysozyme; 11. low pH.
**35.3. Innate Immune Responses** [pp.584–585]
**35.4. Overview of Adaptive Immunity** [pp.586–587]
**35.5. The Antibody-mediated Immune Response**
   [pp.588–589]
1. c; 2. b; 3. a; 4. a; 5. c; 6. b; 7. c; 8. a; 9. e; 10. c; 11. b; 12. d;
13. adaptive; 14. diversity; 15. specificity; 16. memory;
17. self vs. nonself recognition; 18. molecular; 19. one;
20. receptors; 21. remember; 22. d; 23. e; 24. a; 25. c; 26. b;
27. f; 28. a; 29. e; 30. b; 31. c; 32. d; 33. b; 34. c; 35. e; 36. d;
37. a
**35.6. The Cell Mediated Response** [pp.590–591]
**35.7. Remember to Floss** [p.591]
1. dendritic cell; 2. antigen presenting cell; 3. cytokines;
4. effector cytotoxic T cell; 5. activated cytotoxic T cell;
6. effector T helper cell; 7. naïve T helper cell
**35.8. Defenses Enhanced or Compromised**
   [pp.592–593]
**35.9. AIDS Revisited—Immunity Lost** [pp.594–595]
1. DPT; 2. HiB; 3. Pneumonococcal; 4. c; 5. a; 6. e; 7. h;
8. g; 9. f; 10. I; 11. d; 12. b
**Self-Quiz**
1. b; 2. b; 3. c; 4. d; 5. b; 6. e; 7. c; 8. c; 9. a; 10. a
**Chapter Summary**
1. surface barriers, innate immunity, adaptive immunity;
2. innate; 3. adaptive; 4. skin, mucous membranes;
5. innate; 6. phagocytic; 7. antibodies; 8. cytotoxic T;
9. self; 10. allergies, autoimmune disorders

# Chapter 36  Respiration

*Up in Smoke* [pp.598–599]
**36.1. The Nature of Respiration** [p.600]
**36.2. Invertebrate Respiration** [p.601]
1. c; 2. a; 3. c; 4. a; 5. b; 6. b; 7. c; 8. f; 9. e; 10. d; 11. g; 12. a
**36.3. Vertebrate Respiration** [pp.602–603]
**36.4. Human Respiratory System** [pp.604–605]
1. c; 2. a; 3. b; 4. c; 5. b, d; 6. oral cavity (b); 7. pleural
membrane (j); 8. intercostal muscles (f); 9. diaphragm (a);
10. bronchial tree (g); 11. lung (h); 12. trachea (d);
13. larynx (k); 14. epiglottis (i); 15. pharynx (c); 16. nasal
cavity (e)

**36.5. Gas Exchange and Transport** [pp.606–607]

**36.3. Cyclic Reversals in Air Pressure Gradients**
  [pp.608–609]

1. b; 2. a; 3. b; 4. b; 5. a; 6. a; 7. $CO_2$; 8. acidity; 9. chemoreceptors; 10. brain stem; 11. diaphragm; 12. decline; 13. tidal volume

**36.7. Respiratory Diseases and Disorders** [pp.610–611]

**36.8. High Climbers and Deep Divers** [pp.612–613]

1. d; 2. b; 3. e; 4. a; 5. h; 6. f; 7. g; 8. c; 9. i

**Self-Quiz**

1. d; 2.d; 3. e; 4. e; 5. b; 6. a; 7. b; 8. d; 9. b; 10. b

**Chapter Summary**

1. oxygen; 2. carbon dioxide; 3. pH; 4. respiration; 5. tissues; 6. gills; 7. internal; 8. tubes; 9. gills; 10. lungs; 11. lungs; 12. blood; 13. respiratory; 14. interstitial; 15. gas; 16. respiratory; 17. depth; 18. brain; 19. infectious disease; 20. pollutants; 21. cigarette; 22. altitudes; 23. marine

---

# Chapter 37  Digestion and Human Nutrition

*Hominoids, Hips, and Hunger* [pp.616–617]

**37.1. The Nature of Digestive Systems** [pp.618–619]

**37.2. Overview of the Human Digestive System**
  [pp.620–621]

1. organ; 2. incomplete; 3. wastes; 4. gut; 5. pharynx; 6. digested; 7. specialization; 8. complete; 9. mouth; 10. anus; 11. specialized; 12. absorption; 13. salivary glands (j); 14. liver (d); 15. gallbladder (c); 16. pancreas (a); 17. mouth (e); 18. pharynx (h); 19. esophagus (b); 20. stomach (i); 21. small intestine (f); 22. large intestine (k); 23. anus (g); 24. f; 25. g; 26. e; 27. c; 28. a; 29. d; 30. b

**37.3. Food in the Mouth** [p.621]

**37.4. Food Breakdown in the Stomach and Small Intestine** [pp.622–623]

**37.5. Absorption from the Small Intestine** [pp.624–625]

1. polysaccharides; 2. small intestine; 3. pepsins; 4. small intestine; 5. pancreas; 6. intestinal lining; 7. triglycerides; 8. DNA, RNA; 9. nucleotides; 10. b; 11. d; 12. e; 13. c; 14. a; 15. g; 16. f; 17. b; 18. c; 19. f; 20. e; 21. d; 22. a; 23. g

**37.6. The Large Intestine** [p.626]

**37.7. What Happens to Absorbed Organic Compounds?**
  [pp.627–628]

**37.8. Human Nutritional Requirements** [pp.628–629]

**37.9. Vitamins and Minerals** [pp.630–631]

**37.10. Weighty Questions, Tantalizing Answers**
  [pp.632–633]

1. T; 2. T; 3. low; 4. T; 5. 3 cups/week; 6.2 cups/week; 7.3 cups/week; 8. 3 cups/week; 9. 2 cups/day; 10. 3 cups/day; 11. 3 ounces/day 12. 5.5 ounces/day; 13. beta carotene in yellow fruits, fortified milk, egg, liver; 14. inactive form made in sin and activated in the liver; fatty fish, egg yolk; 15. counters free radicals, maintains cell membranes; 16. K; 17. connective tissue formation; 17. coenzyme action; 19. niacin; 20. spinach, tomatoes, potatoes, meats; 21. common in many foods; 22. folate; 23. poultry, fish, red meat; 24. coenzyme in fat, glycogen formation; 25. fruits and vegetables; 26. a; 27. d; 28. b; 29. f; 30. e; 31. c

**Self-Quiz**

1. d; 2. a; 3. a; 4. c; 5. c; 6. e; 7. b; 8. a; 9. c; 10. d

**Chapter Summary**

1. sac; 2. tube; 3. organ systems; 4. nutrients; 5. wastes; 6. mouth; 7. stomach; 8. salivary; 9. pancreas; 10. liver; 11. small intestine; 12. water; 13. wastes; 14. nutrients; 15. vitamins; 16. calories; 17. calories

---

# Chapter 38  The Internal Environment

*Truth in a Test Tube* [pp.636–637]

**38.1. Gains and Losses in Water and Solutes**
  [pp.638–639]

**38.2. Desert Rats** [p.639]

**38.3. Structure of the Urinary System** [pp.640–641]

1. freshwater; 2. marine; 3. f; 4. a; 5. h; 6. g; 7.c; 8. d; 9. e; 10. b; 11. d; 12. a; 13. c; 14. b; 15. kidney; 16. ureter; 17. urinary bladder; 18. urethra; 19. kidney cortex; 20. kidney medulla; 21. renal artery; 22. renal vein; 23. ureter

**38.4. Urine Formation** [pp.642–643]

**38.5. Kidney Disease** [p.644]

**38.6. Acid-Base Balance** [p.644]

1. c; 2. d; 3. f; 4. b; 5. f; 6. e; 7. a; 8. g; 9. c; 10. b; 11. d; 12. a; 13. e; 14. f

**38.7. Heat Gains and Losses** [p.645]

**38.8. Temperature Regulation in Mammals** [pp.646–647]

1. h; 2. m; 3. g; 4. j; 5. a; 6. c; 7. k; 8. e; 9. b; 10. d; 11. f; 12. I; 13. 1

**Self-Quiz**

1. a; 2. b; 3. c; 4. c; 5. d; 6. e; 7. b; 8. e; 8. b; 10. b

**Chapter Summary**

1. wastes; 2. water; 3. volume; 4. extracellular; 5. homeostasis; 6. urinary; 7. kidney; 8. bladder; 9. nephrons; 10. urine; 11. filtration; 12. reabsorption; 13. secretion; 14. ADH; 15. aldosterone; 16. body temperature

---

# Chapter 39  Animal Reproduction and Development

*Mind-Boggling Births* [pp.650–651]

**39.1. Reflections on Sexual Reproduction**
  [pp.652–653]

**39.2. Stages of Reproduction and Development**
  [pp.654–655]

1. meiosis; 2. fertilization; 3. zygote; 4. single; 5. offspring; 6. asexual; 7. genetically; 8. uniformity; 9. environment; 10. variations; 11. packages; 12. animals; 13. resources; 14. sexually; 15. alleles; 16. variation; 17. survive; 18. reproduce; 19. f; 20. b; 21. c; 22. a; 23. e; 24. d; 25.a. mesoderm; b. ectoderm; c. endoderm; d. mesoderm; e. ectoderm; f. mesoderm; g. endoderm; h. mesoderm; i. mesoderm

**39.3. Early Marching Orders** [pp.656–657]

1. i; 2. d; 3. e; 4. g; 5. h; 6. c; 7. j; 8. f; 9. a; 10. b; 11. k

**39.4. Specialized Cells, Tissues, and Organs**
  [pp.658–659]

1. a; 2. c; 3. f; 4. l; 5. i; 6. k; 7. j; 8. b; 9. g; 10. e; 11. d; 12. h; 13. T; 14. T

**39.5. Reproductive System of Human Males**
[pp.660–661]
**39.6. Sperm Formation** [pp.662–663]
1. mitosis; 2. seminiferous; 3. meiosis; 4. sperm;
5. epididymis; 6. vas deferens; 7. urethra; 8. seminal
vesicles; 9. prostate gland; 10. bulbourethral; 11. Leydig;
12. testosterone; 13. testosterone; 14. anterior; 15. hypothalamus; 16. decrease; 17. LH; 18. Sertoli; 19. decreased;
20. elevated; 21. Sertoli; 22. hypothalamus; 23. anterior
pituitary; 24. Sertoli cells; 25. Leydig cells
**39.7. Reproductive System of Human Females**
[pp.664–665]
**39.8. Female Troubles** [p.665]
**39.9. The Menstrual Cycle** [pp.666–667]
1. ovary; 2. oviduct; 3. uterus; 4. urinary bladder;
5. myometrium; 6. endometrium; 7. vagina; 8. labium
major; 9. labium minor; 10. clitoris; 11. urethra; 12. c;
13. d; 14. bI; 15. e; 16. f; 17. g; 18. h; 19. i; 20. a; 21. b; 22. d;
23. c; 24. a
**39.10. FSH and Twins** [p.668]
**39.11. Pregnancy Happens** [pp.668–669]
**39.12. Preventing or Seeking Pregnancy** [pp.670–671]
**39.13. Sexually Transmitted Diseases** [pp.672–673]
1. ejaculation; 2. vagina; 3. ovulation; 4. contractions;
5. hundred; 6. oviduct; 7. zona pellucida; 8. enzymes;
9. digestive; 10. zona pellucida; 11. chemical; 12. oocyte;
13. polar body; 14. ovum; 15. nucleus; 16. nucleus;
17. diploid; 18. j; 19. f; 20. k; 21. d; 22. h; 23. c; 24. l; 25. a;
26. e; 27. I; 28. b; 29. g; 30. m; 31. f; 32. a, c; 33. d, e; 34. f;
35. f; 36. d; 37. c; 38. f; 39. c; 40. f; 41. c; 42. f; 43. d; 44. b;
45. a, b, c, d, e, f
**39.14. Formation of the Early Embryo** [pp.674–675]
**39.15. Emergence of the Vertebrate Body Plan** [p.676]
**39.16. The Function of the Placenta** [p.677]
**39.17. Emergence of Distinctly Human Features**
[pp.678–679]
**39.18. Mother as Provider, Protector, Potential Threat**
[pp.680–681]
**39.19. Birth and Lactation** [p.681]
1. f; 2. g; 3. h; 4. e; 5. d; 6. b; 7. a; 8. c; 9. a; 10. g; 11. c;
12. h; 13. d; 14. e; 15. b; 16. f; 17. c; 18. a; 19. b;
20. maternal; 21. blood vessels; 22. maternal; 23. uterus;
24. fetal; 25. embryonic; 26. umbilical cord;
27. blood-filled; 28. chorionic villus; 29. amniotic;
30. e; 31. c; 32. d; 33. a; 34. b; 35. d; 36. b; 37. f; 38. a;
39. e; 40. c; 41. a; 42. d; 43. c; 44. a; 45. d; 46. b
**39.20. Maturation, Aging, and Death** [p.682]
1. c; 2. b; 3. g; 4. j; 5. h; 6. a; 7. k; 8. l; 9. I; 10. d; 11. f; 12. e
**Self-Quiz**
1. b; 2. a; 3. b; 4. a; 5. d; 6. e; 7. c; 8. e; 9. c; 10. a; 11. b;
12. d; 13. e; 14. d; 15. b; 16. d; 17. e; 18. a; 19. b; 20. c
**Chapter Summary**
1. sexual; 2. asexual; 3. variable; 4. gamete formation;
5. fertilization; 6. cleavage; 7. gastrulation; 8. organ
formation; 9. growth and tissue specialization; 10. sperm;
11. ovaries; 12. hypothalamus; 13. cyclic; 14. sexual
intercourse; 15. pathogens; 16. fertilization; 17. blastocyst;
18. placenta; 19. genetic; 20. cells

# Chapter 40 Population Ecology
*The Numbers Game* [pp.686–687]
**40.1. Population Demographics** [p.688]
**40.2. Elusive Heads to Count** [p.687]
1. population crashes occur when population size
exceeds that allowed by resources; 2. I; 3. f; 4. b; 5. e; 6. g;
7. a; 8. d; 9. k; 10. j; 11. l; 12. c; 13. h; 14. m; 15. 2,500
**40.3. Population Size And Exponential Growth**
[pp.690–691]
**40.4. Limits on Population Growth** [pp.692–693]
1.a. 0.4 births per rat per month, b. 0.1 deaths per rat per
month, c. 0.3 per rat per month; 2. e; 3. j; 4. I; 5. f; 6. b;
7. a; 8. h; 9. c; 10. g; 11. d; 12. b; 13. a; 14. a; 15. b; 16. a; 17. b
**40.5. Life History Patterns** [pp.694–695]
**40.6. Natural Selection and Life Histories** [pp.696–697]
1. e; 2. c; 3. a; 4. b; 5. f; 6. d; 7. F, can; 8. T; 9. T; 10. F,
guppy; 11. F, rapidly; 12. T; 13. F, a genetic; 14. F,
pike-cichlids; 15. F, had; 16. T
**40.7. Human Population Growth** [pp.698–699]
**40.8. Fertility Rates and Age Structure** [pp.700–701]
**40.9. Population Growth and Economic Effects**
[pp.702–703]
**40.10 A No-Growth Society** [p.703]
1. geographic expansion, increased carrying capacity,
sidestepped limiting factors; 2. limited natural resources
(carry capacity); 3. the average number of children born
to a woman during her reproductive years 4; d; 5. c; 6. a;
7. b; 8. c; 9. b; 10. c; 11. c; 12. d; 13. b; 14. d; 15. a; 16. c
**Self-Quiz**
1. d; 2. a; 3. d; 4. b; 5. b; 6. a; 7. c; 8. a; 9. a; 10. b
**Chapter Summary**
1. growth; 2. size; 3. distribution; 4. age; 5. density;
6. reproductive; 7. exponential; 8. carrying capacity;
9. environmental; 10. stabilize; 11. disease; 12. restrict;
13. limiting; 14. history; 15. sidestepped; 16. global;
17. cultural; 18. technological; 19. resources

# Chapter 41 Community Structure
# and Biodiversity
*Fire Ants in the Pants* [pp.706–707]
**41.1. Which Factors Shape Community Structure?** [p.708]
1. *Solenopsis invicta*, one of the Argentine fire ants, can be
attacked and killed in its native habitat by parts of the
life cycle of two flies. Both are parasitoids. Other options
include importation of fungal or protistan pathogens that
will infect *S. invicta* but not native ants; 2. climate;
3. food; 4. adaptive; 5. survive; 6. numbers; 7. community;
8. e; 9. h; 10. g; 11. c; 12. j; 13. a; 14. k; 15. b; 16. f; 17. i;
18. d; 19. commensalism; 20. helpful; 21. harmful;
22. harmful; 23. parasitism
**41.2. Mutualism** [p.709]
**41.3. Competitive Interactions** [pp.710–711]
1.a. The larval stages of the moth grow only in the yucca
plant; they eat only yucca seeds; every plant species of
the genus *Yucca* can be pollinated only by one species of
the yucca moth genus; the yucca moth is the plant's only
pollinator; b. Fungal hyphae obtain sugar molecules

from plant roots; plant roots obtain mineral ions obtained by mycorrhiza; c. Anemone fish are sheltered and protected by nematocyst-laden tentacles of cnidarians; aggressive anemone fish chase away predatory butterflyfishes that can bite off the sea anemone's tentacles; d. Long ago, phagocytic cells engulfed aerobic bacterial cells that tapped host nutrients; host cells obtain ATP produced by the guests (origin of mitochondria and chloroplasts); 2. d; 3. c; 4. a; 5. e; 6. b; 7. a; 8. e

**41.4. Predator–Prey Interactions [pp.712–713]**
**41.5. An Evolutionary Arms Race [pp.714–715]**
1. c; 2. e; 3. b; 4. a; 5. c; 6. f; 7. f; 8. c; 9. f; 10. e; 11. a; 12. b; 13. a; 14. d; 15. d

**41.6. Parasite–Host Interactions [pp.716–717]**
**41.7. Cowbird Chutzpah [p.717]**
1. parasites; 2. parasitic; 3. T; 4. reproductive; 5. T; 6. The agents are adapted to a specific host species and to its habitat; they are good at locating hosts; their population growth rate is high compared to the host's; their offspring are good at dispersing; and they make a type III functional response to prey, without much lag time after shifts occur in the host population size. 7. Parasites can weaken the host, cause sterility, shift the sex ratio, alter birth and death rates, and they may compete with the host.

**41.8. Ecological Succession [p.718]**
1. a; 2. d; 3. a; 4. b; 5. b; 6. a; 7. c; 8. a; 9. a; 10. a; 11. d; 12. physical factors such as soil and climate, chance events, the extent of disturbances

**41.9. Species Interactions and Community Instability [pp.720–721]**
**41.10. Exotic Invaders [pp.722–723]**
1. b; 2. d; 3. c; 4. a; 5. e; 6. b; 7. a; 8. b; 9. c; 10. a; 11. c

**41.11. Biogeographic Patterns in Community Structure [pp.724–725]**
1. equator; 2. tropics; 3. sunlight; 4. resource; 5. evolving; 6. greater; 7. herbivores; 8. herbivore species; 9. predatory; 10. parasitic; 11. reefs; 12. Island C

**41.12. Threats to Biodiversity [pp.726–727]**
**41.13. Sustaining Biodiversity and Human Populations [pp.728–729]**
1. i; 2. d; 3. h; 4. e; 5. a; 6. b; 7. g; 8. c; 9. f

**Self-Quiz**
1. a; 2. a; 3. d; 4. b; 5. d; 6. a; 7. e; 8. e; 9. b; 10. d

**Chapter Objectives/Review Questions**
3. fundamental; 5. obligatory; 10. parasitoids

**Chapter Summary**
1. species; 2. niche; 3. community; 4. commensalism; 5. predation; 6. size; 7. species interactions; 8. disturbances; 9. biogeographers; 10. tropical; 11. species; 12. conservation

## Chapter 42   Ecosystems

*Bye-Bye, Blue Bayou* [pp.732–733]
**42.1. The Nature of Ecosystems [pp.734–735]**
**42.3. Biological Magnification in Food Webs [pp.738–739]**
1. b; 2. d; 3. h; 4. c; 5. k; 6. l; 7. a; 8. e; 9. f; 10. j; 11. g; 12. i; 13. 4; 14. 1, 2, 3; 15. 5, 6, 7, 8; 16. 9, 11; 17. none; 18. 10, 12, 13; 19. 5, 6, 7, 8; 20. 10; 21. In a grazing food web, energy flows from producers to herbivores and

then to carnivores and decomposers. In a detrital food web, energy flows from producers to detritivores and then decomposers. 22. The concentration of a compound in the tissues of organisms at higher trophic levels.

**42.4. Studying Energy Flow Through Ecosystems [pp.740–741]**
**42.5. Biogeochemical Cycles [p.741]**
1. f; 2. h; 3. e; 4. c; 5. b; 6. g; 7. a; 8. d; 9. reservoirs; 10. geochemical; 11. ecosystem; 12. producers; 13. herbivores; 14. detritivores

**42.6. The Water Cycle [pp.742–743]**
1. condenses; 2. precipitation; 3. evaporation; 4. transpiration; 5. watersheds; 6. runoff; 7. aquifers; 8. groundwater

**42.7. Carbon Cycle [pp.744–745]**
**42.8. Greenhouse Gases [pp.746–747]**
1. atmospheric; 2. photosynthesis; 3. aerobic respiration; 4. fossil; 5. combustion; 6. photosynthesis; 7. aerobic respiration; 8. volcanic; 9. The greenhouse gases trap heat radiated from the Earth, thus increasing the temperature of the planet. An increase in greenhouse gases equates to an increase in global temperatures.

**42.9. Nitrogen Cycle [pp.748–749]**
**42.10. Phosphorous Cycle [pp.750–751]**
1. b; 2. c; 3. a; 4. a; 5. a; 6. c; 7. a; 8. b

**Self-Quiz**
1. c; 2. b; 3. b; 4. c; 5. d; 6. d; 7. a; 8. c; 9. c; 10. c

**Chapter Summary**
1. ecosystem; 2. energy; 3. chains; 4. detritivores; 5. webs; 6. biological magnification; 7. productivity; 8. photosynthesis; 9. carbon; 10. nitrogen; 11. reservoirs; 12. human

## Chapter 43   The Biosphere

*Surfers, Seals, and the Sea* [pp.754–755]
**43.1. Global Air Circulation Patterns [pp.756–757]**
**42.2. Circulating Airborne Pollutants [pp.758–759]**
**42.3. The Ocean, Landforms, and Climates [pp.760–761]**
1. j; 2. b; 3. g; 4. k; 5. e; 6. a; 7. i; 8. c; 9. f; 10. d; 11. h; 12. warms; 13. ascends; 14. moisture; 15. descends; 16. moisture; 17. ascends; 18. moisture; 19. descends; 20. easterlies; 21. westerlies; 22. March; 23. December; 24. September; 25. June

**43.4. Biogeographic Realms and Biomes [pp.762–763]**
**43.5. Availability of Sunlight, Soils and Moisture [pp.764–765]**
**43.6. Moisture-Challenged Biomes [pp.766–767]**
1. a; 2. b; 3. c; 4. d; 5. c; 6. d; 7. d; 8. e; 9. a; 10. c; 11. b;

**43.7. More Rain, Broadleaf Forests [p.768]**
**43.8. You and the Tropical Forests [p.769]**
**43.9. Coniferous Forests [p.770]**
**43.10. Brief Summers and Long, Icy Winters [p.772]**
1. c; 2. d; 3. i; 4. h; 5. g; 6. f; 7. e; 8. b; 9. j; 10. a

**43.11. Freshwater Ecosystems [pp.772–773]**
**43.12. "Fresh" Water? [p.774]**
**43.13. Life at Land's End [pp.774–775]**
**43.14. The Once and Future Reefs [pp.776–777]**
**43.15. The Open Ocean [pp.778–779]**
1. g; 2. b; 3. d; 4. a; 5. f; 6. c; 7. e; 8. b; 9. d; 10. a; 11. c; 12. e; 13. b; 14. d; 15. a; 16. benthic; 17. pelagic; 18. intertidal zone; 19. neritic zone; 20. oceanic zone.

**43.16. Climate, Copepods, and Cholera** [pp.780–781]
1. El Niño is characterized by warmer than usual waters in the Pacific, La Niña has cooler than normal waters in the Pacific; 2. fecal contamination of water and the *Vibrio cholerae* bacteria
**Self-Quiz**
1. a; 2. b; 3. c; 4. c; 5. d; 6. b; 7. e; 8. c; 9. a; 10. d
**Chapter Summary**
1. regional; 2. energy; 3. distribution; 4. climates; 5. ocean; 6. circulation; 7. ecosystems; 8. biogeographic; 9. biomes; 10. soil type; 11. 71; 12. light; 13. gases; 14. productivity; 15. ocean; 16. epidemics

# Chapter 44  Behavioral Ecology

*My Pheromones Made Me Do It* [pp.784–785]
**44.1. Behavior's Heritable Basis** [p.786]
**44.2. Instinct and Learning** [p.787]
1. c; 2. e; 3. b; 4. f; 5. a; 6. d; 7. the banana slug; 8. ate; 9. inland; 10. were not; 11. genetic; 12.a. cuckoo birds are social parasites in that adult females lay eggs in the nests of other bird species, b. young toads instinctively capture edible insects with sticky tongues; if a bumblebee is captured and the stings the tongue, the toad learns to leave bumblebees alone.
**44.3. The Adaptive Value of Behavior** [p.788]
**43.4. Communication Signals** [pp.788–789]
**44.5. Mates, Offspring, and Reproductive Success** [pp.790–791]
1. a; 2. c; 3. b; 4. d; 5. a; 6. hangingflies; 7. sage grouse; 8. lions, sheep, elephant seals, bison; 9. crocodiles
**44.6. Living in Groups** [pp.792–793]
**44.7. Why Sacrifice Yourself?** [pp.794–795]
**44.8. Human Behavior** [p.795]
1. d; 2. b; 3. c; 4. a
**Self-Quiz**
1. c; 2. d; 3. b; 4. d; 5. b; 6. b; 7. a
**Chapter Summary**
1. behavior; 2. pheromones; 3. environmental; 4. heritable; 5. natural selection; 6. social; 7. communication; 8. benefits; 9. evolution; 10. human; 11. animals; 12. moral

# Notes

# Notes

# Notes

# Notes

# Notes

# Notes

# Notes